T0320473

AI, Machine Learning and Deep Learning

Today, artificial intelligence (AI) and machine/deep learning (ML/DL) have become the hottest areas in information technology. In our society, many intelligent devices rely on AI/ML/DL algorithms/tools for smart operation. Although AI/ML/DL algorithms/tools have been used in many internet applications and electronic devices, they are also vulnerable to various attacks and threats. AI parameters may be distorted by the internal attacker; the DL input samples may be polluted by adversaries; the ML model may be misled by changing the classification boundary, among many other attacks/threats. Such attacks can make AI products dangerous to use.

While this discussion focuses on security issues in AI/ML/DL-based systems (i.e., securing the intelligent systems themselves), AI/ML/DL models/algorithms can actually also be used for cyber security (i.e., use of AI to achieve security).

Since AI/ML/DL security is a newly emergent field, many researchers and industry people cannot yet obtain a detailed, comprehensive understanding of this area. This book aims to provide a complete picture of the challenges and solutions to related security issues in various applications. It explains how different attacks can occur in advanced AI tools and the challenges of overcoming those attacks. Then, the book describes many sets of promising solutions to achieve AI security and privacy. The features of this book have seven aspects:

1. This is the first book to explain various practical attacks and countermeasures to AI systems.
2. Both quantitative math models and practical security implementations are provided.
3. It covers both "securing the AI system itself" and "using AI to achieve security."
4. It covers all the advanced AI attacks and threats with detailed attack models.
5. It provides multiple solution spaces to the security and privacy issues in AI tools.
6. The differences among ML and DL security/privacy issues are explained.
7. Many practical security applications are covered.

AI, Machine Learning and Deep Learning

A Security Perspective

Edited By
Fei Hu and Xiali Hei

CRC Press
Taylor & Francis Group
Boca Raton London New York

CRC Press is an imprint of the
Taylor & Francis Group, an **informa** business

First edition published 2023
by CRC Press
6000 Broken Sound Parkway NW, Suite 300, Boca Raton, FL 33487-2742

and by CRC Press
4 Park Square, Milton Park, Abingdon, Oxon, OX14 4RN

Library of Congress Cataloging-in-Publication Data
Names: Hu, Fei, 1972– editor. | Hei, Xiali, editor.
Title: AI, machine learning and deep learning : a security perspective / edited by Fei Hu and Xiali Hei.
Description: First edition. | Boca Raton : CRC Press, 2023. |
Includes bibliographical references and index.
Identifiers: LCCN 2022055385 (print) | LCCN 2022055386 (ebook) |
ISBN 9781032034041 (hardback) | ISBN 9781032034058 (paperback) |
ISBN 9781003187158 (ebook)
Subjects: LCSH: Computer networks–Security measures. |
Machine learning–Security measures. | Deep learning (Machine learning)–Security measures. |
Computer security–Data processing. | Artificial intelligence.
Classification: LCC TK5105.59 .A39175 2023 (print) |
LCC TK5105.59 (ebook) | DDC 006.3/1028563–dc23/eng/20221223
LC record available at https://lccn.loc.gov/2022055385
LC ebook record available at https://lccn.loc.gov/2022055386

ISBN: 9781032034041 (hbk)
ISBN: 9781032034058 (pbk)
ISBN: 9781003187158 (ebk)

DOI: 10.1201/9781003187158

Typeset in Times
by Newgen Publishing UK

Contents

PART III   Using AI/ML Algorithms for Cyber Security

PART IV   Applications

Preface

Today, artificial intelligence (AI) and machine/deep learning (ML/DL) have become the hottest areas in information technology. In our society, many intelligent devices rely on AI/ML/DL algorithms/tools for smart operation. For example, a robot uses a reinforcement learning algorithm to automatically adjust its motions; an autonomous vehicle uses AI algorithms to recognize road conditions and control its movements and direction; an UAV uses Bayesian learning to recognize weather/air patterns for automatic velocity adjustments, and so on.

Although AI/ML/DL algorithms/tools have been used in many internet applications and electronic devices, they are also vulnerable to various attacks and threats. AI parameters may be distorted by the internal attacker; the DL input samples may be polluted by adversaries; the ML model may be misled by changing the classification boundary, among many other attacks/threats. Such attacks can make AI products dangerous to use. For example, the autonomous car may deviate from the normal trajectory under the wrong blockage image classification algorithm (e.g., thinking a STOP sign is GO sign); the robot may hurt human beings under distorted Bayesian learning hyperparameters, and so on.

While this discussion focuses on security issues in AI/ML/DL-based systems (i.e., securing the intelligent systems themselves), AI/ML/DL models/algorithms can actually also be used for cyber security (i.e., use of AI to achieve security). The typical examples include the use of classification algorithms for intrusion detection and attack identification on the internet, the use of pattern recognition to recognize malicious behaviors in database access, the use of DL for radio fingerprint recognition to detect enemies' RF devices, and so on.

Since the AI/ML/DL security is a newly emergent field, many researchers and industry people cannot yet obtain a detailed, comprehensive understanding of this area. This book aims to provide a complete picture of the challenges and solutions to related security issues in various applications. It explains how different attacks can occur in advanced AI tools and the challenges of overcoming those attacks. Then, the book describes many sets of promising solutions to achieve AI security and privacy.

The features of this book have seven aspects:

1. This is the first book to explain various practical attacks and countermeasures to AI systems.
2. Both quantitative math models and practical security implementations are provided.
3. It covers both "securing the AI system itself" and "use of AI to achieve security."
4. It covers all the advanced AI attacks and threats with detailed attack models.
5. It provides multiple solution spaces to the security and privacy issues in AI tools.
6. The differences among ML and DL security/privacy issues are explained.
7. Many practical security applications are covered.

This book will have a big impact in AI-related products since today so many AI/ML/DL products have been invented and implemented, albeit with very limited security supports. Many companies have serious concerns about the security/privacy issues of AI systems. Millions of AI engineers/researchers want to learn how to effectively prevent, detect, and overcome such attacks. To the best of our knowledge, we have not seen a book that comprehensively discusses these critical areas. This book is more than purely theoretical description. Instead, it has many practical designs and implementations for a variety of security models. It has detailed hardware and software design strategies.

This book has invited worldwide AI security/privacy experts to contribute to the complete picture of AI/ML/DL security and privacy issues.

First, the book explains the evolution process from conventional AI to various ML algorithms (Bayesian learning, SVN, HMM, etc.) to the latest DL models. It covers various applications based on those algorithms, from robots to automatic vehicles, from smart city to big data applications.

Then it explains how an adversary can possibly attack an intelligent system with built-in AI/ML/DL algorithms/tools. The attack models are described with different environment settings and attack goals. Then the book compares all those attacks from perspectives of attack severity, attack cost, and attack pattern. The differences between internal and external attacks are explained, and intentional attacks and privacy leak issues are also covered.

Then the book provides different sets of defense solutions to those attacks. The solutions may be based on various encryption/decryption methods, statistical modeling, distributed trust models, differential privacy methods, and even the ML/DL algorithms.

Next, it discusses the use of different AI/ML/DL models and algorithms for cyber security purposes. Particularly, it explains the use of pattern recognition algorithms for the detection of external attacks on a network system.

Finally, it discusses some practical applications that can utilize the discussed AI security solutions. Concrete implementation strategies are also explained, and the application scenarios with different external attacks are compared.

The targeted audiences of this book include people from both academia and industry. Currently, more than 75 million people in the world are involved in the R&D process of AI products. This book will be highly valuable to those AI/ML/DL engineers and scientists in the industry of vehicles, robots, smart cities, intelligent transportation, etc. In addition, many academic people (faculty/students, researchers) who are working on intelligent systems will also benefit from this book.

About the Editors

Fei Hu is a professor in the department of Electrical and Computer Engineering at the University of Alabama. He has published more than ten technical books with CRC Press. His research focus includes cyber security and networking. He obtained his PhD degrees at Tongji University (Shanghai, China) in the field of Signal Processing (in 1999) and at Clarkson University (New York, USA) in Electrical and Computer Engineering (in 2002). He has published over 200 journal/conference papers and books. Dr. Hu's research has been supported by the US National Science Foundation, Cisco, Sprint, and other sources. In 2020, he won the school's President's Faculty Research Award (<1% faculty are awarded each year).

Xiali (Sharon) Hei is an assistant professor in the School of Computing and Informatics at the University of Louisiana at Lafayette. Her research focus is cyber and physical security. Prior to joining the University of Louisiana at Lafayette, she was an assistant professor at Delaware State University from 2015 to 2017 and Frostburg State University, 2014–2015. Sharon received her PhD in computer science from Temple University in 2014, focusing on computer security.

Contributors

Darine Ameyed
École de technologie supèrieure, Montreal, Canada

Pallavi Arora
IK Gujral Punjab Technical University, Punjab, India

Sikha Bagui
University of West Florida, USA

Lauren Burgess
Towson University, Maryland, USA

Riadh ben Chaabene
École de technologie supèrieure, Montreal, Canada

Yen-Hung Chen
College of Informatics, National Chengchi University, Taipei, Taiwan

Mohamed Cheriet
École de technologie supèrieure, Montreal, Canada

Nicole do Vale Dalarmelina
University of São Paulo, São Carlos, SP, Brasil

Long Dang
University of South Florida, USA

Regina Eckhardt
University of West Florida, USA

E. Bijolin Edwin
Karunya Institute of Technology and Sciences, Tamil Nadu, India

Tugba Erpek
Intelligent Automation, Inc., Maryland, USA

Bryse Flowers
University of California, USA

Nur Imtiazul Haque
Florida International University, USA

Linsheng He
University of Alabama, USA

William Headley
Virginia Tech National Security Institute, USA

Xiali Hei
University of Louisiana at Lafayette, USA

Diego Heredia
Escuela Politécnica Nacional, Quito, Ecuador

Fei Hu
University of Alabama, USA

Yuh-Jong Hu
National Chengchi University, Taipei, Taiwan

Mu-Tien Huang
National Chengchi University, Taipei, Taiwan

Fehmi Jaafar
Quebec University at Chicoutimi, Quebec, Canada

Krzysztof Jagiełło
Warsaw School of Economics, Poland

Brian Jalaian
Joint Artificial Intelligence Center, Washington, DC, USA

G. Jaspher
Karunya Institute of Technology and Sciences, Tamil Nadu, India

Gabriel Kabanda
University of Zimbabwe, Harare, Zimbabwe

W. Kathrine
Karunya Institute of Technology and Sciences,
 Tamil Nadu, India

Baljeet Kaur
Guru Nanak Dev Engineering College, Punjab,
 India

Joseph Layton
University of Alabama, USA

Hengshuo Liang
Towson University, Maryland, USA

Weixian Liao
Towson University, Maryland, USA

Jing Lin
University of South Florida, USA

Zhuo Lu
University of South Florida, USA

Rodolfo Ipolito Meneguette
University of São Paulo, São Carlos, SP, Brasil

Mahbub Rahman
University of Alabama, USA

Mohamed Rahouti
Fordham University, New York USA

Mohammad Ashiqur Rahman
Florida International University, USA

Yalin E. Sagduyu
Intelligent Automation, Inc., Maryland, USA

Yi Shi
Intelligent Automation, Inc., Maryland, USA

George Stantchev
Naval Research Laboratory, Washington, DC, USA

Jerzy Surma
Warsaw School of Economics, Poland

Marcio Andrey Teixeira
Federal Institute of Education, Science, and
 Technology of Sao Paulo, SP, Brasil

D. Roshni Thanka
Karunya Institute of Technology and Sciences,
 Tamil Nadu, India

Qianlong Wang
Towson University, Maryland, USA

Kaiqi Xiong
University of South Florida, USA

Wei Yu
Towson University, Maryland, USA

Jiamiao Zhao
University of Alabama, USA

PART I

Secure AI/ML Systems: Attack Models

Machine Learning Attack Models

1

Jing Lin,[1] Long Dang,[2] Mohamed Rahouti,[3] and Kaiqi Xiong[1]

[1]*ICNS Lab and Cyber Florida, University of South Florida, Tampa, FL*

[2]*ICNS Lab & Computer Science and Engineering, University of South Florida, Tampa, FL*

[3]*Computer and Information Sciences, Fordham University, Bronx, NY*

Contents

DOI: 10.1201/9781003187158-2

3

1.1 INTRODUCTION

Machine learning (ML) is a research field focusing on the theory, properties, and performance of learning algorithms and systems. The ML field is considered highly interdisciplinary as it is built on the top of various disciplines, including but not limited to statistics, information theory, cognitive science, and mathematics (e.g., optimization theory). ML techniques have been intensively studied for many years. Notably, the achievements in many recent research studies related to smart city applications have been made possible using ML advances, where AI safety is directly relevant to ML security. As another example, ML implementations in an intelligent transportation system (ITS) can be deployed to analyze data generated by the different parts of the system (e.g., roads, number of passengers, and commute mode). The analysis of collected data is consequently used for future planning and decision making within a transportation scheme [1].

Moreover, deep learning (DL) is another vital component of the ML field. Unlike many traditional learning techniques based on shallow-structured learning paradigms, DL is based upon deep architectures, where unsupervised and/or supervised learning strategies are mainly used to learn hierarchical representations autonomously. DL architectures (deep architectures) are often enabled to capture more hierarchically launched statistical patterns (complicated input patterns) in comparison to shallow-structured learning architectures [2].

Many state-of-the-art ML models have outperformed humans in various tasks, such as image classification [3]. With such outstanding performance, ML models are widely used today. However, the existence of adversarial attacks and data poisoning attacks really places the robustness of ML models in question. For instance, Engstrom [4] demonstrated that state-of-the-art image classifiers could be easily fooled by a small rotation on an arbitrary image. As ML systems are being increasingly integrated into safety- and security-sensitive applications [5], adversarial attacks and data poisoning attacks pose a considerable threat. This chapter focuses on the two broad and important areas of ML security: adversarial attacks and data poisoning attacks.

In *adversarial attacks*, attackers attempt to perturb a data point x to an adversarial data point x' so that x' is misclassified by an ML model with high confidence, even though x' is visually indistinguishable from its original data point x by humans. A visual illustration of an adversarial attack is shown in Figure 1.1,

  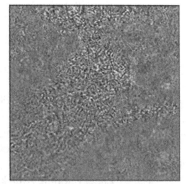

Original Image:
Egyptian cat (9.48%)

Image + Noise:
Egyptian cat (0.00%)
bookcase (99.83%)

Amplified Noise

FIGURE 1.1 Example of L_∞-norm-based FGSM attacks using a perturbation magnitude $\epsilon = 3$. Left: Base image (classified as a cat correctly by an Inception-v3 network [11]). Center: Adversarial image generated by the FGSM attacks (mislabeled as a bookcase with high confidence of 99.83%). Right: (Amplified) adversarial noise.

which presents an L_∞-norm-based fast gradient sign method (FGSM) attack [6]. Adversarial attacks can be conducted in different application domains, such as audio [7], text [8], network signals [9], and images [10]. They can be further classified based upon the knowledge level that adversaries maintain for a target model. The attack model here can be specified as either *black-box*, *white-box*, or *gray-box*. *Black-box* attacks refer to the case when an attacker has no access to the trained ML model. Instead, the attacker only knows the model outputs (logit, confidence score, or label). Black-box attacks are commonly observed in online learning techniques used in anomaly detection systems that effectively retrain the detector when new data is available. In a *gray-box* attack, attackers may have some incomplete knowledge about the overall structure of a target model. Finally, in a *white-box* threat model, attackers have complete knowledge about a target model, thereby facilitating the task of generating adversarial examples and crafting poisoned input data. Even though there are more white-box attacks among the three types of attacks, black-box attacks are more practical for ML systems because essential model information is often confidential and protected from a normal user interacting with a system. Furthermore, due to the phenomenon of attack transferability, white-box attacks can be utilized to attack black-box models (transfer attacks). In this chapter, we focus on the two primary threat models: white-box and black-box attacks.

In *data poisoning attacks*, adversaries try to manipulate training data in an attempt to

- decrease the overall performance (i.e., accuracy) of an ML model,
- induce misclassification to a specific test sample or a subset of the test sample, or
- increase training time.

If an attacker has a specified target label to which a specific test sample is misclassified, the attack is called a *targeted data poisoning attack*; otherwise, it is called an *untargeted data poisoning attack*. Adversaries are assumed either to be able to contribute to the training data or to have control over the training data. For instance, crowdsourcing is regarded as a vital data source for various smart cities, such as intelligent transportation systems and traffic monitoring [12]. Hence, such systems are vulnerable to data poisoning threats [13, 14] as data poisoning attacks are often difficult to detect [15, 16].

This chapter presents adversarial attacks and data poisoning attacks in both white-box and black-box settings. These security attack classes are comprehensively discussed from different adversarial capabilities. The main objective of this chapter is to help the research community gain the insights and implications of existing adversarial attacks and data poisoning attacks and to increase awareness of potential adversarial threats when researchers develop future learning algorithms. The rest of this chapter is organized as follows. In section 1.2, we first provide some background information on support vector machines (SVMs) and neural networks. Then we introduce important definitions and typical white-box adversarial attacks in section 1.3. In section 1.4, we further discuss black-box adversarial attacks. Moreover, section 1.5 covers data poisoning attacks, along with their adversarial aspects. Finally, section 1.6 provides concluding remarks.

1.2 BACKGROUND

In this section, we introduce our notation before providing background information on the SVMs and neural networks.

1.2.1 Notation

Table 1.1 presents a list of notations used throughout the chapter.

TABLE 1.1   Notations Used in the Chapter

SYMBOL	MEANING	SYMBOL	MEANING
Z_+	Set of positive integers	C	Classifier whose output is a label
R	Set of real numbers	∇	Gradient
K	Number of classes	$\|\cdot\|_p$	L_p-norm
x	Original (clean, unmodified) input	δ	Perturbation $\delta = x' - x$
x'	Adversarial image (perturbed input image)	ε	Some application-specific maximum perturbation
y	Input x's ground truth label	g_i	The discriminant function for class i
t	Label of the targeted class in an adversarial attack	f	Neural network (NN) model whose output is a confidence or probability score vector: $f(x) \in R^K$
L	The loss function of a NN	Z	Logits
Θ	Parameters off		

1.2.2 Support Vector Machines

SVMs, known as large margin classifiers, are supervised learning models for both regression and classification problems. Formally, SVMs are ML methods to construct a hyperplane such that the distance between the hyperplane and the nearest training sample is maximized.

As an example, we focus here on the hard-margin SVM for a linearly separable binary classification problem. Given a training dataset of n points written as $(x_1, y_1), \ldots, (x_n, y_n)$, where y_i is the class to which x_i belongs. Each data point x_i is a p-dimensional real vector, and $y_i \in \{-1, 1\}$. The goal of SVMs is to search for the maximum-margin hyperplane that separates the group of points x_i for which $y_i = 1$ from the group of points x_i for which $y_i = -1$. The SVM, proposed by Cortes and Vapnik [17], finds the maximum-margin hyperplane by solving the following quadratic programming (QP) problem :

$$\frac{1}{2}\|w\|^2 , \tag{1.2.1}$$

subject to the constraints, $y_i(w^T x_i + b) \geq 1$, for all $i = 1, 2, \ldots, n$, where w is the normal vector of the maximum-margin hyperplane $y = w^T x + b$. This QP problem can be reformatted as an unconstrained problem using Lagrange multipliers $\alpha_i \geq 0$, for $i = 1, 2, \ldots, n$, as follows:

$$L_p = \frac{1}{2}\|w\|^2 - \sum_{i=1}^{n} \alpha_i \left[y_i \left((w^T x_i + b) \right) - 1 \right]. \tag{1.2.2}$$

Taking the derivative of the preceding expression with respect to w, we have

$$w = \sum_{i=1}^{n} \alpha_i x_i y_i. \tag{1.2.3}$$

This indicates that the normal vector w of the hyperplane depends on α_i, x_i and y_i. Similarly, taking the derivative of the preceding expression with respect to w_0, we obtain

$$\sum_{i=1}^{n} \alpha_i y_i = 0. \tag{1.2.4}$$

Substituting equations (1.2.2) and (1.2.3) to L_p, we maximize the resulting L_p with respect to α_i subject to equation (1.2.4) by applying a quadratic optimization method. Then we can easily solve for α_i. Once solved, most α_i are equal to zero. The set of data points (x_i, y_i) for which the corresponding $\alpha_i \neq 0$ are called support vectors. Since the normal vector $w = \sum_{i=1}^{n} \alpha_i x_i y_i$, the resulting hyperplane depends on the support vectors only.

However, if two classes are nonlinearly separable, a soft-margin SVM with a kernel function can be utilized to learn a nonlinear decision boundary (linear in transformed space).

1.2.3 Neural Networks

In this subsection, we discuss neural networks, focusing on image-recognition systems since most existing adversarial examples introduced in the literature are related to image recognition.

Let $X = R^{hwc}$ be an input image space, where $h \in Z_+, w \in Z_+$, and $c \in Z_+$. For instance, a grayscale image $x \in X$ has a color channel $c = 1$, and a colored RGB image $x \in X$ has color channel $c = 3$. The element $x_{i,j,k}$ of vector x denotes the value of the pixel located at (i, j, k). Each pixel value indicates how bright a pixel is, and the pixel value is an integer between 0 and 255. However, the pixel value is usually rescaled to be in the range of [0, 1].

A feed-forward neural network can be written as a function $f : X \rightarrow R^K$ that maps an input image $x \in X$ to an output $y \in R^K$, where $K \in Z_+$ is the number of classes, and R is a set of real numbers. Each element y_j of y can be considered as the probability or likelihood that input x belongs to class j, where j is an integer between 1 and K. Each element y_j of y satisfies $y_j \in [0,1]$ and $\sum_{j=1}^{K} y_j = 1$. The output vector y can be considered a probability distribution over the discrete label set $\{1,2,\ldots,K\}$. Furthermore, let $C : X \rightarrow \{1,2,\ldots,K\}$ be the corresponding classifier that maps an input image $x \in X$ to $\{1,2,\ldots,K\}$. The relationship between C and the output vector y is as follows:

$$C(x) = \underset{i=1,2,\ldots,K}{\operatorname{argmax}} y_i. \tag{1.2.5}$$

A feed-forward neural network f with $L \in Z_+$ layers can be considered a composition function such that for each $x \in X$,

$$f(x) = f_L\left(f_{L-1}\left(f_{L-2}\cdots\left(f_1(x)\right)\right)\right), \tag{1.2.6}$$

where each component function $f_i(x_i) = \sigma_i(\theta_i \cdot x_i + b_i), \sigma_i$ is an activation function at layer i, θ_i is a matrix of weight parameters at layer i, and b_i is a bias unit at layer i. Both $\theta = (\theta_1, \theta_2, \ldots, \theta_L)$ and $b = (b_1, b_2, \ldots, b_L)$ form model parameters $\Theta = (\theta, b)$. To keep things simple, we do not explicitly state the preceding function's dependence on Θ. If an attack model explicitly depends on it, we would write it as $f(x \mid \Theta) = y$. To find model parameters Θ such that the classifier C can predict most instances in X correctly, a stochastic gradient descent method is often utilized to minimize a loss function $L : X \times \{1,2,\ldots,K\} \times P \rightarrow R^+$, where P is parameter space. To keep things simple, we sometimes simply use $L : X \times \{1,2,\ldots,K\} \rightarrow R^+$ when there is no confusion.

Let us further discuss activation function σ_i as it is essential in expanding a linear hypothesis space to a nonlinear hypothesis space. Rectified Linear Unit (ReLU) is widely used in ML/DL applications, especially in an image classification task. $ReLU : R \rightarrow R$ is defined on a real number set R into a real number set R such that $ReLU(x) = \max(0, x)$ for each $x \in R$. Therefore, the $ReLU$ is differentiable at all the points except $x = 0$. Usually, data scientists are interested in a nonlinear activation function that is bounded and has a smooth gradient. Boundedness is an important property in preventing the return values of the activation function from becoming overly large. A smooth gradient is desirable since it makes back-propagation possible. However, $ReLU$ is neither upper-bounded nor nondifferentiable at $x = 0$. Thus, why is it so popular and useful compared to traditional nonlinear activation functions (such as sigmoid and hyperbolic tangent functions) that are bounded and have a smooth gradient? Existing studies have not yet laid a theoretical foundation to answer this question. However, the only nondifferentiable point for $ReLU$ is when $x = 0$ and an input value of exactly zero for a $ReLU$ activation is not likely at any stage of the calculation, that so back-propagation is working, even though $ReLU$ is not everywhere differentiable.

Many variations of $ReLU$ have been developed over the past few years, such as the Exponential Linear Unit (Exponential LU or simply ELU), the Gaussian Error Linear Unit ($GELU$), and Leaky $ReLU$. For instance, ELU transforms the negative input through an exponential function instead of assigning all negative input to be zero. $GELU$ [18], an empirical improvement of $RELU$ and ELU activations, is defined as follows:

$$GELU(x) = \left(\frac{x}{2}\right)\left[1 + \tanh\left(\sqrt{\frac{2}{\pi}}\left(x + 0.044715x^3\right)\right)\right].$$ (1.2.7)

$GELU$ has been the state-of-the-art activation function in Natural Language Processing for the past few years. Moreover, the Leaky $ReLU$ [19] scales down the negative input value through multiplying it by a small positive constant $\alpha \in R_+$, e.g., $\alpha = 0.01$ instead of returning zero when $x < 0$.

Lastly, we discuss the softmax function, a popular activation function for the last layer of a multiclassification model. The softmax function is a generalization of the logistic function, or sigmoid function, and can be defined as follows: $Softmax : R^K \rightarrow [0,1]^K$ such that the ith element of the output for an input $z \in R^K$ is

$$Softmax_i(z) = \frac{\exp(z_i)}{\sum_{j=1}^{K}\exp(z_j)}.$$ (1.2.8)

1.3 WHITE-BOX ADVERSARIAL ATTACKS

We start with a reasonably comprehensive definition of adversarial attacks based on L_p-norm.

Definition 1 (Adversarial Attack): Let $x \in R^n$ be a legitimate input data that is correctly classified as class y by an ML classifier f. A target class t such that $t \neq y$ is given. An adversarial attack is a mapping $\alpha : R^n \rightarrow R^n$ such that the adversarial example $\alpha(x) = x'$ is misclassified as class t by f, whereas the difference between x' and x is trivial; i.e., $\|x' - x\|_p < \epsilon$ for some small value ϵ.

The L_p-norm $\|x' - x\|_p$ used in Definition 1 measures the magnitude of perturbation generated by an adversarial attack α. The L_p-norm of a vector v is defined as

$$\|v\|_p = (\sum_{i=1}^n |v_i|^p)^{\frac{1}{p}}, \tag{1.2.10}$$

where v_i is the ith component of v. Here are three frequently used norms in literature.

- L_0-norm:

$$\|x' - x_0\| = \sum_{i=1}^n J[x' \neq x], \text{ where } J[x' \neq x] = \begin{cases} 1, & x' \neq x \\ 0, & otherwise \end{cases},$$

  counts the number of coordinates i such that $x_i \neq x_i'$, where x_i and x_i' are the ith component of x and x', respectively. The norm is useful when an attacker wants to limit the number of attack pixels without limiting the size of the change to each pixel.

- L_2-norm:

$$\|x' - x\|_2 = \sqrt{\sum_{i=1}^n (x_i' - x_i)^2}$$

  measures the Euclidean distance between x' and x

- L_∞-norm or the Chebyshev Distance:

$$\|x' - x\|_\infty = \max \left(|x_1' - x_1|, |x_2' - x_2|, ..., |x_n' - x_n| \right)$$

  measures the maximum absolute change to any of the coordinates of x.

Definition 1 is recognized as "targeted adversarial attacks." The "targeted" attacks comprise a target class t, and a function α aims at identifying a legitimate input data x' (called adversarial example) such that the classifier f misclassifies x' as an instance of class t. In some scenarios, an attacker may not have a target class in mind. The goal is simply to mislead the classifier f to misclassify an adversarial example x' to a class other than its original class y.

This section mainly focuses on white-box attacks, assuming that an attacker has complete knowledge about the trained ML model. When attackers have access to this essential information, they can modify any arbitrary clean input using adversarial attack models. White-box attacks are the worst-case scenario because attackers have access to a model architecture and model weights.

1.3.1 L-BGFS Attack

Adversarial examples are often generated based on either first-order gradient information (such as the FGSM and DeepFool attack [20]) or gradient approximations. We first discuss some classic gradient-based attacks. In this case, the goal of an attacker is either to minimize the norm of the added perturbation that is needed to cause misclassification or to maximize the loss function model f with respect to an input data.

Szegedy et al. [21] demonstrated the first targeted adversarial attack on deep neural networks. They found a minimal distorted adversarial example x' by solving the following optimization problem:

$$\min_{x'} \| x' - x \|_p, \tag{1.3.1}$$

subject to the constraints, $C(x')=t$ and $x' \in [0,1]^n$, where $\| . \|_p$ is the L_p-norm. Finding a precise solution to this problem is very challenging since the constraint $C(x')=t$ is nonlinear. Szegedy et al. replaced the constraint $C(x')=t$ with the continuous loss function L, such as the cross-entropy for classification. That is, they instead solved the following optimization problem:

$$\min_{x'} c \| x' - x \|_p + L(x',t), \tag{1.3.2}$$

subject to the constraint $x' \in [0,1]^n$, where the constant $c > 0$. By first adaptively choosing different values of the constant c and then iteratively solving the problem (1.3.2) using the L-BFGS algorithm [22] for each selected c, an attacker can obtain an adversarial example x' that has the minimal distortion to x and the $C(x')=t$. The L-BGFS attack is formally defined next.

Definition 2 (L-BGFS Attack by Szegedy et al. 2014): Let $x \in R^n$ be legitimate input data that is correctly classified as class y by an ML classifier C. Given a loss function L, the L-BFGS attack generates an adversarial example $x' \in R^n$ by minimizing the following objective function:

$$c \| x' - x \|_p + L(x',t),$$

subject to the constraint $x' \in [0,1]^n$, where x' is misclassified as class $t \neq y$ by C.

1.3.2 Fast Gradient Sign Method

Unlike the L-BFGS attack, the fast gradient sign method (FGSM) proposed by Goodfellow et al. [6] focuses on finding an adversarial perturbation limited by the L_∞-norm efficiently rather than on producing an optimal adversarial example. In the training of an ML model, a given loss function is minimized to find an optimal parameter set Θ for a classifier C so that C classifies most training data correctly. In contrast, FGSM maximizes the loss function as it tries to make the classifier perform poorly on the adversarial example x'. Therefore, the FGSM perturbed image for an untargeted attack is constructed by solving the following maximization problem:

$$\max_{x' \in [0,1]^n} L(x',y), \tag{1.3.3}$$

subject to the constraint $\| x' - x \|_\infty \leq \varepsilon$, where y is the ground-truth label of the original image x, and ε is application-specific maximum perturbation.

Applying the first-order Taylor series approximation, we have

$$L(x', y) = L(x + \delta, y) \approx L(x, y) + \nabla_x L(x, y)^T \cdot \delta, \tag{1.3.4}$$

where $\delta = x' - x$ is an adversarial perturbation, and $\nabla_x L(x,y)^T$ refers to the gradient of the loss function L with respect to input x, and it can be computed quickly by a back-propagation algorithm.

The objective function in (1.3.3) is rewritten as:

$$\min_{\delta} -L(x,y) - \nabla_x L(x,y)^T \cdot \delta, \tag{1.3.5}$$

subject to the constraint $\| \delta \|_\infty \leq \varepsilon$, where we convert the maximization problem to the minimization problem by flipping the signs of the two components. Note that $L(x,y)$ and $\nabla_x L(x,y)^T$ are constant and already known because the attack is in the white-box attack setting.

Furthermore, since $0 \leq \| \delta \|_\infty \leq \varepsilon$, we have

$$\delta = \varepsilon \cdot sign\left(\nabla_x L(x,y)\right), \tag{1.3.6}$$

Where the *sign* function outputs the sign of its input value.

For the targeted attack setting, an attacker tries to minimize the loss function $L(x,t)$, where t is the target class. In this case, we can show that

$$\delta = -\varepsilon \cdot sign\left(\nabla_x L(x,t)\right). \tag{1.3.7}$$

Formally, we can define FGSM with L_∞-norm bounded perturbation magnitude next.

Definition 3 (FGSM Adversarial Attack by Goodfellow et al. 2014): Let $x \in R^n$ be legitimate input data that is correctly classified as class y by an ML classifier f. The FGSM with L_∞-norm bounded perturbation magnitude generates an adversarial example $x' \in R^n$ by maximizing the loss function $L(x',y)$ subject to the constraint $\| x' - x \|_\infty \leq \varepsilon$. That is,

$$x' = \begin{cases} x + \varepsilon \cdot sign\left(\nabla_x L(x,y)^T\right), untargeted \\ x - \varepsilon \cdot sign\left(\nabla_x L(x,t)^T\right), targeted\ on\ t. \end{cases}$$

Furthermore, we can easily generalize the attack to other L_p-norm attacks[1]. For example, we can use the Cauchy–Schwarz inequality, a particular case of the Holder's inequality with $p = q = 2$, to find the lower bound of the objective function in problem (1.3.5) subject to the constraint $\| x' - x \|_2 \leq \varepsilon$ and to obtain the perturbation for an untargeted FGSM for L_2-norm-based constraint

$$\delta = \frac{\varepsilon \nabla_x L(x,y)}{\left\| \nabla_x L(x,y) \right\|_2}. \tag{1.3.8}$$

Unless otherwise specified, we focus on the attacks with the L_∞-norm bounded perturbation magnitude in the rest of this chapter.

1.3.3 Basic Iterative Method

The basic iterative method proposed by Kurakin et al. [23] is an iterative refinement of the FGSM. BIM uses an iterative linearization of the loss function rather than the one-shot linearization in FGSM. Furthermore, multiple smaller but fixed step sizes α are taken instead of a single larger step size ε in the direction of the gradient sign. For instance, in an untargeted attack with L_∞-norm bounded perturbation magnitude, the initial point $x^{(0)}$ is set to the original instance x, and in each iteration, the perturbation $\delta^{(k)}$ and $x^{(k+1)}$ are calculated as follows:

$$\delta^{(k)} = \alpha \times sign\left(\nabla_x L(x^{(k)},y)\right) \tag{1.3.9}$$

$$x^{(k+1)} = Clip_{x,\varepsilon}\left\{x^{(k)} + \delta^{(k)}\right\}, \tag{1.3.10}$$

where $\varepsilon > 0$ denotes the size of the attack, and $Clip_{x,\varepsilon}(x')$ is a projection operator that projects the value of x' onto the intersection of the box constraint of x (for instance, if x is an image, then a box constraint of x can be the set of integer values between 0 and 255) and the ε neighbor ball of x. This procedure ensures that the produced adversarial examples are within ε bound of the input x. Using a reduced perturbation magnitude α limits the number of active attack pixels and thus prevents a simple outlier detector from detecting the adversarial examples.

Definition 4 (BIM Adversarial Attack by Kurakin et al. 2016): The BIM attack generates an L_∞-norm bounded adversarial example $x' \in R^n$ by iteratively maximizing the loss function $L(x', y)$, subject to the constraint $\| x' - x \|_\infty \le \varepsilon$. For iteration k, $x^{(k+1)}$ is calculated as follows:

$$\delta^{(k)} = \alpha \times sign\left(\nabla_x L\left(x^{(k)}, y\right)\right)$$

$$x^{(k+1)} = Clip_{x,\varepsilon}\left\{x^{(k)} + \delta^{(k)}\right\},$$

where $x^{(0)} = x, \alpha < \varepsilon$ is a smaller but fixed perturbation magnitude in each iteration, and the number of iterations is chosen by the user.

A popular variant of BIM is the projected gradient descent (PGD) attack, which uses a uniformly random noise as an initialization instead of setting $x^{(0)} = x$ [24]. The random initialization is used to explore the input space. Later, Croce and Hein [24] improved PGD by adding a momentum term for a gradient update and utilizing the exploration vs. exploitation concept for optimization. They called this approach the auto-projected gradient descent (AutoPGD) and showed that AutoPGD is more effective than PGD [24]. It is further improved with AutoAttack [24], an ensemble of various attacks. Indeed, AutoAttack combines three white-box attacks with one black-box attack. The four attack components are two extensions of AutoPGD attacks (one maximizes cross-entropy loss, and another maximizes the difference of logistic ratio loss) and two existing supporting attacks (a fast adaptive boundary [FAB] attack [25] and a square attack [26], respectively). It has been shown that using the ensemble can improve the attack effectiveness over multiple defense strategies [25]. An attack is considered successful for each test example if at least one of the four attack methods finds an adversarial example.

Furthermore, AutoAttack can run entirely without predefined user inputs across benchmark datasets, models, and norms. Therefore, it can provide a reliable, quick, and parameter-free evaluation tool when a researcher develops adversarial defenses [24]. We will introduce fast adaptive boundary attacks in section 1.3.5 and square attacks in section 1.4.4.

1.3.4 DeepFool

The DeepFool [20] attack is an untargeted white-box attack. Like Szegedy et al. [21], Moosavi-Dezfooli et al. [20] studied the minimally distorted adversarial example problem. However, rather than finding the gradient of a loss function, DeepFool searches for the shortest distance from the original input to the nearest decision boundary using an iterative linear approximation of the decision boundary/hyperplane and the orthogonal projection of the input onto the approximated decision boundary.

Moosavi-Dezfooli et al. searched the shortest distance path for an input x to cross the decision boundary and get misclassified. Formally, the DeepFool attack can be defined as follows.

Definition 5 (DeepFool Adversarial Attack by Moosavi-Dezfooli et al. 2016): The DeepFool L_2-norm attack generates adversarial example $x' \in R^n$ by solving the following optimization problem:

$$\min_{x'} \|x' - x\|_2^2 , \tag{1.3.11}$$

subject to the constraint $x' \in g$, where g is the decision boundary separating class c_i from class $c_j \, (j \neq i)$.

To find such a path, an attacker first approximates the decision boundary using the iterative linear approximation method and then calculates the orthogonal vector from x to that linearized decision boundary. Furthermore, we present the solution to the binary classification case and extend the solution to the multiclassification case later.

Given a nonlinear binary classifier with discriminant functions g_1 and g_2, the decision boundary between class 1 and class 2 is $g = \{x : g_1(x) - g_2(x) = 0\}$. A per-iteration approximate solution x'_{i+1} can be derived by approximating the decision boundary/hyperplane g based on the first-order Taylor expansion:

$$g(x') \approx g(x_i) + \nabla_x g(x_i)^T (x' - x_i), \tag{1.3.12}$$

where x_i is the ith iterate of the solution and $x_0 = x$. Then we solve the following optimization problem:

$$\min_{x'} \|x'_{i+1} - x_i\|, \tag{1.3.13}$$

subject to $g(x_i) + \nabla_x g(x_i)^T (x' - x_i) = 0$ by finding the saddle node of the Lagrangian given by:

$$\frac{1}{2}\|x'_{i+1} - x_i\|_2^2 + \lambda \left(g(x_i) + \nabla_x g(x_i)^T (x' - x_i) \right). \tag{1.3.14}$$

The optimization problem can be solved to obtain the saddle node

$$x_{i+1} = x_i - \left(\frac{g(x_i)}{\|\nabla_x g(x_i)^T\|_2} \right) \frac{\nabla_x g(x_i)}{\|\nabla_x g(x_i)\|_2}.$$

This saddle node can be considered the orthogonal projection of x_i onto the decision hyperplane g at iteration i, and the iteration stops when x_{i+1} is misclassified; i.e., $sign(g(x_{i+1})) \neq sign(g(x_0))$.

The multiclass (in one-vs.-all scheme) approach of the DeepFool attack follows the iterative linearization procedure in the binary case, except that an attacker needs to determine the closest decision boundary l to the input x. The iterative linear approximation method used to find the minimum perturbation is a greedy method that may not guarantee the global minimum. Moreover, the closest distance from the original input to the nearest decision boundary may not be equivalent to the minimum difference observed by human eyes. In practice, DeepFool usually generates small unnoticeable perturbations.

So far, we have only considered untargeted DeepFool attacks. A targeted version of DeepFool can be achieved if the input x is pushed toward the boundary of a target class $t \neq y$ subject to $C(x'_{i+1}) = t$. Furthermore, the DeepFool attack can also be adapted to find the minimal distorted perturbation for any L_p-norm, where $p \in [1, \infty]$. If interested, see [28] for details.

1.3.5 Fast Adaptive Boundary Attack

The fast adaptive boundary attack, proposed by Croce and Hein [25], is an extension of the DeepFool attack [20]. Specifically, the computation of x'_i for a FAB attack follows the DeepFool algorithm closely.

The main differences between the two types of attacks are that a DeepFool attacker projects x_i' onto the decision hyperplane g only, $Proj_g\left(x_i'\right)$, whereas a FAB attacker projects x_i' onto the intersection of $[0,1]^n$ and the approximated decision hyperplane g, $Proj_{g\cap[0,1]^n}\left(x_i'\right)$, with the following additional steps.

- Add a momentum term α to regulate the influence of the additional term $x + \eta * Proj_{g\cap[0,1]^n}\left(x\right)$ on the modified image $x_i' + \eta * Proj_{g\cap[0,1]^n}\left(x_i'\right)$. Thus x_i' is biased toward x, and hence the x_i' generated by the FAB attack method is closer to the input x than one generated by the DeepFool attack. The momentum term α is updated for each iteration i as follows. First, a FAB attacker computes the minimal perturbations $\delta_i = Proj_{g\cap[0,1]^n}\left(x_i'\right)$ and $\delta_0 = Proj_{g\cap[0,1]^n}\left(x_0\right)$ as the current best point x_i', and the inputs x are projected onto the intersection of $[0,1]^n$ and the approximated linear decision boundary, respectively. Then, α is chosen as follows:

$$\alpha = \min\left\{\frac{\left|Proj_{g\cap[0,1]^n}\left(x_i'\right)\right|}{\left|Proj_{g\cap[0,1]^n}\left(x_i'\right)\right| + \left|Proj_{g\cap[0,1]^n}\left(x\right)\right|}, \alpha_{max}\right\} \in [0,1], \tag{1.3.15}$$

where $\alpha_{max} \in [0,1]$ is a hyperparameter that provides an upper limit value for α. α is defined in such a way that the per-iteration x_i' might go too far away from x.
- Add a backward step at the end of each iteration to further increase the quality of the adversarial example. The DeepFool stops when x_i' is successfully misclassified. However, the FAB attack has this additional step for enhancement. The backward step is simply a movement of x_i' closer to the input x by calculating the linear combination of x and x_i' with a hyperparameter $\beta \in (0,1)$. Croce and Hein [25] used $\beta = 0.9$ for all their experiments.

$$x_i' = (1-\beta)x + \beta x_i' = x + \beta\left(x_i' - x\right). \tag{1.3.16}$$

- Add random restarts to widen the search space for adversarial examples. That is, instead of initializing $x_0' = x$, a FAB attacker sets x_0' as a random sample in a ϵ'-neighborhood of x, where $\epsilon' = \min(\left\|x_i' - x\right\|_p, \epsilon)$.
- Add a final search step to further reduce the distance between the adversarial example and its original input. This step uses a modified binary search on x_i' and x to find a better adversarial example within a few iterations; e.g., Croce and Hein [25] set the number of iterations to 3 for their experiments. For details of the final search step and the random restarts, see [25].

1.3.6 Carlini and Wagner's Attack

Carlini and Wagner's (C&W) attack [27] aims to find the minimally disturbed perturbation to solve the box-constrained optimization problem defined in section 1.3.1. Specifically, they converted the box-constrained optimization problem into an unconstrained optimization problem, which can then be solved through standard optimization algorithms instead of the L-BFGS algorithm. Carlini and Wagner investigated three different methods to get rid of the box-constraint $x' \in [0,1]^n$: projected gradient descent, clipped gradient descent, and change of variables. They concluded that the method of the change of variables is the most effective in generating adversarial examples fast. That is, they introduced a set of new variables $w_i \in R, i = 1,2,...,n$ such that

$$x_i' = \frac{1}{2}\left(\tanh\left(w_i\right)+1\right),\tag{1.3.17}$$

although $x_i' \in [0,1]$, w_i can be any real number so that the box-constraint is eliminated.

Furthermore, the constraint $C(x')=t$ is highly nonlinear. Carlini and Wagner [29] considered seven objective candidate functions g to replace the constraint $C(x')=t$. Each g satisfies the condition that $C(x')=t$ if and only if $g(x')\leq 0$ (for targeted attacks) and $C(x') \neq y$ if and only if $g(x')\leq 0$ (for untargeted attacks). For instance, C&W's L_2-norm attack has g defined as follows:

$$g(x') = \begin{cases} max\left\{\max_{j\neq t} Z_j(x') - Z_t(x'), -\kappa\right\}, & target\ on\ t \\ max\left\{Z_y(x') - \max_{j\neq y} Z_j(x'), -\kappa\right\}, & untargeted, \end{cases}\tag{1.3.18}$$

where Z is the logits, and Z_j is the jth element of the logits; $\kappa \geq 0$ is a parameter that controls the strength of adversarial examples. The default value is zero for experiments in [29]. However, increasing κ can generate adversarial examples with a higher transfer success rate.

Instead, Carlini and Wagner [27] solved the optimization problem

$$\min_{x'} \| x' - x \|_p + c \cdot g(x'),\tag{1.3.19}$$

where $c > 0$ is a regularization parameter that controls the relative importance of the L_p-norm perturbation over g. If a large value of c is chosen, the attacker generates an adversarial example that is further away from the base example but is misclassified with high confidence and vice versa. Carlini and Wagner [27] used a modified binary search to find the smallest value of c for which the solution x' to the optimization problem (1.3.19) satisfies the condition $g(x')\leq 0$. As an alternative to the modified binary search, a line search can also be used to find the adversarial example with the minimal L_p distance from x.

Furthermore, Carlini and Wagner considered three types of attacks, L_0-norm attack, L_2-norm, and L_∞-norm attacks. The formal definition of the C&W attack is presented next.

Definition 6 (C&W Adversarial Attack by Carlini et al. 2017): The C&W attack generates an adversarial example $x' \in R^n$ by solving the following optimization problem:

$$\min_{x'} \| x' - x \|_2^2 + c \cdot g(x'),$$

where $x_i' = \frac{1}{2}\left(\tanh\left(w_i\right)+1\right), c > 0$ is the regularization term, and

$$g(x') = \begin{cases} max\left\{\max_{j\neq t} Z_j(x') - Z_t(x'), -\kappa\right\}, & target\ on\ t \\ max\left\{Z_y(x') - \max_{j\neq y} Z_j(x'), -\kappa\right\}, & untargeted. \end{cases}$$

Note that the standard gradient descent algorithm can be used to find the solution of this minimization problem.

Carlini and Wagner [27] showed that the C&W attack is very powerful, and ten proposed defense strategies cannot withstand C&W attacks constructed by minimizing defense-specific loss functions [28]. Furthermore, the C&W attack can also be used to evaluate the efficacy of potential defense strategies since it is one of the strongest adversarial attacks [27].

In the next two subsections, we discuss some adversarial attacks other than L_p-norm-based attacks.

1.3.7 Shadow Attack

In this subsection, we introduce a class of adversarial attacks, called "semantic" adversarial attacks, on image datasets that usually generate adversarial examples through a larger perturbation on an image in terms of L_p-norm but that are still semantically similar to its original image. A typical example of the attack is simply a small color shift, rotation, shearing, scaling, or translation of a natural image to fool a state-of-the-art image classifier [4, 29]. The attack is successful due to differences in the way the ML model recognizes an image and the human visual recognition system. A semantic adversarial example can be formally defined next.

Definition 7 (Semantic Adversarial Attack): Let $x \in R^n$ be a legitimate input data that is correctly classified as class y by an ML classifier. A semantic adversarial attack is a mapping $\alpha : R^n \to R^n$ such that the adversarial example $\alpha(x) = x'$ is misclassified as class $t \neq y$ by the classifier and $\beta(x) = \beta(x')$, where β is the human visual recognition system.

Semantic adversarial attacks can be either white-box or black-box. While we discuss one of the white-box semantic adversarial attacks in detail here, subsection 1.4.3 is concerned with a spatial transformation attack, a box–box semantic adversarial attack.

The shadow attack [30] is a type of "semantic" adversarial attack that is different from those introduced in subsections 1.3.1–1.3.6, as its targets is not only the classifier's output label but also the certifiable defenses [31, 32]. This attack reveals that certifiable defenses are not inherently secure. Furthermore, shadow attacks can also be considered as a generalization of PGD attacks with three penalty terms that minimize the perceptibility of perturbations. Formally, shadow attacks are defined as follows, with three penalty terms added to the loss term.

Definition 8 (Shadow Attack by Ghiasi et al. 2020): The shadow attack generates the adversarial example by solving the following optimization problem:

$$\max_{\delta} L(x+\delta,\theta) - \lambda_{tv} TV(\delta) - \lambda_c C(\delta) - \lambda_s D(\delta), \tag{1.3.20}$$

where θ is the model weights, x is an arbitrary natural image, δ is the perturbation added to image x, L is a loss function, $\lambda_c > 0, \lambda_{tv} > 0$, and $\lambda_s > 0$ are scalar penalty weights. $TV(\delta)$ is the total variation of δ, $C(\delta) = \left\|Avg\left(\left|\delta_R\right|\right)\right\|_p + \left\|Avg\left(\left|\delta_B\right|\right)\right\|_p + \left\|Avg\left(\left|\delta_G\right|\right)\right\|_p$ restricts the change in the mean of each color channel, and $D(\delta) = \left\|\delta_R - \delta_B\right\|_p + \left\|\delta_R - \delta_G\right\|_p + \left\|\delta_G - \delta_B\right\|_p$ promotes even perturbation between the color channels.

Note that the total variation $TV(\delta)$ is calculated element-wise as $TV(\delta_{i,j}) = anisotropic - TV(\delta_{i,j})^2$ [30], where $anisotropic - TV(\delta_{i,j})^2$ estimates the L_1-norm of the gradient of $\delta_{i,j}$.

1.3.8 Wasserstein Attack

Different from most adversarial attacks that focus on the L_p-norm, the Wasserstein attack uses the Wasserstein distance (also known as the optimal transport, Kantorovich distance, or Earth mover's distance) to generate Wasserstein adversarial examples. L_p-norms, though used extensively in adversarial attacks, do not capture the image transforms such as rotation, translation, distortion, flipping, and reflection.

Definition 9 (Wasserstein Distance between Two Inputs x and y): Let $x \in R_+^n$ and $y \in R_+^n$ be two inputs that are normalized with $\sum_{i=1}^n x_i = 1$ and $\sum_{j=1}^n x_j = 1$. Furthermore, let $C \in R^{n \times n}$ be a cost matrix such that each element $C_{i,j}$ of C measures the cost of moving mass from x_i to y_j (i.e., the Euclidean distance between the pixel x_i, and the pixel y_j, if x and y are images). Then the Wasserstein distance between two inputs x and y is:

$$d_W(x,y) = \langle T,C \rangle, \tag{1.3.21}$$

subject to $T1 = x$, $T^T1 = y$, where $1 \in R^{n \times 1}$ is a column vector of ones, $T \in R^{n \times n}$ is a transport plan whose element $T_{i,j} > 0$ encodes how the mass moves from x_i to y_j and $\langle T,C \rangle$ denotes the sum of the matrix–matrix element-wise multiplication. That is, $\langle T,C \rangle = \sum_{i=1}^n \sum_{j=1}^n T_{ij} * C_{ij}$.

Probabilistically, you can think of a transport plan as a joint probability of x and y [33]. Although the Wasserstein distance has many theoretical advantages, calculating a Wasserstein distance between two images is computationally expensive as it requires solving a linear programming problem with $n \times n$ number of variables. Wong et al. [34] developed a fast approximate projection algorithm, called the projected Sinkhorn iteration, that modifies the standard PGD attack by projecting onto a Wasserstein ball. Formally, Wong et al. proposed to solve the following objective function.

Definition 10 (Wasserstein Adversarial Example by Wong et al. 2019). The Wasserstein attack generates a Wasserstein adversarial example $x' \in R^n$ by minimizing the following objective function:

$$\|x - x'\|_2^2 + \frac{1}{\lambda} \sum_{ij} T_{ij} \log \log (T_{ij}), \tag{1.3.22}$$

subject to the Wasserstein ball constraints, $T1 = x$, $T^T1 = x'$, and $d_W(x,y) < \epsilon$, where $\lambda \in R$.

Note that an entropy-regularized term, $\sum_{ij} T_{ij} \log\log(T_{ij})$, on the transport plan T, is included in the objective function so that projection is approximated; however, this entropy-regularized term speeds up the computation (near-linear time). To further reduce the computational cost, the local transport plan restricts the movement of each pixel to within the $k \times k$ region of its original location. For instance, Wong et al. [34] set $k = 5$ for all their experiments.

1.4 BLACK-BOX ADVERSARIAL ATTACKS

Recent research on the adversarial attack has shown particular interest in black-box settings. Although many attacks require the attacker to have full access to the network architecture and weight, this information is often protected in commercialized products such as the vision system in modern self-driving vehicles. As a result, the black-box attack, in which an attacker has no knowledge of the network and the training set except the ability to query the probability/confidence score for each class (score-based black-box attack) or the label (decision-based black-box attack), is increasingly important to investigate. Furthermore, since black-box attack models do not require the knowledge of gradients as in most white-box attacks introduced in section 1.3, the defense strategies based on the masking gradient or nondifferentiability do not work with the attacks introduced in this section.

In section 1.4.1, we introduce a transfer attack, a popular black-box attack that utilizes the transferability of adversarial examples. This attack can even be applied to no-box settings, in which an attacker does not even have access to the output of a model. In section 1.4.2, we present two score-based black-box attacks, zeroth-order-optimization- (ZOO-) based attack and square attack. Then we discuss three popular decision-based black-box attacks in section 1.4.3.

1.4.1 Transfer Attack

Szegedy et al. [6] highlight the transferability of adversarial examples through the generalization power of DNNs. That is, the same adversarial image can be misclassified by a variety of classifiers with different architectures or trained on different training data. This property is useful in black-box settings. When an attacker does not have knowledge of model architecture, they can generate a substitute model to imitate the target model and then apply white-box attack methods such as a L-BGFS attack to the obtained substitute model in order to generate an adversarial example. This type of attack is called a transfer attack.

To do a transfer attack, an attacker needs to have a labeled training set first. With free query access, an attacker can feed the substitute training set to the target classifier in order to obtain the label for these instances. In the extreme case of limited query access, attackers may have to do the labeling on their own or find a substitute training set that is already labeled. However, with the power of free query access, the generated model is usually more representative of the target model. With a labeled substitute training set, attackers can select a substitute model that has a similar structure as the target model if they have any knowledge of the underlying structure. For instance, an image classifier usually has multiple CNN layers, whereas a sequential model may have some sort of RNN layers. Then they can train the selected substitute model with the substitute training set. This trained substitute model serves as an attack surrogate. Finally, white-box attack methods introduced in section 3.3 can be used to generate adversarial examples. In order to have a high transfer attack success rate, FGSM, PGD, C&W, or other attack methods with high transferability are selected to generate adversarial examples. Furthermore, the adversarial perturbation required is usually somewhat larger in order to ensure the success of a transfer attack.

1.4.2 Score-based Black-box Attacks

In this section, we discuss two popular score-based attacks: ZOO and square attacks. The ZOO attack is to find adversarial examples based on zero-order optimization, whereas the square attack generates adversarial examples by using derivative-free optimization (DFO). The detailed discussions of the two attacks follow.

ZOO Attack

Instead of using transfer attacks to exploit the transferability of adversarial images, Chen et al. [35] proposed a ZOO attack to directly approximate the gradients of the target model using confidence scores. Therefore, the ZOO attack is considered a score-based black-box attack, and it does not need to train a substitute model. The ZOO attack is as effective as the C&W attack and markedly surpasses existing transfer attacks in terms of success rate. Chen et al. [35] also proposed a general framework utilizing capable gradient-based white-box attacks for generating adversarial examples in the black-box setting.

The ZOO attack finds the adversarial example by also solving the optimization problem (1.3.19). Motivated by the attack loss functions (1.3.6.2) used in the C&W attack, a new hinge-like loss function [35] based on the log probability score vector $p = f(x')$ of the model f, instead of Z, is proposed as follows:

$$g(x') = \begin{cases} max\left\{ \max_{j \neq t} \left[f(x') \right]_j - \left[f(x') \right]_t, -\kappa \right\}, & target\ on\ t \\ max\left\{ \left[f(x') \right]_y - \max_{j \neq y} \left[f(x') \right]_j, -\kappa \right\}, & untargeted, \end{cases}$$

(1.4.1)

where $\left[f(x') \right]_j$ is the jth element of the probability score vector, and the parameter $\kappa \geq 0$ ensures a constant gap between the log probability score of the adversarial example classified as class t and all remaining classes. The log probability score is used instead of a probability score since well-trained DNNs yield a significant high confidence score for a class compared to other classes, and the log function lessens this dominance effect without affecting the order of confidence score. The ZOO attack is defined as follows.

Definition 11 (ZOO Adversarial Attack by Chen et al. 2017): A ZOO attack is a score-based black-box attack that generates the adversarial example x' by solving the following optimization problem using the zeroth-order stochastic coordinate descent with a coordinate-wise ADAM.

$$\min_{x'} \| x' - x \|_p + c \cdot g(x')$$

subject to the constraint $x' \in [0,1]^n$.

In the white box setting, finding an adversarial example requires taking the gradient of the model function $\partial f(x)$. However, in the black-box setting, the gradient is inadmissible, and one can use only the function evaluation $f(x)$, which makes it a zeroth-order optimization problem. Chen et al. [35] provided a method for estimating the gradient information surrounding the victim sample x by watching variations in prediction confidence $f(x)$ when the coordinate values of x are adjusted [36]. Chen et al. [35] used the following symmetric difference quotient to approximate the gradient:

$$\frac{\partial f(x)}{\partial x_i} \approx \frac{f(x + he_i) - f(x - he_i)}{2h}$$

(1.4.2)

where h is a small constant, and e_i is a standard basis vector. For networks with a large input size n, e.g., Inception-v3 network [11] with $n = 299 \times 299 \times 3$, the number of model queries per gradient evaluate is $2 \times n$ (two function evaluations per coordinate-wise gradient estimation). This is very query inefficient.

To overcome this inefficiency, Chen et al. [29] proposed the five acceleration techniques.

1. Use zeroth-order stochastic coordinate descent with a coordinate-wise ADAM to minimize the objective function.

2. Evaluate the objective function in batches instead of one by one to take advantage of the GPU.
3. Reduce the attack space and perform the gradient estimation from a dimension-reduced space R^m, where $m < n$, to improve computational efficiency: specifically, a dimension reduction transformation $D : R^m \rightarrow R^n$ such as $D(\delta') = x' - x$, where $\delta' \in R^m$ is used to reduce the attack space. For instance, D can be an upscale operator that scales up an image, such as the bilinear interpolation method, discrete cosine transformation (DCT), or an autoencoder for attack dimension reduction [37]. Although this attack space dimension reduction method reduces the computation costs for high-resolution image datasets (such as ImageNet [38]), the reduced attack space also limits the ability to find a valid attack.
4. Use hierarchical attack to overcome the limited search space issue. Instead of just one transformation, a series of transformations $D_1, D_2, ..., D_v$ with dimensions $m_1, m_2, ..., m_v$, where $m_1 < m_2 < m_v$ are used to find the adversarial example. Starting with the small attack space m_1, attackers try to find an adversarial example. If they are unsuccessful, then they increase the attack space to m_2. Otherwise, the process stops.
5. Choose which coordinates to update based on their "importance." For instance, the pixel near the edge or corner of an image is less important than a pixel near the main object (usually near center) for an image classifier. Chen et al. [35] divided an image into 8×8 regions and defined the "importance" as the absolute pixel values change in each region.

Note that acceleration techniques 3–5 are not required when n is small. For instance, Chen et al. [35] did not use techniques 3–5 for the MNIST and CIFAR10 dataset. Although the attack success rate is comparable to the success rate of the C&W attack, the number of required queries is large for gradient estimation despite the proposed acceleration techniques.

Square Attack

Andriushchenko et al. [26] proposed a square attack (SA), a query-efficient attack on both L_∞ and L_2-bounded adversarial perturbations. The SA is based on a random search (RS), an iterative technique in DFO methods. Therefore, the SA is a gradient-free attack [39], and it is resistant to gradient masking. SA improves the query efficiency and success rate by employing RS and a task-specific sampling distribution. In some cases, SA even competes with white-box attacks' performance.

Compared to an untargeted attack, SA is targeted to find a L_p-norm bounded adversarial example by solving the following box-constrained optimization problem:

$$\min_{x'} L\big(f(x',y)\big) \tag{1.4.3}$$

subject to $\| x' - x \|_p \le \epsilon$ where $p \in \{2, \infty\}$, $x' \in [0,1]^n$ and $L\big(f(x',y)\big) = \big[f(x')\big]_y - \max_{j \ne y} \big[f(x')\big]_j$ is a margin-based loss function measuring the level of confidence with which the model f labels an adversarial input x' *as* the ground-truth class y over other classes.

In a targeted attack with target class t, an attacker could simply minimize the loss function

$$L\big(f(x',t)\big) = \max_{j \ne t} \big[f(x')\big]_j - \big[f(x')\big]_t . \tag{1.4.4}$$

However, a cross-entropy loss of the target class t

$$L\big(f(x', t)\big) = -f_t(x') + \log\left(\sum_{i=1}^{K} e^{f_i(x)}\right) \tag{1.4.5}$$

is used to make the targeted attack more query efficient.

This optimization problem can be solved by using the classical RS method. In general, the RS method can be described as follows. Given an objective function g to minimize, a starting point x_0, and a sampling distribution D, the RS algorithm outputs an estimated solution x_{out} after N iterations. For each iteration i, the RS algorithm starts with a random update $\delta \sim D(x_i)$ and then adds this update to the current iteration; i.e., $x_{new} = x_i + \delta$. If the objective function g_{new} evaluated at x_{new} is less than the best g^* obtained so far, we update x_i and g^*. That is, if $g_{new} < g^*$, then $x_i = x_{new}$ and $g^* = g_{new}$. Otherwise, the update δ is discarded [40]. This process stops after a user-specified number of iterations has been reached. Thus using RS to optimize g does not rely on the gradient of g.

Since Moon et al. [41] showed that when one runs the white-box PGD attacks until convergence on CIFAR10 and ImageNet datasets, the successful adversarial perturbation δ is mainly found on vertices (corners) of L_∞-norm ball [26]. Based on this observation, the SA attack only searches over the boundaries of the L_∞ or L_2-ball to obtain the potential adversarial perturbation of the current iterate δ_i. Hence, the perturbation for each pixel (before projection) is either $-2\epsilon, 0$, or 2ϵ for the L_∞-norm attack, where the perturbation of zero means no perturbation for a given pixel. This critical design makes the SA mechanism different from the standard RS algorithm, and it significantly improves the query efficiency.

Another key difference is that at each step the SA attack limits the modification of the current iterate x_i' by updating only a small fraction (a square of side length v_i for each color channel) of the neighboring pixels of x_i'. This helps reduce the dimension of the search space, especially when the input space $[0,1]^n$ is high-dimensional, e.g., $n = 290 \times 290 \times 3$ in the ImageNet dataset [42].

In summary, the SA via random search consists of the following main steps for each iteration i[40]:

1. Find the side length v_i of the square by determining the closest positive integer to $\sqrt{p * w^2}$, where $p \in [0,1]$ is the percentage of pixels of the original image x that can be perturbed and w is the width of an image. p gradually decreases with the iterations, but $v_i \geq 3$ for the L_2-norm SA. This mimics the step size reduction in gradient-based optimization, in which we begin with initial large learning rates to quickly shrink the loss value [40].
2. Find the location of the square with side length v_i for each color channel by uniformly random selection. The square denotes the set of pixels that can be modified.
3. Uniformly assign all pixels' values in the square to either $-2 \times \epsilon$ or $+2 \times \epsilon$ for each color channel c.
4. Add the square perturbation generated in step 3 to the current iterate x_i to obtain the new point x_{i+1}.
5. Project x_{i+1} onto the intersection of $[0,1]^n$ and the L_∞-norm ball of radius ϵ to obtain x_{new}.
6. If the new point x_{new} attains a lower loss than the best loss so far, the change is accepted, and the best loss is updated. Otherwise, it is discarded.

The iteration continues until the adversarial example is found. For those who are interested, see [26].

1.4.3 Decision-based Attack

The decision-based attack uses only the label of an ML output for generating adversarial attacks, and this makes it easily applicable to real-world ML attacks. In this section, we discuss three popular decision-based black-box attacks: a boundary attack [43], a HopSkipJump attack [44], and a spatial transformation attack [4].

Boundary Attack

Relying neither on training data nor on the assumption of transferability, the boundary attack uses a simple rejection sampling algorithm with a constrained independent and identically distributed Gaussian distribution as a proposed distribution and a dynamic step-size adjustment inspired by Trust Region methods to generate minimal perturbation adversarial samples. The boundary attack algorithm is given as follows. First, a data point is sampled randomly from either a maximum entropy distribution (for an untargeted attack) or a set of data points belonging to the target class (for a targeted attack). The selected data point serves as a starting point. At each step of the algorithm, a random perturbation is drawn from a proposed distribution such that the perturbed data still lies within the input domain, and the difference between the perturbed image and the original input is within the specified maximum allowable perturbation ϵ. Newly perturbed data is used as a new starting point if it is misclassified for an untargeted attack (or misclassified as the target class for a targeted attack). The process continues until the maximum number of steps is reached.

The boundary attack is conceptually simple, requires little hyperparameter tuning, and performs as well as the state-of-the-art gradient attacks (such as the C&W attack) in both targeted and untargeted computer vision scenarios without algorithm knowledge [43]. Furthermore, it is robust against common deceptions such as gradient obfuscation or masking, intrinsic stochasticity, or adversarial training. However, the boundary attack has two main drawbacks. First, the number of queries for generating an adversarial sample is large, making it impractical for real-world applications [44]. Instead of a rejection sampling algorithm, Metropolis-Hastings's sampling may be a better option since it does not simply discard the rejected sample. Secondly, it only considers L_2-norm.

HopSkipJump Attack

Conversely, the HopSkipJump attack is a family of query-efficient algorithms that generate both targeted and untargeted adversarial examples for both L_2 and L_∞-norm distances. Furthermore, the HopSkipJump attack is more query efficient than the Boundary attack [45], and it is a hyperparameter-free iterative algorithm. The HopSkipJump attack is defined as follows.

Definition 12 (HopSkipJump Attack by Chen et al. 2020): A HopSkipJump attack is a decision-based attack that generates the adversarial example x' by solving the following optimization problem:

$$\min_{x'} \left\| x' - x \right\|_p, \tag{1.4.6}$$

subject to the constraint $\phi_x(x') = 1$, where $p = \{2, \infty\}$,

$$\phi_x(x') = sign\left(S_{x^*}(x')\right) = \begin{cases} 1, & S_{x^*}(x') > 1 \\ -1, & otherwise, \end{cases} \tag{1.4.7}$$

and

$$S_{x^*}(x') = \begin{cases} \left[f(x') \right]_t - \max_{j \neq t} \left[f(x') \right]_j, & target\ on\ t \\ \max_{j \neq y} \left[f(x') \right]_y - \left[f(x') \right]_y, & untargeted. \end{cases} \tag{1.4.8}$$

Note that S_{x^*} is similar to (1.3.18) with $\kappa = 0$. Let us discuss a HopSkipJump L_2-based target attack in detail. Interested readers can consult [44] for untargeted attacks or L_2 based attacks.

The HopSkipJump L_2-based target attack is an iterative algorithm consisting mainly of an initialization step (step 1) and three iterative steps (steps 2–4, which are repeated until the maximum number of iterations specified by an attacker is reached):

1. Select a random data point $\tilde{x}_0$ from a target class (similar to the initialization step of a boundary attack).
2. Approach the decision boundary/hyperplane by using a binary search algorithm. That is, move $\tilde{x}_t$ toward the input x and the decision hyperplane by interpolating between the input x and $\tilde{x}_t$ to get a new data point x_t such that $\phi_x(x_t) = 1$; i.e.,

$$x_t = \alpha_t x + (1 - \alpha_t) \tilde{x}_t, \tag{1.4.9}$$

where α_t is obtained via a binary search between 0 and 1 and stops at x_t such that $\phi_x(x_t) = 1$.
3. Estimate the gradient direction using the method similar to FGSM (1.3.8); i.e.,

$$v_t = \frac{\widehat{\nabla S}(x_t, \delta_t, B_t)}{\left\| \widehat{\nabla S}(x_t, \delta_t) \right\|_2} \tag{1.4.10}$$

where $\delta_t = \dfrac{\left\| x_t - x \right\|_2}{n}$ depends on the distance between x_t and x, the image size n, $\{u_b\}_{b=1}^{B_t}$ is a set of independent and identically distributed uniform random noise, $B_t = B_0 \sqrt{t}$ is the batch size of $\{u_b\}_{b=1}^{B_t}$, the initial batch size B_0 is set to 100 in [53], and

$$\widehat{\nabla S}(x_t, \delta_t, B_t) = \frac{1}{B_t - 1} \sum_{b=1}^{B_t} [\phi_x(x_t + \delta_t u_b) - \frac{1}{B_t} \sum_{b=1}^{B_t} \phi_x(x_t + \delta_t u_b)] u_b \tag{1.4.11}$$

4. Estimate the step size ξ_t by geometric progression, and then update the sample point $\tilde{x}_t$. That is, starting with $\xi_t = \dfrac{\left\| x_t - x \right\|_2}{\sqrt{t}}$, the step size is reduced by half until $\phi_x(\tilde{x}_t) = 1$, where

$$\tilde{x}_t = x_t + \xi_t v_t = x_t + \xi_t \frac{\widehat{\nabla S}(x_t, \delta_t, B_t)}{\left\| \widehat{\nabla S}(x_t, \delta_t) \right\|_2}. \tag{1.4.12}$$

Spatial Transformation Attack

In contrast to the boundary attack and the HopSkipJump attack that are based on L_p bounded adversarial perturbations, the spatial transformation attack [4] is a black-box "semantic" adversarial attack that reveals a small random transformation, such as a small rotation on an image, and can easily fool a state-of-the-art image classifier. This attack really questions the robustness of these state-of-the-art image classifiers. Furthermore, the primary disadvantage of the ZOO attack is the need to probe the classifiers thousands of times before the adversarial examples are found [27]. This is not a case for spatial transformation attacks. Engstrom et al. [4] showed that the worst-of-10 ($k = 10$) is able to reduce the model accuracy significantly with just 10 queries. Basically, it rotates or translates a natural image slightly to cause misclassification. This attack reveals the vulnerability of the current state-of-the-art ML models.

Definition 13 (Spatial Transformation Attack): Let $x \in R^n$ be a legitimate input image that is correctly classified as class y by an image classifier. A spatial transformation attack generates an adversarial example $x' \in R^n$ by finding a mapping $\alpha : R^n \to R^n$ such that the adversarial example $\alpha(x) = x'$, where the relationship between x' and x is as follows:

Each pixel (a,b,c) in x is translated $(\delta a, \delta b, 0)$ pixels to the right and rotate γ degrees counterclockwise around the origin to obtain the pixel (a',b',c) in x', where $a' = a * \mathrm{coscos}(\gamma) - b * \mathrm{sinsin}(\gamma) + \delta a$ and $b' = a * \mathrm{sinsin}(\gamma) + b * \mathrm{coscos}(\gamma) + \delta b$.

To find the mapping α, an attacker needs to solve the following optimization problem:

$$\max_{a,b,c} L\big(\alpha(x), y\big), \tag{1.4.13}$$

where L is the loss function.

Engstrom et al. [4] proposed three different methods to solve (1.4.13): (1) a first-order method to maximize the loss function, (2) a grid search on a discretized parameter space, and (3) a worst-of-k (i.e., randomly sample k different set of attack parameters, and choose one that the model performs worst). The first-order method requires a gradient of the loss function, which is not possible in the black-box setting. Even if it is possible, Engstrom et al. [4] showed that the first-order method performs worst due to the nonconcavity of loss function for spatial perturbation.

1.5 DATA POISONING ATTACKS

While adversarial attacks cannot change the training process of a model and can only modify the test instance, data poisoning attacks, on the contrary, can manipulate the training process. Specifically, in data poisoning attacks, attackers aim to manipulate the training data (e.g., poisoning features, flipping labels, manipulating the model configuration settings, and altering the model weights) in order to influence the learning model. It is assumed that attackers have the capability to contribute to the training data or have control over the training data itself. The main objective of injecting poison data is to influence the model's learning outcome.

Recent studies on adversarial ML have demonstrated particular interest in data poisoning attack settings. This section discusses a few of the data poisoning attack models. We start with briefly going over label flipping attacks in subsection 1.5.1 and then focus on clean label data poisoning attacks in subsection 1.5.2 since they are stealthy. To that end, we introduce backdoor attacks.

1.5.1 Label Flipping Attacks

A simple and effective way to attack a training process is to simply flip the labels of some training instances. This type of data poisoning attack is called label flipping attacks. Mislabeling can be done easily in crowdsourcing, where an attacker is one of the annotators, for example. In this subsection, we discuss some common label flip attacks against ML models. These label flip attacks could be either model independent or model dependent. For instance, a random label flipping attack is model independent. It simply selects a subset of training instances and flips their labels. The attacker does not need to have any knowledge of the underlying target ML model. Even though this random strategy looks simple, it is capable of reducing classification accuracy significantly depending on the type of dataset under attack, the training set size, and the portion of the training labels that are flipped. The random label flip attack can be mathematically defined as follows.

Definition 14 (Random Label Flipping Attack): Given a training set $\left\{x_i, y_i\right\}_{i=1}^n$, where $x_i \in X$ and $y_i \in \{-1,1\}$, the attack is called a random label flipping attack if an attacker with the ability to change the $p \in (0,1)$ fraction of the training label can randomly select np training labels and flip the labels.

This random label flipping attack can further be divided into two groups: targeted and untargeted. In an untargeted random label flipping attack, an attacker may select some instances from class 1 to misclassify as class -1 and some instances from class -1 to misclassify as class 1. In contrast, an attacker misclassifies one class as another consistently in a targeted random label flipping attack. The targeted random label flipping attack is more severe compared to the untargeted one as the targeted attack consistently misleads the learning algorithm to classify a specific class of instances as another specific class.

Rather than random label flipping, an attacker can also utilize label flip attacks that are model dependent. For instance, in subsection 1.2.2, we showed that SVM constructs a decision hyperplane using only the support vectors. Existing studies presented a few label flipping attacks based on this characteristic. For example, a Farfirst attack [46] is such a label flipping attack, where a training instance far away from the margin of an SVM is flipped. This attack effectively changes many nonsupport vector training instances (training instances that are far from the margin and are correctly classified by an untainted SVM) to support vectors and significantly alter the decision boundary consequently.

Formally, an optimal label flipping attack can be considered a bilevel optimization problem defined as follows.

Definition 15 (Optimal Label Flipping Attack): Suppose that training set $D = \left\{x_i, y_i\right\}_{i=1}^n$ and a test set $T = \left\{\tilde{x}_i, \tilde{y}_i\right\}_{i=1}^m$, where $x_i \in X, \tilde{x}_i \in X,\ y_i \in \{-1,1\}$, and $\tilde{y}_i \in \{-1,1\}$. Let l be the number of training labels that an attacker has the ability to modify, and let $I \in \{-1,1\}^n$ be an indicator vector such that its ith element is I_i equal to -1 if the ith training label is flipped and 1 otherwise. Then $|I| = l$ and the poisoned training set $D' = \left\{x_i, y_i'\right\}_{i=1}^n$, where $y_i' = y_i I_i$. It is called an optimal label flipping attack if an attacker can find an optimal I by solving the following optimization problem:

$$O(\min_{\Theta} L(D',\Theta), T), \tag{1.5.1}$$

where O is an objective function specified by the attacker. A common objective function is to maximize the test error rate, in which case $O(\Theta^*, T) = \dfrac{1}{m}\sum_{i=1}^m J\left[C(\tilde{x}_i) = \tilde{y}_i\right]$, where J is an indicator function such that

$$J\left[C(\tilde{x}_i) = \tilde{y}_i\right] = \begin{cases} 1, & C(\tilde{x}_i) = \tilde{y}_i \\ 0, & otherwise \end{cases}. \tag{1.5.2}$$

However, solving such a bilevel optimization problem is NP-hard. Here, we introduce a simple greedy algorithm that is suboptimal but tractable. At iteration $t = 1$, an attacker initially flips the label of the first training instance to obtain a poisoned training set D'. Then the attacker trains an ML model using D' and evaluates the performance p_1 of the poisoned model. Next, the attacker flips the label of the second training instance while keeping other labels untainted. Similarly, the attacker trains the ML model using the poisoned dataset and evaluates its performance p_2 on the test set. The attacker repeats this process for each training label to obtain $\{p_1, p_2, \ldots, p_n\}$. Now, the attacker can determine which training label is actually flipped by simply finding the training label corresponding to the worst performance. After that label is flipped, the attacker can select the next training label to flip out of the remaining $n-1$ training labels by following the same process as iteration 1. For each iteration, one additional training label is flipped. The process stops until l training labels have been flipped.

1.5.2 Clean Label Data Poisoning Attack

Security of DL networks can be degraded by the emergence of clean label data poisoning attacks. This is achieved by injecting legitimately labeled poison samples into a training set. Although poison samples seem normal to a user, they indeed comprise illegitimate characteristics to trigger a targeted misclassification during the inference process.

In this subsection, an attacker deploys "clean-labels" data poisoning attacks without the need to control the labeling process. Furthermore, such adversarial attacks are often "targeted" such that they allow for controlling the classifier's behavior on a particular test instance while maintaining the overall performance of the classifier [47]. For instance, an attacker can insert a seemingly legitimate image, i.e., a correctly labeled image, into a training dataset for a facial identification engine and thus control the chosen person's identity during the test time. This attack is severe since the poison instance can be placed online and waits to be scraped by a bot for data collection. The poison instance is then labeled correctly by a trusted annotator but is still able to subvert the learning algorithm.

Here, we introduce two clean-label data poisoning attacks against both transfer learning and an end-to-end training.

Feature Collision Attack

The first one is the feature collision attack introduced by Shafahi et al. [47] who assumed that an attacker has

- no knowledge of the training data,
- no ability to influence the labeling process,
- no ability to modify the target instance during inference, and
- knowledge of the feature extractor's architecture and its parameters.

An attacker tries to find a poison instance that collides with a given target instance in a feature space while maintaining its indistinguishability with a base instance from class c other than target class $t \neq c$. Hence, the generated poison instance looks like a base instance, and an annotator labels it as an instance from class c. However, that poison instance is close to the target instance in a feature space, and the ML model is likely classifying it as an instance from class t. This causes the targeted misclassification, and only one such poison instance is needed to poison a transfer learning model. That is why this attack is sometimes called a one-shot attack as well, and its formal definition is as follows.

Definition 16 (Feature Collision Attack by Shafahi et al. 2018): Given a feature extractor f, a target instance x_t from class t, and a base instance x_b that belongs to a targeted class b such that $b \neq t$, it is called a feature collision attack if an attacker finds a poison instance x_p as follows:

$$x_p = \operatorname*{argmin}_{x} \left\| f(x) - f(x_t) \right\|_2^2 + \beta \left\| x - x_b \right\|_2^2 , \tag{1.5.3}$$

where β is a similarity parameter that indicates the importance of the first component $\left\| f(x) - f(x_t) \right\|_2^2$ (the first component measures the similarity of the poison instance and the target instance in the feature space created by f) over the second component $\left\| x - x_b \right\|_2^2$, which measures the similarity of the poison instance and the base instance in the input space.

The base instance can be selected randomly from any classes other than the target class. However, some base instances may be easier for an attacker to find a poison instance than others. The coefficient β must be tuned by attackers in order for them to make the poison instance seem indistinguishable from a base instance. Shafahi et al. [47] solved the optimization problem by using a forward-backward-splitting iterative procedure [48].

This attack has a remarkable attack success rate (e.g., 100% in one experiment presented in [47]) against transfer learning. Nevertheless, we want to point out that such an impressive attack success rate reported in [47] is due to the overfitting of the victim model to the poison instance. The data poisoning attack success rate drops significantly on end-to-end training and in black-box settings, and Shafahi et al. [47] proposed to use a watermark and multiple poison instances to increase the attack success rate on end-to-end training. However, one obvious drawback is that the pattern of the target instance sometimes shows up in the poison instances.

Convex Polytope Attack and Bullseye Polytope Attack

Although in many ML-based attacks, an attacker is required to obtain full access to the network architecture and weight of a model (e.g., a feature collision attack), such a privilege is likely protected in commercialized products such as perception systems in smart cars. In this subsection, we introduce a transferable and scalable clean label targeted data poisoning attack against transfer learning and end-to-end training in a black-box setting, where attackers do not have access to the target model. Aghakhani et al. [49] proposed such a clean label data poisoning attack model, named the bullseye polytope attack, to create poison instances that are similar to a given target instance in a feature space. The bullseye polytope attack achieves a higher attack success rate compared to a feature collision attack in a black-box setting and increases the speed and success rate of the targeted data poisoning attack on end-to-end training compared to the convex polytope attack.

The convex polytope attack [50] is very similar to the bullseye polytope attack; both of them are extensions of a feature collision attack into a black-box setting. Let us define the convex polytope attack formally as follows before we discuss the bullseye polytope attack, a scalable alternative to the convex polytope attack.

Definition 17 (Convex Polytope Attack by Zhu et al. 2019): Given a set of feature extractors $\left\{ f^i \right\}_{i=1}^m$, a target instance x_t from class t and a set of base instances $\left\{ x_b^j \right\}_{j=1}^k$ that belong to class b such that $b \neq t$, the attack is called a convex polytope attack if an attacker finds a set of poison instances $\left\{ x_p^j \right\}_{j=1}^k$ by solving the following optimization problem:

$$\min_{\substack{\{x_p^j\}_{j=1}^k \\ \{c^i\}_{i=1}^m}} \frac{1}{2} \sum_{i=1}^m \left(\frac{\left\| f^i(x_t) - \sum_{j=1}^k c_j^i f^i\left(x_p^{(j)}\right) \right\|^2}{\left\| f^i(x_t) \right\|^2} \right), \tag{1.5.4}$$

subject to the constraints $\sum_{j=1}^k c_j^i = 1, c_j^i \geq 0, \forall i = 1,2,...,m, \forall j = 1,2,...,k,$ and $\left\| x_p^j - x_b^j \right\|_\infty \leq \epsilon, \forall j = 1,2,...,k,$ where $c^i \in R^k$ is a coefficient vector consisting of elements c_j^i.

Without the knowledge of a victim model, an attacker can utilize the attack transferability to attack the victim model by substituting feature extractors $\left\{f^i\right\}_{i=1}^m$ (similar to transfer attacks in section 1.4.1). This is the main idea behind both the convex polytope attack and the bullseye polytope attack. The main difference between the two attacks is that there is no incentive to refine the target deeper inside the convex hull of poisons ("attack zone") for the convex polytope attack, whereas the bullseye polytope attack instead pushes the target into the center of the convex hull by solving the optimization problem (1.5.5). Consequently, the target is often near the convex hull boundary in the convex polytope attack. The optimization problem (1.5.4) requires finding a set of coefficient vectors $\left\{c^i\right\}_{i=1}^m$ in addition to finding a set of poisons $\left\{x_p^j\right\}_{j=1}^k$ that minimize its objective function. Hence the convex polytope attack has a much higher computation cost than that of the bullseye polytope attack. As shown next, the bullseye polytope attack simply sets $c_j^i = \frac{1}{k}, \forall i = 1, 2, \ldots, m, \forall j = 1, 2, \ldots, k$, to assure that the target remains near the center of the poison samples.

Definition 18 (Single-target Mode Bullseye Polytope Attack by Aghakhani et al. 2020): Given a set of feature extractors $\left\{f^i\right\}_{i=1}^m$, a target instance x_t from class t, and a set of base instances $\left\{x_b^j\right\}_{j=1}^k$ that belong to class b such that $b \neq t$, the attack is called a single-target mode bullseye polytope attack if an attacker finds a set of poison instances $\left\{x_p^j\right\}_{j=1}^k$ by solving the following optimization problem:

$$\min_{\left\{x_p^j\right\}_{j=1}^k} \frac{1}{2} \sum_{i=1}^m \left(\frac{\left\| f^i(x_t) - \frac{1}{k} \sum_{j=1}^k f^i\left(x_p^{(j)}\right) \right\|^2}{\left\| f^i(x_t) \right\|^2} \right), \tag{1.5.5}$$

subject to the constraints $\left\| x_p^j - x_b^j \right\|_\infty \leq \epsilon, \forall j = 1, 2, \ldots, k,$ where ϵ determines the maximum allowed perturbation.

The bullseye polytope attack can be further extended to a more sophisticated target mode (called the multitarget mode) by constructing a set of poison images that targets multiple instances enclosing the same object from different angles or lighting conditions. Specifically, if n_t target instances of the same object are used in a multitarget mode attack, $f^i(x_t)$ in (1.5.5) is replaced by the average target feature vectors $\sum_{v=1}^{n_t} f_v^i(x_t)$. Such a model extension can remarkably enhance the transferability of the attack to unseen instances comprising the same object without having to use additional poison samples [49].

1.5.3 Backdoor Attack

In this last subsection, we introduce backdoor attacks, a type of data poisoning attacks/causative attacks that contain a backdoor trigger (an adversary's embedded pattern). Specifically, we consider backdoor attacks on an outsourcing scenario [51], a transfer learning scenario [51], and a federated learning scenario [52]. Here are the three attack scenarios:

- *Outsourced training*: In this attack scenario, a user aims to train the parameters Θ of a network f_Θ by using a training dataset. For this purpose, the user sends the model description to the trainer who will then return trained parameters Θ'. The user herein verifies the trained model's parameters by checking the classification accuracy on a "held-out" validation dataset, D_{valid}, as the trainer cannot be fully trusted. The model will be accepted only if its accuracy fulfills a target accuracy, $a*$, on the validation set. Building on this setting, a malicious trainer will return an illegitimate backdoored model $f_{\Theta_{adv}}$ to the user such that $f_{\Theta_{adv}}$ meets the target accuracy requirement while $f_{\Theta_{adv}}$ misclassifies the backdoored instances.

- *Transfer learning*: In this attack scenario, a user unintentionally downloads an illegitimately trained model $f_{\Theta_{adv}}$ and the corresponding training and validation set D_{valid} from an online repository. Then the user develops a transfer learning model $f_{\Theta_{adv},TL}$ based on it. The main attacker's goal is similar to his or her goal in the adversarial outsourced training scenario. However, both $f_{\Theta_{adv}}$ and $f_{\Theta_{adv},TL}$ must have reasonably high accuracy on D_{valid} and the user's private validation set for the new domain of application.

- *Federated learning*: This attack scenario is especially vulnerable to a backdoor attack since, by design, the central server has no access to the participants' private local data and training. A malicious participant or a compromised participant can manipulate the joint model by providing a malicious interned update to the central server based on the techniques such as contrain-and-scale [52]. As a result, the joint model would behave in accordance with the attacker's goal as long as the input encloses the backdoor features while the model can maintain the classification accuracy on the clean test instances.

An attacker can generate an illegitimately trained DL network, also known as a backdoored neural network (BadNet), for both targeted or untargeted misclassification on backdoor instances by modifying the training procedure, such as injecting poisoned instances with a backdoor trigger superimposed and the label altered, manipulating the model configuration settings, or altering model parameters directly. Subsequently, the malicious learning network behaves wrongly on backdoor instances but has high classification accuracy on the user's private validation set.

Although various detection mechanisms have been elaborated on adversarial backdoors detection, e.g., using statistical differences in latent representations between clean input data and attacker's input samples in a poisoned model, we introduce an adversarial backdoor embedding attack [53] that can bypass several detection mechanisms (e.g., Feature Pruning [54] and Dataset Filtering by Spectral Signatures [55]) simultaneously in this section. Notably, Shokri et al. [53] developed an adversarial training mechanism, which creates a poisoned model that can mitigate either a specific defense or multiple defenses at the same time by utilizing an objective function that penalizes the difference in latent representations between the clean inputs and backdoored inputs.

The proposed mechanism maximizes the indistinguishability of both clean and poisoned data's hidden representations by utilizing the idea similar to a generative adversarial network. The poisoned model decouples the adversarial inputs comprising the backdoor triggers from clean data by developing a discriminative network. The conventional backdoor detection mechanisms may succeed in recognizing the backdoor triggers only if the attacker naively trains the poisoned model in a way that leaves a remarkable difference in the distribution of latent representations between the clean data and backdoor instances. Therefore, a knowledgeable attacker can make the model more robust against existing backdoor detection mechanisms by minimizing the difference in latent representations of two. As a result, the attacker can attain high accuracy of classification by the poisoned model (the behavior of the poisoned model remains unmodified on clean data), while fooling the backdoor detection mechanism.

1.6 CONCLUSIONS

As ML systems have been dramatically integrated into a broad range of decision-making-sensitive applications for the past years, adversarial attacks and data poisoning attacks have posed a considerable threat against these systems. For this reason, this chapter has focused on the two important areas of ML security: adversarial attacks and data poisoning attacks. Specifically, this chapter has studied the technical aspects of these two types of security attacks. It has comprehensively described, discussed, and scrutinized adversarial attacks and data poisoning attacks with regard to their applicability requirements and adversarial capabilities. The main goal of this chapter is to help the research community gain insights into and the implications of existing adversarial attacks and data poisoning attacks, as well as to increase the awareness of potential adversarial threats when developing learning algorithms and applying ML methods to various applications in the future.

Researchers have developed many adversarial attack and data poisoning attack approaches over the years, but we are not able to cover all of them due to the space limit. For instance, Biggio et al. [56] proposed a set of poisoning attacks against SVMs, where attackers aim to increase the SVM's test error in 2012. In this study, Biggio et al. also demonstrated that an attacker could predict the SVM's decision function change, which can then be used to construct malicious training data. Furthermore, Li et al. [57] developed a spatial transformation robust backdoor attack in 2020. Moreover, many defense mechanisms have been proposed as well. Those who are interested may see [28, 50, 54, 57–64].

ACKNOWLEDGMENT

We acknowledge National Science Foundation (NSF) for partially sponsoring the work under grants #1633978, #1620871, #1620862, and #1636622, and BBN/GPO project #1936 through an NSF/CNS grant. We also thank the Florida Center for Cybersecurity (Cyber Florida) for a seed grant. The views and conclusions contained herein are those of the authors and should not be interpreted as necessarily representing the official policies, either expressed or implied of NSF.

NOTE

1 Lecture note: Chan, Stanley H. "Adversarial Attack." Machine Learning ECE595, April 2019, Purdue University. Portable document format (PDF) [23].

REFERENCES

[1] K. Ahmad, M. Maabreh, M. Ghaly, K. Khan, J. Qadir, and A. I. Al-Fuqaha, "Developing future human-centered smart cities: critical analysis of smart city security, interpretability, and ethical challenges," *arXiv preprint arXiv:2012.09110*, 2020.

[2] J. Qiu, Q. Wu, G. Ding, Y. Xu, and S. Feng, "A survey of machine learning for big data processing," *EURASIP Journal on Advances in Signal Processing*, vol. 2016, p. 85, 2016, doi: 10.1186/s13634-016-0382-7

[3]   K. He, X. Zhang, S. Ren, and J. Sun, "Delving deep into rectifiers: surpassing human-level performance on ImageNet classification," *Proceedings of the IEEE International Conference on Computer Vision (ICCV)*, pp. 1026–1034, 2015, doi: 10.1109/ICCV.2015.123

[4]   L. Engstrom, B. Tran, D. Tsipras, L. Schmidt, and A. Madry, "Exploring the landscape of spatial robustness," *Proceedings of the 36th International Conference on Machine Learning (ICML)*, vol. 97, pp. 1802–1811, 2019, doi: 10.1109/EuroSP.2016.36

[5]   H. Hosseini, Y. Chen, S. Kannan, B. Zhang, and R. Poovendran, "Blocking transferability of adversarial examples in black-box learning systems," *arXiv preprint arXiv:1703.04318*, 2017.

[6]   I. J. Goodfellow, J. Shlens, and C. Szegedy, "Explaining and harnessing adversarial examples," *arXiv preprint arXiv:1412.6572*, 2014.

[7]   N. Carlini and D. A. Wagner, "Audio adversarial examples: targeted attacks on speech-to-text," *IEEE Symposium on Security and Privacy (SP)*, pp. 1–7, 2018, doi: 10.1109/SPW.2018.00009

[8]   M. Sato, J. Suzuki, H. Shindo, and Y. Matsumoto, "Interpretable adversarial perturbation in input embedding space for text," *Proceedings of the 27th International Joint Conference on Artificial Intelligence (IJCAI)*, pp. 4323–4330, 2018, doi: 10.24963/ijcai.2018/601

[9]   I. Corona, G. Giacinto, and F. Roli, "Adversarial attacks against intrusion detection systems: Taxonomy, solutions and open issues," *Information Sciences*, vol. 239, pp. 201–225, 2013, doi: 10.1016/j.ins.2013.03.022

[10] N. Papernot, P. McDaniel, S. Jha, M. Fredrikson, Z. B. Celik, and A. Swami, "The limitations of deep learning in adversarial settings," *IEEE European Symposium on Security and Privacy (EuroS&P)*, pp. 372–387, 2016, doi: 10.1109/EuroSP.2016.36

[11] C. Szegedy, V. Vanhoucke, S. Ioffe, J. Shlens, and Z. Wojna, "Rethinking the Inception architecture for computer vision," *Proceedings of the IEEE Conference on Computer Vision and Pattern Recognition (CVPR)*, pp. 2818–2826, 2016, doi: 10.1109/CVPR.2016.308

[12] O. Alvear, C. T. Calafate, J.-C. Cano, and P. Manzoni, "Crowdsensing in smart cities: overview, platforms, and environment sensing issues," *Sensors*, vol. 18, p. 460, 2018, doi: 10.3390/s18020460

[13] M. Li, Y. Sun, H. Lu, S. Maharjan, and Z. Tian, "Deep reinforcement learning for partially observable data poisoning attack in crowdsensing systems," *IEEE Internet of Things*, vol. 7, pp. 6266–6278, 2019, doi: 10.1109/JIOT.2019.2962914

[14] Z. Huang, M. Pan, and Y. Gong, "Robust truth discovery against data poisoning in mobile crowdsensing," *Proceeedings of the IEEE Global Communications Conference (GLOBECOM)*, pp. 1–6, 2019, doi: 10.1109/GLOBECOM38437.2019.9013890

[15] C. Miao, Q. Li, H. Xiao, W. Jiang, M. Huai, and L. Su, "Towards data poisoning attacks in crowd sensing systems," *Proceedings of the 19th ACM International Symposium on Mobile Ad Hoc Networking and Computing (MobiHoc)*, pp. 111–120, 2018, doi: 10.1145/3209582.3209594

[16] C. Miao, Q. Li, L. Su, M. Huai, W. Jiang, and J. Gao, "Attack under disguise: an intelligent data poisoning attack mechanism in crowdsourcing," *Proceedings of the World Wide Web Conference*, pp. 13–22, 2018, doi: 10.1145/3178876.3186032

[17] C. Cortes and V. Vapnik, "Support-vector networks," *Machine Learning*, vol. 20, pp. 273–297, 1995.

[18] D. Hendrycks and K. Gimpel, "Bridging nonlinearities and stochastic regularizers with Gaussian error linear units," *arXiv preprint arXiv:1606.08415*, 2016.

[19] A. L. Maas, A. Y. Hannun, and A. Y. Ng, "Rectifier nonlinearities improve neural network acoustic models," *Proceedings of the International Conference on Machine Learning (ICML)*, vol. 30, p. 3, 2013.

[20] S.-M. Moosavi-Dezfooli, A. Fawzi, and P. Frossard, "DeepFool: a simple and accurate method to fool deep neural networks," *Proceedings of the IEEE Conference on Computer Vision and Pattern Recognition (CVPR)*, pp. 2574–2582, 2016, doi: 10.1109/CVPR.2016.282

[21] C. Szegedy *et al.*, "Intriguing properties of neural networks," *arXiv preprint arXiv:1312.6199*, 2013.

[22] D. C. Liu and J. Nocedal, "On the limited memory BFGS method for large scale optimization," *Mathematical Programming*, vol. 45, pp. 503–528, 1989, doi: 10.1007/BF01589116

[23] A. Kurakin, I. J. Goodfellow, and S. Bengio, "Adversarial examples in the physical world," *Proceedings of the 5th International Conference on Learning Representations (ICLR)*, 2016, doi: 10.1201/9781351251389-8

[24] F. Croce and M. Hein, "Reliable evaluation of adversarial robustness with an ensemble of diverse parameter-free attacks," *Proceedings of the 37th International Conference on Machine Learning (ICML)*, pp. 2206–2216, 2020.

[25] F. Croce and M. Hein, "Minimally distorted adversarial examples with a fast adaptive boundary attack," *Proceedings of the 37th International Conference on Machine Learning (ICML)*, vol. 119, pp. 2196–2205, 2020.

[26] M. Andriushchenko, F. Croce, N. Flammarion, and M. Hein, "Square Attack: a query-efficient black-box adversarial attack via random search," *Proceedings of the 16th European Conference of Computer Vision (ECCV)*, vol. 12368, pp. 484–501, 2020, doi: 10.1007/978-3-030-58592-1_29

[27] N. Carlini and D. Wagner, "Towards evaluating the robustness of neural networks," *IEEE Symposium on Security and Privacy (SP)*, 2017, doi: 10.1109/SP.2017.49

[28] N. Carlini and D. Wagner, "Adversarial examples are not easily detected: bypassing ten detection methods," *Proceedings of the 10th ACM Workshop on Artificial Intelligence and Security (AISec)*, pp. 3–14, 2017, doi: 10.1145/3128572.3140444

[29] L. Engstrom, B. Tran, D. Tsipras, L. Schmidt, and A. Madry, "A rotation and a translation suffice: fooling CNNs with simple transformations," *arXiv preprint arXiv:1712.02779*, 2017.

[30] A. Ghiasi, A. Shafahi, and T. Goldstein, "Breaking certified defenses: semantic adversarial examples with spoofed robustness certificates," *arXiv preprint arXiv:2003.08937*, 2020.

[31] S. Gowal *et al.*, "On the effectiveness of interval bound propagation for training verifiably robust models," *arXiv preprint arXiv:1810.12715*, 2018.

[32] J. Cohen, E. Rosenfeld, and Z. Kolter, "Certified adversarial robustness via randomized smoothing," *Proceedings of the 36th International Conference on Machine Learning (ICML)*, pp. 1310–1320, 2019.

[33] G. Peyré and M. Cuturi, "Computational optimal transport: With applications to data science," *Foundations and Trends in Machine Learning*, vol. 11, no. 5–6, pp. 355–607, 2019.

[34] E. Wong, F. R. Schmidt, and J. Z. Kolter, "Wasserstein adversarial examples via projected Sinkhorn iterations," *Proceedings of the 36th International Conference on Machine Learning (ICML)*, vol. 97, pp. 6808–6817, 2019.

[35] P.-Y. Chen, H. Zhang, Y. Sharma, J. Yi, and C.-J. Hsieh, "Zoo: zeroth order optimization based black-box attacks to deep neural networks without training substitute models," *Proceedings of the 10th ACM workshop on artificial intelligence and security (AISec)*, vol. 17, pp. 15–26, 2017, doi: 10.1145/3128572.3140448

[36] H. Xu *et al.*, "Adversarial attacks and defenses in images, graphs and text: A review," *International Journal of Automation and Computing*, vol. 17, pp. 151–178, 2020.

[37] C.-C. Tu *et al.*, "Autozoom: Autoencoder-based zeroth order optimization method for attacking black-box neural networks," *Proceedings of the 33rd AAAI Conference on Artificial Intelligence*, vol. 33, pp. 742–749, 2019, doi: 10.1609/aaai.v33i01.3301742

[38] O. Russakovsky *et al.*, "Imagenet large scale visual recognition challenge," *International Journal of Computer Vision (IJCV)*, vol. 115, pp. 211–252, 2015, doi: 10.1007/s11263-015-0816-y

[39] B. C. Kim, Y. Yu, and Y. M. Ro, "Robust decision-based black-box adversarial attack via coarse-to-fine random search," *Proceedings of the IEEE International Conference on Image Processing (ICIP)*, pp. 3048–3052, 2021, doi: 10.1109/ICIP42928.2021.9506464

[40] F. Croce, M. Andriushchenko, N. D. Singh, N. Flammarion, and M. Hein, "Sparse-RS: a versatile framework for query-efficient sparse black-box adversarial attacks," *arXiv preprint arXiv:2006.12834*, vol. abs/2006.1, 2020.

[41] S. Moon, G. An, and H. O. Song, "Parsimonious black-box adversarial attacks via efficient combinatorial optimization," *Proceedings of the 36th International Conference on Machine Learning (ICML)*, vol. 97, pp. 4636–4645, 2019.

[42] G. Ughi, V. Abrol, and J. Tanner, "An empirical study of derivative-free-optimization algorithms for targeted black-box attacks in deep neural networks," *Optimization and Engineering*, pp. 1–28, 2021, doi: 10.1007/S11081-021-09652-W

[43] W. Brendel, J. Rauber, and M. Bethge, "Decision-based adversarial attacks: reliable attacks against black-box machine learning models," *Proceedings of the 6th International Conference on Learning Representations (ICLR)*, 2018.

[44] J. Chen, M. I. Jordan, and M. J. Wainwright, "HopSkipJumpAttack: a query-efficient decision-based attack," *IEEE Symposium on Security and Privacy (SP)*, pp. 1277–1294, 2020, doi: 10.1109/SP40000.2020.00045

[45] Y. Xue, M. Xie, and U. Roshan, "Towards adversarial robustness with 01 loss neural networks," *Proceedings of the 19th IEEE International Conference on Machine Learning and Applications (ICMLA)*, pp. 1304–1309, 2020, doi: 10.1109/ICMLA51294.2020.00204

[46] S. Weerasinghe, T. Alpcan, S. M. Erfani, and C. Leckie, "Defending support vector machines against data poisoning attacks," *IEEE Transactions on Information Forensics and Security*, vol. 16, pp. 2566–2578, 2021, doi: 10.1109/TIFS.2021.3058771

[47] A. Shafahi *et al.*, "Poison frogs! targeted clean-abel poisoning attacks on neural networks," *Advances in Neural Information Processing Systems,* vol. 31, pp. 6103–6113, 2018.

[48] T. Goldstein, C. Studer, and R. Baraniuk, "A field guide to forward-backward splitting with a FASTA implementation," *arXiv preprint arXiv:1411.3406,* 2014.

[49] H. Aghakhani, D. Meng, Y.-X. Wang, C. Kruegel, and G. Vigna, "Bullseye polytope: a scalable clean-label poisoning attack with improved transferability," *arXiv preprint arXiv:2005.00191,* 2020.

[50] C. Zhu, W. R. Huang, H. Li, G. Taylor, C. Studer, and T. Goldstein, "Transferable clean-label poisoning attacks on deep neural nets," *Proceedings of the 36th International Conference on Machine Learning (ICML),* pp. 7614–7623, 2019.

[51] T. Gu, K. Liu, B. Dolan-Gavitt, and S. Garg, "BadNets: evaluating backdooring attacks on deep neural networks," *IEEE Access,* vol. 7, pp. 47230–47244, 2019, doi: 10.1109/ACCESS.2019.2909068

[52] E. Bagdasaryan, A. Veit, Y. Hua, D. Estrin, and V. Shmatikov, "How To backdoor federated learning," *Proceedings of the 23rd International Conference on Artificial Intelligence and Statistics, (AISTATS) Online,* vol. 108, pp. 2938–2948, 2020.

[53] T. J. L. Tan and R. Shokri, "Bypassing backdoor detection algorithms in deep learning," *IEEE European Symposium on Security and Privacy (EuroS&P),* pp. 175–183, 2020, doi: 10.1109/EuroSP48549.2020.00019

[54] B. Wang *et al.*, "Neural cleanse: identifying and mitigating backdoor attacks in neural networks," *IEEE Symposium on Security and Privacy (SP),* pp. 707–723, 2019, doi: 10.1109/SP.2019.00031

[55] B. Tran, J. Li, and A. Madry, "Spectral signatures in backdoor attacks," *Advances in Neural Information Processing Systems,* pp. 8011–8021, 2018.

[56] B. Biggio, B. Nelson, and P. Laskov, "Poisoning attacks against support vector machines," *arXiv preprint arXiv:1206.6389,* 2012.

[57] Y. Li, J. Lin, and K. Xiong, "An adversarial attack defending system for securing in-vehicle networks," *Proceedings of the IEEE 18th Annual Consumer Communications Networking Conference (CCNC),* pp. 1–6, 2021, doi: 10.1109/CCNC49032.2021.9369569

[58] N. Papernot, P. McDaniel, X. Wu, S. Jha, and A. Swami, "Distillation as a defense to adversarial perturbations against deep neural networks," *IEEE Symposium on Security and Privacy (SP),* pp. 582–597, 2016, doi: 10.1109/SP.2016.41

[59] J. Lin, R. Luley, and K. Xiong, "Active learning under malicious mislabeling and poisoning attacks," *arXiv preprint arXiv:210100157,* vol. abs/2101.0, 2021.

[60] Y. Li, T. Zhai, B. Wu, Y. Jiang, Z. Li, and S. Xia, "Rethinking the trigger of backdoor attack," *arXiv preprint arXiv:2004.04692,* 2020.

[61] C. Guo, M. Rana, M. Cissé, and L. van der Maaten, "Countering adversarial images using input transformations," *arXiv preprint arXiv:1711.00117,* 2017.

[62] R. Feinman, R. R. Curtin, S. Shintre, and A. B. Gardner, "Detecting adversarial samples from artifacts," *arXiv preprint arXiv:1703.00410,* vol. abs/1703.0, 2017.

[63] B. Chen *et al.*, "Detecting backdoor attacks on deep neural networks by activation clustering," *arXiv preprint arXiv:1811.03728,* vol. 2301, 2018.

[64] J. Buckman, A. Roy, C. Raffel, and I. J. Goodfellow, "Thermometer encoding: one hot way to resist adversarial examples," *Proceedings of the 6th International Conference on Learning Representations (ICLR),* 2018.

Adversarial Machine Learning: A New Threat Paradigm for Next-generation Wireless Communications

Yalin E. Sagduyu,[1] Yi Shi,[1] Tugba Erpek,[1]
William Headley,[2] Bryse Flowers,[3] George Stantchev,[4]
Zhuo Lu,[5] and Brian Jalaian[6]

[1]*Intelligent Automation, Inc., Rockville, MD*

[2]*Virginia Tech National Security Institute, Blacksburg, VA*

[3]*University of California, San Diego, CA*

[4]*Naval Research Laboratory, Washington, DC*

[5]*University of South Florida, Tampa, FL*

[6]*Joint Artificial Intelligence Center, Washington, DC*

Contents

DOI: 10.1201/9781003187158-3

2.1 INTRODUCTION

The demand for high-performance (e.g., high-rate, high-reliability, and low-latency) communications is ever increasing over the generations of wireless communications systems. There is a growing need to support diverse services for a massive number of smart edge devices (such as in the Internet of Things [IoT]) that are connected in a heterogeneous network setting including nonterrestrial segments. Communication among wireless devices is subject to various effects of channel, traffic, interference, jamming, and mobility. Consequently, next-generation communications need to operate with rich representations of high-dimensional and dynamic spectrum data. The massive numbers of sensors involved in next-generation communications increase the availability of (training and test) data and raise the need for data-driven approaches.

In 4G and earlier communications systems, expert knowledge and analytical solutions were often sufficient for communication design to meet users' needs. However, these traditional approaches cannot capture complex waveforms, channels, and resources used by 5G and later communications systems. In response, machine learning has emerged as a viable tool to learn from the spectrum data and to solve the underlying optimization problems in order to improve the performance of next-generation communications.

The application of machine learning in the wireless domain, commonly referred to as radio frequency machine learning (RFML), has grown robustly in recent years to solve various problems in the areas of wireless communications, networking, and signal processing. Examples include resource allocation (e.g., for network slicing) and spectrum sensing and signal classification (e.g., for spectrum coexistence) [1, 2]. Machine learning has found applications in wireless security such as user equipment (UE) authentication, intrusion detection, and detection of conventional attacks such as jamming and eavesdropping. On the other hand, wireless systems have become vulnerable to machine-learning-based security and privacy attack vectors that have recently been considered in other modalities such as computer vision and natural language processing (NLP). With the emerging commercial applications driven by 5G and later communications, it is timely to study the security and privacy concerns of machine learning systems in the area of RFML.

2.1.1 Scope and Background

Machine learning has found diverse and far-reaching applications ranging from computer vision and NLP to cyber security and autonomous navigation. Over time, machine learning capabilities have evolved from engineering intelligent machines and programs, to learning without being explicitly programmed, and now to learning using deep neural networks, as illustrated in Figure 2.1. Recently, machine learning has been also applied to the wireless domain. This emerging research area is colloquially referred to as RFML. As wireless technologies evolve over time (as depicted in Figure 2.2) to meet the ever growing demand of supporting high rates, high reliability, and low latency with smart edge devices in heterogeneous networks, machine learning has provided an automated means to learn from and adapt to spectrum data and solve complex tasks in the wireless domain. Empowered by the ever increasing hardware accelerations for computing, machine learning systems based on advanced deep learning architectures have been considered for various wireless applications in the emerging landscape of next-generation communications systems (see Figure 2.3). These applications include but are not limited to spectrum sensing applications (such as signal detection, estimation, classification, and identification), waveform design (such as beam optimization, power control, and channel access), channel estimation, emitter identification, cognitive jamming, and antijamming. These research thrusts are expected to play an increasingly prevalent role in the development of next-generation machine-learning-empowered wireless communications [3].

Recent research in the more established machine learning domains has indicated that machine learning is vulnerable to adversarial manipulations, forming the basis for the emerging field of adversarial machine

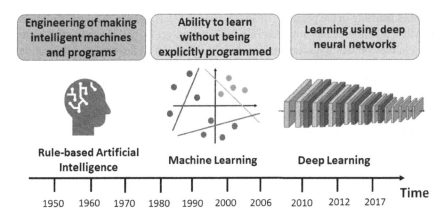

FIGURE 2.1 Evolution of machine learning over time.

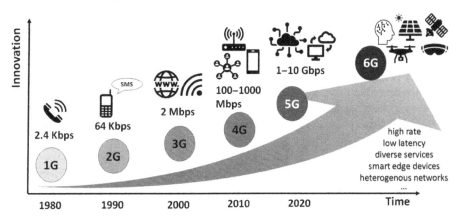

FIGURE 2.2 Wireless technologies evolving over time.

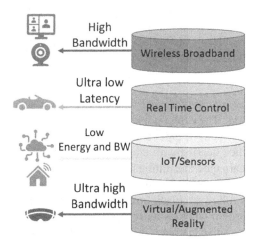

FIGURE 2.3 Wireless applications and objectives that will benefit from machine learning.

learning that studies learning in the presence of adversaries [4–7]. However, the impact of machine learning security on RFML technologies has not been fully understood yet [8–10]. Therefore, it is timely to study wireless security in the context of adversarial machine leaning and understand its implications for 5G and later communications systems. While adversarial machine learning can take a variety of shapes and forms, our focus is on adversarial techniques in the context of deep neural networks since they currently represent the majority of commonly used RFML architectures.

Research efforts that are needed to address machine learning for wireless security are diverse, including but not limited to the following directions:

- *Security attacks built upon adversarial machine learning*: evasion attacks (adversarial examples), poisoning (causative) attacks, Trojan (backdoor or trapdoor) attacks, and generative adversarial learning for signal spoofing
- *Privacy threats for machine learning solutions*: membership inference attacks, model inversion, and physical layer privacy
- *Machine learning hardening techniques*: privacy-preserving learning, secure learning, hardware and software implementations, edge computing, testbeds, experiments, and datasets
- *Expansion of machine learning applications for wireless security*: device identification, spectrum monitoring, RF fingerprinting, smart jamming, smart eavesdropping, localization, covert communications, authentication, anonymity, intrusion detection, and IoT security

The remainder is organized as follows. Section 2.2 presents different types of adversarial machine learning attacks and describes the vulnerabilities of machine learning systems used in the wireless domain. Section 3.3 discusses challenges and gaps for machine learning applications in wireless communications and security. Section 2.4 makes concluding remarks with recommendations for future research directions.

2.2 ADVERSARIAL MACHINE LEARNING

Different types of attacks built upon adversarial machine learning are illustrated in Figure 2.4. Adversarial machine learning studies the attacks launched on the training and/or testing of machine learning models. Recently, these attacks have been extended to wireless applications and have been shown to be stealthier and more energy efficient compared to conventional attacks such as brute-force jamming of data transmissions [11,12] in the presence of different levels of wireless network uncertainty [13–15]. In this section, we discuss exploratory (inference) attacks, evasion attacks (adversarial examples), poisoning (causative) attacks, and Trojan (backdoor) attacks as applied to the wireless domain:

- *Exploratory (inference) attack*: The goal of the exploratory attack is to learn how a machine learning algorithm works. The exploratory attack can be launched to build a surrogate (shadow) model that has the same types of input and outputs as the victim system and provides the same (or similar) outputs when queried with the same input [16, 17]. Exploratory attacks have been studied in [18, 19] to learn the transmitting behavior of a dynamic spectrum access (DSA) system and to develop jamming strategies accordingly. The exploratory attack can be considered the initial step for various follow-up black-box attacks without requiring any knowledge about the machine learning algorithm (such as its type, its hyperparameters, and the underlying training dataset). In this context, active learning [20] can be applied to reduce the number of data samples needed by the exploratory attack [21].
- *Evasion (adversarial) attack*: The goal of the evasion attack is to tamper with the input data in test time and fool a machine learning algorithm into making errors in their decisions. Figure 2.5 shows two examples of the evasion attack, one from computer vision and another from

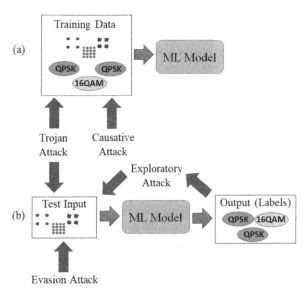

FIGURE 2.4 Taxonomy of adversarial machine learning: (a) in training time, (b) in test time. Arrows indicate which part of the machine learning model each attack type is supposed to manipulate.

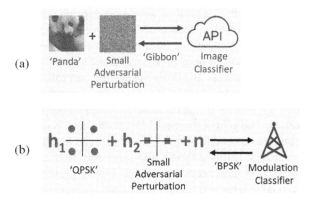

FIGURE 2.5 Examples of evasion (adversarial) attack from (a) computer vision [4], (b) wireless signal classification.

wireless signal classification. Note that the evasion attacks in the wireless domain require the over-the-air manipulation of input data and need to account for channel effects (for both the victim system and the adversary). Most of the studies on evasion attacks in the wireless domain have considered the modulation classifier as the victim system [22–32]. Additional constraints on communication performance at one receiver have been imposed in [33–35], along with the evasion attack launched on another receiver. This setting also extends to adversarial machine-learning-driven covert communications, where the objective is to hide communications from eavesdroppers equipped with machine learning algorithms for spectrum sensing [36]. Various defense mechanisms have been proposed against evasion attacks [27, 37–40], such as statistical outlier detection and certified defense. Evasion attacks have been also considered for other machine learning applications including spectrum sensing [41–43], end-to-end communications with an autoencoder (that consists of an encoder at the transmitter and a decoder at the receiver [44, 45], multiuser power control [46, 47], initial access for directional transmissions [48], and DSA systems [49].

- *Poisoning (causative) attack*: The goal of the poisoning attack is to tamper with the (re)training process of a machine learning algorithm such that the machine learning algorithm makes errors in test time [50]. The poisoning attacks have been considered for spectrum sensing [42, 43] and cooperative spectrum sensing [51–53] as the machine learning extension of spectrum sensing data falsification attacks [54, 55]. As spectrum environments change over time, it is useful to update the machine learning models. To that end, causative attacks can be launched when the wireless system retrains the machine learning models over time.
- *Trojan (backdoor) attack*: The goal of the Trojan attack is to insert Trojans (i.e., backdoors or triggers) into some training data samples in training time and then activate them in test time such that the machine learning makes errors in test time only for selected input samples. For example, [56] investigated input samples with Trojans while the machine learning model keeps high accuracy for the input samples that are not poisoned with Trojans. The Trojan attack was considered in [57] against a wireless signal classifier. For that purpose, the adversary modifies the phases of some signal samples and changes the corresponding labels in training time. In test time, the adversary transmits signals with the same phase shift that was added as a trigger, and thus the wireless signal classifier provides wrong results only for those signals, while the classifier keeps performing well for other input signals.

In addition to security attacks, adversarial machine learning has been applied to study various privacy issues, including the following:

- *Membership inference attack*: The goal of a membership inference attack is to infer whether or not a data sample of interest has been used in the training data of a target machine learning algorithm. This private information can be used to launch follow-up attacks on the machine learning algorithm. In the wireless domain, the membership inference attack can reveal information regarding waveform, channel, and device characteristics [58, 59], and, if leaked, this information could be exploited by an adversary to identify vulnerabilities of the underlying machine learning model (e.g., to infiltrate the physical layer authentication).
- *Model inversion attack*: The goal of the model inversion attack is to infer additional private information by observing the inputs and outputs of the machine learning model when the adversary has access to the machine learning model and some private information [60]. In the wireless domain, the adversary may have access to some information, such as waveform characteristics, and may then aim to infer some additional information such as radio hardware characteristics.

Attacks on reinforcement learning: Attacks built upon adversarial machine learning have been also launched against reinforcement learning algorithms with various applications starting in computer vision [61]. Recently, these attacks have been extended to wireless communications. Adversarial attacks on channel access have been considered in [62–64]. In addition, these attacks have been applied in [65, 66] to launch attacks on the use of reinforcement learning in 5G network slicing [67, 68]. The main goal is to manipulate the inputs to the reinforcement learning algorithms such that it takes a long time for the underlying performance to recover (as the reinforcement learning algorithm adapts to changes) even after the attack stops.

Attacks on machine learning applications in 5G: Adversarial machine learning has been extended to various use cases in 5G communications systems, including but not limited to:

- *Attack on 5G network slicing*: Network slicing provides the capability to multiplex virtualized networks on the same physical network and allocate communication and computational resources dynamically to secure and prioritize different services. Adversaries may send fake network slice requests as a flooding attack [65] or jam the resource blocks [66] to tamper with the reinforcement learning engine used for admission control and resource allocation for 5G network slices.

- *Attack on 5G spectrum coexistence*: The Federal Communications Commission (FCC) has enabled the spectrum coexistence of both federal and nonfederal users in the 3.5 GHz Citizens Broadband Radio Service (CBRS) band. The adversary may attack the spectrum coexistence between the incumbent radar system and the 5G communication network. As there should not be any harmful interference from the 5G network to the radar system, Environmental Sensing Capability (ESC) sensors need to detect the radar signals. For that purpose, deep learning can be used as a powerful means to detect signals from the I/Q samples captured over the air. The adversary may try to affect the decision-making process of the ESC sensor by transmitting small perturbation signals over the air to manipulate the input signal to the ESC sensor's deep learning algorithm [49].
- *Attack 5G UE authentication*: Massive number of heterogeneous devices connecting to 5G network for 5G-enabled services raise the need for physical layer authentication at the signal level. Deep learning algorithms can be used for signal classification based on RF fingerprints that take waveform, channel, and radio hardware characteristics into account. An adversary can train a generative adversarial network (GAN) to spoof wireless signals and bypass the authentication process.
- *Attack to enable covert 5G communications*: Adversarial machine learning can be used to enable covert 5G communications by fooling spectrum sensors even when they are employed with deep neural networks. Either the transmitter adds a small perturbation to its signal before transmitting it, or the perturbation signal is added over the air by a cooperative jammer. While this perturbation fools the spectrum sensor, it should not have a significant effect on the bit error rate of 5G communications [36].
- *Attack on 5G power control*: The 5G base station needs to allocate the transmit power to serve multiple UEs and to maximize the aggregate rate across all UEs. Deep learning can be used for a low-complexity solution to this optimization problem. In the meantime, an adversary can craft adversarial perturbations to manipulate the inputs to the underlying deep neural network. This adversary may be an external transmitter interfering with the pilot signals that are transmitted to measure the channel gain or a rogue UE sending fabricated channel estimates back to the base station [46, 47].
- *Attack on 5G initial access*: Operation on wide mmWave and THz bands achieves high data rates by using narrow directional beams. When a UE connects to the 5G network for the first time, it needs to establish the initial access with the base station. For that purpose, the base station transmits pilot signals with different narrow beams, and the UE computes the received signal strengths (RSSs) for all beams to find the most suitable beam. However, sweeping all beams is a long process. Instead, the base station can transmit with a small number of narrow beams, and the UE can compute the RSSs for these beams and run a deep neural network to predict the best beam from a larger set of beams [69, 70]. In the meantime, an adversary can transmit adversarial perturbations over the air during the initial access. This way, the adversary can manipulate the RSSs and fool the deep neural network into choosing the beams with small RSSs.

Studies of adversarial machine learning in the wireless domain are still in their early stages [71], and more work is needed to fully characterize the attack and defense spaces by accounting for unique features of wireless applications such as differences in features and labels observed by adversaries and defenders due to different channel and interference effects.

2.3 CHALLENGES AND GAPS

With the increasing use of machine learning in the wireless domain, a fundamental question arises whether to trust machine learning for RFML applications. Challenges regarding trust in RFML are illustrated in

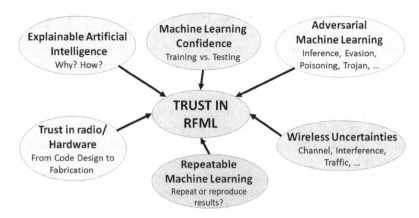

FIGURE 2.6   Trust in machine learning for RFML.

Figure 2.6. Some of these challenges (such as explainability and confidence) apply to all data domains, and therefore they are universal. However, other challenges are specific to the wireless domain, such as the characteristics of the radio hardware used and the inherent uncertainties associated with the wireless medium, including channel, traffic, mobility, and interference effects.

2.3.1 Development Environment

The first challenge that we consider concerns the development environment for RFML applications. Next-generation communications systems rely on complex ecosystems. For example, the 5G ecosystem consists of various components, including:

- *Radio access network (RAN)*: Connects the network and the UE as the final link.
- *Multi-access edge computing (MEC)*: Provides services/computing functions for edge nodes.
- *Core network*: Controls the networks.
- *Transport*: Links RAN components (fronthaul/midhaul) and links RAN and core (backhaul).

Our focus is on the challenges regarding the RFML applications for the air interface and RAN. As RFML solutions rely on various waveform, channel, and radio hardware characteristics, a multilevel development environment is ultimately needed, as illustrated in Figure 2.7.

A starting point for the RFML development is based on simulation tests. While these tests mainly represent virtual hardware and channel effects, they fall short of reflecting real-world scenarios. Therefore, there is a need to support the initial simulations with high-fidelity emulation tests that use actual radios to generate real network traffic and make real transmissions that propagate over emulated (virtual) channels [72, 73]. Instead of over-the-air transmissions in the dynamic wireless medium, emulation tests provide the flexibility to control channel effects such as path loss, fading, and delay [74, 75] and to repeat the tests (e.g., by comparing different network protocols under the same channel effects). On the other hand, test beds with the over-the-air transmissions provide the opportunity to test the system with real channels along with real hardware [76].

One key need for the development of RFML is the availability of necessary training and test datasets. In addition, it is important to account for potential changes between training and test conditions (e.g., training under indoor channel effects and testing under outdoor channel effects). See section 2.3.2 for a more detailed discussion of RFML datasets. Note that RFML applications may need to support edge computing, and therefore it is often the case that RFML applications are deployed on embedded platforms such as software-defined radios (SDRs). See section 2.3.4 for a more detailed discussion of embedded implementation of RFML applications.

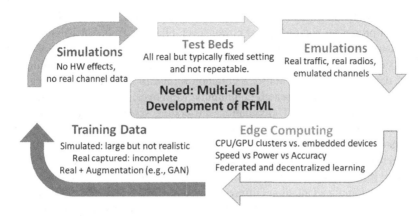

FIGURE 2.7   Multilevel development environment for RFML.

As RFML is heavily used in next-generation communications systems, the openness of the software development ecosystem for these systems poses another challenge for RFML applications. An example is Open RAN (O-RAN) that provides the virtualized RAN capability and builds on an open software development environment supported by the 5G community. For example, various RFML applications such as UE identification and network slicing can be implemented through the RAN Intelligent Controller (RIC) of O-RAN. While this open software paradigm aims to accelerate and unify efforts across different technology vendors, adversaries may abuse this openness and get access to various network functions including machine learning (such as training data and models) and exploit the underlying vulnerabilities (such as launching attacks built on adversarial machine learning).

2.3.2 Training and Test Datasets

The success of machine learning applications heavily hinges on the availability of relevant data for both training and testing. Specific to the wireless domain, various channel, waveform, radio, interference, and adversary (e.g., jammer) characteristics should be accounted for when data samples are collected and curated. One approach is to rely on simulations to generate RFML datasets. This is a controlled process that can be configured to the needs of the machine learning problem at hand. However, whether simulated data has fidelity for RFML applications is questionable [77]. The reason lies in the difficulty of modeling channel and radio hardware effects. Therefore, real datasets collected through emulated or over-the-air transmissions are ultimately needed for machine learning problems in the wireless domain.

It is a common practice for other data domains like computer vision and NLP to publish datasets used for machine learning. Recently, the wireless domain has also observed a trend of releasing datasets (e.g., see [78]). However, there is still a lack of real datasets to accommodate various wireless scenario needs, e.g., spectrum dynamics subject to intelligent jammers. Therefore, the RFML domain has faced the risk that the availability of datasets for a particular application drives the research efforts. For example, a plethora of research has focused on modulation recognition for which representative datasets are publicly available [79]. On the other hand, other wireless signal classification tasks (especially the ones focused on wireless security) are lagging behind due to the absence of relevant datasets. Therefore, the RFML research area needs publicly released real datasets for different wireless applications of machine learning across the network protocol stack (not limited to the physical layer). Ultimately, a wireless security database similar to the National Vulnerability Database [80] in the cyber domain is needed to foster the RFML research efforts.

Unlike other data domain, e.g., computer vision, there is no obvious universal data format for RFML data. One example of the signal metadata format is SigMF that provides a standard way to describe data recordings [81]. Based on different applications in the wireless domain, machine learning may consider different signal properties such as signal power level, phase shift, on/off behavior, or raw IQ data. Although raw IQ data could be used to establish a universal RFML data format, it is not easy to determine the sampling rate since a low sampling rate may lose information while a high sampling rate may yield a large amount of data. Thus more efforts for a common format for RFML data are needed to foster the adoption of datasets by the RFML research community. Moreover, different applications may consider different network settings, in terms of modulation and coding scheme, traffic, network topology, and mobility, further increasing the challenge to build datasets for RFML.

One way to satisfy the large data needs of machine learning algorithms, in particular deep learning, is synthetic data generation. Various methods have been suggested for other domains such as computer vision [82] and NLP [83]. Built on deep neural networks, the GAN has emerged as a viable approach for data augmentation and domain adaptation. Different versions of GANs have been effectively used to generate synthetic data samples for computer vision [84], for NLP [85], and to support various attacks on the victim machine-learning-based systems [86]. However, efforts to generate synthetic samples of wireless signal data are lagging (see [77, 87–91]). To account for rich spectrum characteristics, synthetic data generation is critical for various wireless applications and serve important purposes such as

1. data augmentation to expand the training dataset (e.g., when there is limited time for spectrum sensing or the sampling rate is low),
2. domain adaptation of test and training data samples (e.g., when there is a change in spectrum conditions such as channel, interference, and traffic effects from training time to test time),
3. defense against adversarial attacks (e.g., when an adversary manipulates the input data to a wireless signal classifier such as the one used for user authentication), and
4. signal spoofing (e.g., when an adversary generates and transmit signals to infiltrate an access control system or to set up decoys/honeypots).

As illustrated in Figure 2.8, the GAN structure consists of two neural networks, one for the generator and one for the discriminator. The generator takes noise as the input and generates synthetic data samples. In the meantime, the discriminator receives either a real or synthetic data sample as the input and tries to classify each sample. The classification result is sent as the feedback to the discriminator. Note that, in a wireless system, the generator and the discriminator may not be colocated (e.g., the generator is located at a transmitter, and the discriminator is located at another receiver), and they may communicate over noisy channels. As the spectrum data is subject to various channel, interference, and traffic effects, it is challenging to train the GAN subject to issues regarding stability and mode collapse. Note that the GAN can be also used in conjunction with adversarial attacks, as shown in [92] for the purpose of evading device identification.

The use of pretrained machine learning models is a common practice in wireless security applications. The continuation of attack and defense optimization in test time can be ensured through online machine learning update mechanisms to adapt to spectrum dynamics. One promising update mechanism for wireless security is reinforcement learning that can learn from the spectrum environment and take actions to secure wireless communications in the presence of adversaries such as jammers and eavesdroppers

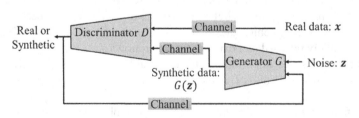

FIGURE 2.8   Generative adversarial network (GAN) subject to channel effects.

[93]. One aspect that differentiates machine learning security in the wireless domain from its counterpart in other data domains is the extent of performance measures. To that end, machine learning can be applied to optimize the quality of communications measured by various performance metrics including but not limited to throughput and delay subject to power consumption requirements. As these metrics are often coupled with one another, their joint optimization is needed [94].

Channel, interference, waveform, and traffic effects jointly shape the spectrum [95]. Since traffic flows and links in a wireless network are often dynamic and random, packet queues at individual nodes should be stabilized such that queue lengths remain bounded [96–99] while optimizing the performance via machine learning. In addition to conventional performance measures such as throughput and delay (and its extension to deadline [100]), the Age of Information (AoI) has raised interest in measuring information freshness (the time elapsed since the last received update) [101, 102]. Machine learning is promising for the optimal determination of the AoI, which is a complex process in a wireless network setting [103].

2.3.3 Repeatability, Hyperparameter Optimization, and Explainability

In addition to the release of the datasets, RFML research can benefit from the open software as also experienced in 5G and later ecosystems. This approach can help with the rapid extensions of the proposed solutions to other wireless security problems or the comparison of different solutions. Therefore, one critical task for the RFML research is to develop a common library of algorithmic solutions to machine-learning-based wireless attacks. Empowered by recent algorithmic and computational advances, deep learning finds rich applications in wireless security as it can better capture physical signal characteristics compared to conventional machine learning algorithms.

Before deep learning can be practically applied in RFML applications, four major issues need to addressed:

1. The complex structures of deep neural networks require large datasets for training purposes. As discussed earlier, the GAN can be used to augment datasets. In addition, other methods can be adopted from computer vision, NLP, and other domains, or new ones can be developed to meet this need.

2. Deep neural networks involve many hyperparameters to optimize, including the number of layers and the number of neurons per layer. To meet the low-latency and low-complexity processing needs of the edge computing applications of RFML, novel techniques of hyperparameter optimization can be applied to reduce the computational complexity and runtime of parameter searches. One example of hyperparameter optimization is Hyperband [104], which iteratively selects and evaluates the performance of hyperparameter configurations with increasing runtimes. The alternative approach of exhaustive search becomes easily intractable as the neural network structures become deeper and cannot match the need of edge computing applications of RFML.

3. A common practice of deep learning is to use the deep neural networks as a black box. However, justification of the operational functionality and performance is essential to foster the wide adoption of RFML solutions based on deep learning [105]. In this context, the preprocessing of spectrum data for feature extraction is needed through novel statistical, temporal, or spectral methods. Another point that requires further attention is that the wireless domain further increases the discrepancy between test and training data samples due to
   - spectrum dynamics (channel, traffic, interference, and radio effects change over time),
   - potential changes in the model (data labels [e.g., signal types] may change over time),
   - noisy data (features, such as I/Q samples hidden under noise, fading, and interference effects),
   - restrictions due to distributed computing (training and inference may be distributed at edge devices), and

TABLE 2.1 Performance of Different Embedded Computation Platforms

MEASURE\PLATFORM	ARM	EMBEDDED GPU	FPGA
Latency	High	Medium	Low
Power Consumption	Medium	High	Low
Programmability	High	Medium	Low

- adversarial machine learning (wireless signals are vulnerable to over-the-air attacks built on adversarial machine learning).
4. In a dynamic spectrum environment, it is likely that the types of signals (e.g., jammers) change over time. When wireless signals are set as inputs to a machine learning classifier, these changes translate to changes in output labels. One option is to retrain the classifiers with only new signal types, leading to catastrophic forgetting. Instead, it is possible to strike a balance between old and new tasks to learn through continual learning. To that end, elastic weight consolidation (EWC) can be used to slow down the learning process on selected machine learning model parameters in order to remember previously learned tasks [106]. EWC finds applications in wireless signal classification when wireless signal types such as modulation types change over time [107].

2.3.4 Embedded Implementation

Other data domains such as computer vision and NLP often rely on offline computation using central processing unit (CPU) or graphics processing unit (GPU) resources for machine learning tasks. However, wireless security applications may require embedded computing for online decision making. In this case, machine learning algorithms need to run on radio platforms with size, weight, and power (SWaP) constraints. Note that computational resources such as ARM processors, embedded GPUs, and field-programmable gate arrays (FPGA) are already available for embedded implementation. However, RFML research efforts supported by embedded platform implementations are lagging behind and cannot match the low-latency and low-power requirements of wireless security. As examples of embedded implementation of machine learning solutions, signal authentication was implemented on embedded GPU in [108], jammer detection was implemented on SDRs in [109], and modulation classification was implemented on the Android smartphone in [110]. Since embedded platforms have limited memory, there is a need to compress the deep neural networks [111] before deploying them on embedded platforms. In addition, embedded platforms such as embedded GPU and FPGA operate with finite precision. Therefore, RFML applications need to account for quantization effects on machine learning models [112].

In particular, initial studies have indicated that FPGAs provide major gains in latency and energy consumption for machine-learning-based RF signal classification tasks compared to embedded GPUs [113, 114]. Since FPGAs are hard to program, novel techniques that can convert software codes of machine learning solutions to the FPGA implementation are ultimately needed. A qualitative comparison of expected latency, power consumption, and programmability of different embedded computing resources is shown in Table 2.1, motivating the need for more efforts to bridge the gap between theory and practice by implementing the proposed solutions on embedded platforms.

2.4 CONCLUSIONS AND RECOMMENDATIONS

We draw the following conclusions built on the challenges and gaps that we identified in the previous section:

- Machine learning has emerged as a critical means to secure wireless systems. However, the attack vectors are not yet fully understood. Therefore, we need more efforts to identify the vulnerabilities of wireless systems against adversaries that may apply machine learning techniques. In particular, there is a need to characterize clearly defined machine-learning-based attack models as the precondition to formally analyze machine-learning-related wireless security.

- Defense schemes in the wireless domain need to become more agile and adaptive as adversaries are getting smarter. Machine learning can be used as a powerful means to detect and mitigate wireless attacks.

- Adversarial machine learning has emerged as a major threat for wireless systems, given its potential to disrupt the machine learning process. As the wireless medium is open and shared, it provides new means for the adversaries to manipulate the test and training processes of machine learning algorithms. To that end, novel techniques are needed to safeguard the wireless attack surface against adversarial machine learning.

- As machine learning leads the duel between the adversary and the defender as an evolving, interactive process, there is a need to understand the dynamic evolution of this process by quantifying the performance metrics for the adversary and the defender.

- More efforts are needed to make more high-fidelity and diverse datasets publicly available for the wide adoption of machine learning in wireless security. These datasets are needed not only for back-end applications but also for embedded implementation. In this context, it is crucial to meet latency, power, and computational complexity requirements of machine-learning-based attacks and defenses in SWaP-constrained environments.

- Last but not least, it is important to recognize that the robustness of RFML classifiers to adversarial machine learning attacks can, in general, depend on the classifier type. In particular, discriminative classifiers tend to be more vulnerable than generative classifiers [115], such as GANs or deep Bayesian models. Many of the classifiers currently used in RFML are of the discriminative type, thereby posing the following dilemma: Is it more efficient to fortify existing classifiers using state-of-the-art defensive techniques or to opt for generative classifiers with commensurable performance characteristics? To answer this question, it will be necessary to study and quantify the trade-off between the robustness and accuracy of discriminative and generative classifiers in the RFML domain, so that informed decisions can be made, depending on the specific application.

REFERENCES

[1]   T. Erpek, T. O'Shea, Y. E. Sagduyu, Y. Shi, and T. C. Clancy, "Deep Learning for Wireless Communications," *Development and Analysis of Deep Learning Architectures*, Springer, 2019.

[2]   L. J. Wong, W. H. Clark, B. Flowers, R. M. Buehrer, W. C. Headley and A. J. Michaels, "An RFML Ecosystem: Considerations for the Application of Deep Learning to Spectrum Situational Awareness," *IEEE Open Journal of the Communications Society*, vol. 2, pp. 2243–2264, 2021.

[3]   K. B. Letaief, W. Chen, Y. Shi, J. Zhang, Y.A. Zhang, "The Roadmap to 6G: AI Empowered Wireless Networks." *IEEE Communications Magazine*, Aug. 2019.

[4]   I. Goodfellow, J. Shlens, and C. Szeged, "Explaining and Harnessing Adversarial Examples." *International Conference on Learning Representations (ICLR)*, 2015.

[5]   A. Kurakin, I. Goodfellow, and S. Bengio, "Adversarial Machine Learning at Scale," arXiv preprint arXiv:1611.01236, 2016.

[6]   Y. Vorobeychik and M. Kantarcioglu, "Adversarial Machine Learning," *Synthesis Lectures on Artificial Intelligence and Machine Learning*, Aug. 2018.

[7]   C. Szegedy, W. Zaremba, I. Sutskever, J. Bruna, D. Erhan, and I. Goodfellow, and R. Fergus, "Intriguing Properties of Neural Networks," *arXiv preprint arXiv:1312.6199*, 2013.

[8]   D. Adesina, C-C Hsieh, Y. E. Sagduyu, and L. Qian, "Adversarial Machine Learning in Wireless Communications using RF Data: A Review," *arXiv preprint arXiv:2012.14392*, 2020.

[9]   L. Pajola, L. Pasa, and M. Conti, "Threat is in the Air: Machine Learning for Wireless Network Applications," *ACM Workshop on Wireless Security and Machine Learning (WiseML)*, 2019.

[10] Y. E. Sagduyu, Y. Shi, T. Erpek, W. Headley, B. Flowers, G. Stantchev, and Z. Lu, "When Wireless Security Meets Machine Learning: Motivation, Challenges, and Research Directions," arXiv preprint arXiv:2001.08883.

[11] W. Xu, W. Trappe, Y. Zhang, and T. Wood, "The Feasibility of Launching and Detecting Jamming Attacks in Wireless Networks," *ACM International Symposium on Mobile Ad Hoc Networking and Computing (MobiHoc)*, 2005.

[12] Y. E. Sagduyu and A. Ephremides, "A Game-Theoretic Analysis of Denial of Service Attacks in Wireless Random Access," *Wireless Networks*, July 2009.

[13] Y. E. Sagduyu, R. Berry, and A. Ephremides, "MAC Games for Distributed Wireless Network Security with Incomplete Information of Selfish and Malicious User Types," *IEEE International Conference on Game Theory for Networks (GameNets)*, 2009.

[14] Y. E. Sagduyu, R. Berry and A. Ephremides, "Wireless Jamming Attacks under Dynamic Traffic Uncertainty," *IEEE International Symposium on Modeling and Optimization in Mobile, Ad Hoc, and Wireless Networks* (WIOPT), 2010.

[15] Y. E. Sagduyu, R. Berry, and A. Ephremides, "Jamming Games in Wireless Networks with Incomplete Information," *IEEE Communications Magazine*, Aug. 2011.

[16] F. Tramer, F. Zhang, A. Juels, M. K. Reiter, and T. Ristenpart, "Stealing Machine Learning Models via Prediction APIs," *USENIX Security Symposium*, 2016.

[17] Y. Shi, Y. E. Sagduyu, and A. Grushin, "How to Steal a Machine Learning Classifier with Deep Learning," *IEEE Symposium on Technologies for Homeland Security (HST)*, 2017.

[18] Y. Shi, Y. E Sagduyu, T. Erpek, K. Davaslioglu, Z. Lu, and J. Li, "Adversarial Deep Learning for Cognitive Radio Security: Jamming Attack and Defense Strategies," *IEEE International Conference on Communications (ICC) Workshop on Promises and Challenges of Machine Learning in Communication Networks*, 2018.

[19] T. Erpek, Y. E. Sagduyu, and Y. Shi, "Deep Learning for Launching and Mitigating Wireless Jamming Attacks," *IEEE Transactions on Cognitive Communications and Networking*, 2019.

[20] B. Settles, "Active Learning Literature Survey," *Science*, vol. 10, no. 3, pp. 237–304. 1995.

[21] Y. Shi, Y. E. Sagduyu, K. Davaslioglu, and J. H Li, "Active Deep Learning Attacks under Strict Rate Limitations for Online API Calls," *IEEE Symposium on Technologies for Homeland Security (HST)*, 2018.

[22] M. Sadeghi and E. G. Larsson, "Adversarial Attacks on Deep-learning Based Radio Signal Classification," *IEEE Wireless Communications Letters*, Feb. 2019.

[23] B. Flowers, R. M. Buehrer, and W. C. Headley, "Evaluating Adversarial Evasion Attacks in the Context of Wireless Communications," *arXiv preprint, arXiv:1903.01563*, 2019.

[24] B. Flowers, R. M. Buehrer, and W. C. Headley, "Communications Aware Adversarial Residual Networks," *IEEE Military Communications Conference (MILCOM)*, 2019.

[25] S. Bair, M. Delvecchio, B. Flowers, A. J. Michaels, and W. C. Headley, "On the Limitations of Targeted Adversarial Evasion Attacks Against Deep Learning Enabled Modulation Recognition," *ACM Workshop on Wireless Security and Machine Learning (WiseML)*, 2019.

[26] B. Kim, Y. E. Sagduyu, K. Davaslioglu, T. Erpek, and S. Ulukus, "Over-the-air Adversarial Attacks on Deep Learning based Modulation Classifier over Wireless Channels," *Conference on Information Sciences and Systems (CISS)*, 2020.

[27] B. Kim, Y. E. Sagduyu, K. Davaslioglu, T. Erpek, and S. Ulukus, "Channel-aware Adversarial Attacks against Deep Learning-based Wireless Signal Classifiers," *arXiv preprint arXiv:2005.05321*, 2020.

[28] B. Kim, Y. E. Sagduyu, T. Erpek, K. Davaslioglu, S. Ulukus, "Adversarial Attacks with Multiple Antennas against Deep Learning-based Modulation Classifiers," *IEEE GLOBECOM Open Workshop on Machine Learning in Communications*, 2020.

[29] B. Kim, Y. E. Sagduyu, T. Erpek, K. Davaslioglu, and S. Ulukus, "Channel Effects on Surrogate Models of Adversarial Attacks against Wireless Signal Classifiers," *IEEE International Conference on Communications (ICC)*, 2021.

[30] Y. Lin, H. Zhao, Y. Tu, S. Mao, and Z. Dou, "Threats of Adversarial Attacks in DNN-based Modulation Recognition," *IEEE Conference on Computer Communications (INFOCOM)*, 2020.

[31] F. Restuccia, S. D'Oro S, A. Al-Shawabka, B. C. Rendon, K. Chowdhury, S. Ioannidis, and T. Melodia, "Generalized Wireless Adversarial Deep Learning," *ACM Workshop on Wireless Security and Machine Learning (WiseML)*, 2020.

[32] R. Sahay, C. G. Brinton, and D. J. Love, "A Deep Ensemble-based Wireless Receiver Architecture for Mitigating Adversarial Attacks in Automatic Modulation Classification," *IEEE Transactions on Cognitive Communications and Networking*, 2021.

[33] M. Z. Hameed, A. Gyorgy, and D. Gunduz, "Communication without Interception: Defense Against Deep-learning-based Modulation Detection," *IEEE Global Conference on Signal and Information Processing (GlobalSIP)*, 2019.

[34] M. Z. Hameed, A. Gyorgy, and D. Gunduz, "The Best Defense is a Good Offense: Adversarial Attacks to Avoid Modulation Detection," *IEEE Transactions on Information Forensics and Security*, vol. 16, 2021.

[35] M. DelVecchio, V. Arndorfer, and W. C. Headley, "Investigating a Spectral Deception Loss Metric for Training Machine Learning-based Evasion Attacks," *ACM Workshop on Wireless Security and Machine Learning (WiseML)*, 2020.

[36] B. Kim, Y. E. Sagduyu, K. Davaslioglu, T. Erpek, and S. Ulukus, "How to make 5G Communications 'Invisible': Adversarial Machine Learning for Wireless Privacy," *Asilomar Conference on Signals, Systems, and Computers*, 2020.

[37] S. Kokalj-Filipovic and R. Miller, "Adversarial Examples in RF Deep Learning: Detection of the Attack and its Physical Robustness," *IEEE Global Conference on Signal and Information Processing (GlobalSIP)*, 2019.

[38] S. Kokalj-Filipovic, R. Miller, N. Chang, and C. L. Lau, "Mitigation of Adversarial Examples in RF Deep Classifiers Utilizing Autoencoder Pre-training," *arXiv preprint arXiv:1902.08034*, 2019.

[39] S. Kokalj-Filipovic, R. Miller, and J. Morman, "Targeted Adversarial Examples against RF Deep Classifiers," *ACM Workshop on Wireless Security and Machine Learning (WiseML)*, 2019.

[40] M. DelVecchio, B. Flowers and W. C. Headley, "Effects of Forward Error Correction on Communications Aware Evasion Attacks," *IEEE International Symposium on Personal, Indoor and Mobile Radio Communications*, 2020

[41] Y. Shi, T. Erpek, Y. E Sagduyu, and J. Li, "Spectrum Data Poisoning with Adversarial Deep Learning," *IEEE Military Communications Conference (MILCOM)*, 2018.

[42] Y. E. Sagduyu, Y. Shi, and T. Erpek, "IoT Network Security from the Perspective of Adversarial Deep Learning," *IEEE International Conference on Sensing, Communication and Networking (SECON) Workshop on Machine Learning for Communication and Networking in IoT*, 2019.

[43] Y. E. Sagduyu, Y. Shi, and T. Erpek, "Adversarial Deep Learning for Over-the-Air Spectrum Poisoning Attacks," *IEEE Transactions on Mobile Computing*, 2019.

[44] M. Sadeghi and E. G. Larsson, "Physical Adversarial Attacks Against End-to-end Autoencoder Communication Systems," *IEEE Communications Letters*, Feb. 2019. 847–850.

[45] T. J. O'Shea and J. Hoydis, "An Introduction to Deep Learning for the Physical Layer," *IEEE Transactions on Cognitive Communications and Networking*, Dec. 2017.

[46] B. Kim, Y. Shi, Y. E. Sagduyu, T. Erpek, and S. Ulukus, "Adversarial Attacks against Deep Learning Based Power Control in Wireless Communications," *IEEE Global Communications Conference (GLOBECOM) Workshops*, 2021.

[47] B. Manoj, M. Sadeghi, and E. G. Larsson, "Adversarial Attacks on Deep Learning Based Power Allocation in a Massive MIMO Network," *arXiv preprint arXiv:2101.12090*, 2021.

[48] B. Kim, Y. E. Sagduyu, T. Erpek, and S. Ulukus, "Adversarial Attacks on Deep Learning Based mmWave Beam Prediction in 5G and Beyond," *IEEE Statistical Signal Processing (SSP) Workshop*, 2021.

[49] Y. E. Sagduyu, T. Erpek, and Y. Shi, "Adversarial Machine Learning for 5G Communications Security," *Game Theory and Machine Learning for Cyber Security, IEEE*, pp.270–288, 2021.

[50] Y. Shi and Y. E Sagduyu, "Evasion and Causative Attacks with Adversarial Deep Learning," *IEEE Military Communications Conference (MILCOM)*, 2017.

[51] Z. Luo, S. Zhao, Z. Lu, J. Xu, and Y. E. Sagduyu, "When Attackers Meet AI: Learning-empowered Attacks in Cooperative Spectrum Sensing," *arXiv preprint arXiv:1905.01430*.

[52] Z. Luo, S. Zhao, Z. Lu, Y. E. Sagduyu, and J. Xu, "Adversarial machine learning based partial-model attack in IoT," *ACM Workshop on Wireless Security and Machine Learning (WiseML)*, 2020.

[53] Z. Luo, S. Zhao, R. Duan, Z. Lu, Y. E. Sagduyu, and J. Xu, "Low-cost Influence-Limiting Defense against Adversarial Machine Learning Attacks in Cooperative Spectrum Sensing," *ACM Workshop on Wireless Security and Machine Learning (WiseML)*, 2021.

[54] F. R. Yu, H. Tang, M. Huang, Z. Li and P. C. Mason, "Defense Against Spectrum Sensing Data Falsification Attacks in Mobile Ad Hoc Networks with Cognitive Radios," *IEEE Military Communications Conference*, 2009.

[55] Y. E. Sagduyu, "Securing Cognitive Radio Networks with Dynamic Trust against Spectrum Sensing Data Falsification," *IEEE Military Communications Conference (MILCOM)*, 2014.

[56] Y. Liu, S. Ma, Y. Aafer, W.-C. Lee, J. Zhai, W. Wang, and X. Zhang, "Trojaning Attack on Neural Networks," *Network and Distributed System Security Symposium*, 2018.

[57] K. Davaslioglu and Y. E. Sagduyu, "Trojan Attacks on Wireless Signal Classification with Adversarial Machine Learning," *IEEE Workshop on Data-Driven Dynamic Spectrum Sharing of IEEE DySPAN*, 2019.

[58] Y. Shi, K. Davaslioglu, and Y. E. Sagduyu, "Over-the-Air Membership Inference Attacks as Privacy Threats for Deep Learning-based Wireless Signal Classifiers," *ACM Workshop on Wireless Security and Machine Learning (WiseML)*, 2020.

[59] Y. Shi and Y. E. Sagduyu, "Membership Inference Attack and Defense for Wireless Signal Classifiers with Deep Learning," *arXiv preprint arXiv:2107.12173, 2021.*

[60] M. Fredrikson, S. Jha, and T. Ristenpart, "Model Inversion Attacks that Exploit Confidence Information and Basic Countermeasures," *ACM SIGSAC Conference on Computer and Communications Security*, 2015.

[61] S. Huang, N. Papernot, I. Goodfellow, Y. Duan, and P. Abbeel, "Adversarial Attacks on Neural Network Policies," arXiv preprint arXiv:1702.02284, 2017.

[62] F. Wang, C. Zhong, M. C. Gursoy, and S. Velipasalar, "Adversarial Jamming Attacks and Defense Strategies via Adaptive Deep Reinforcement Learning," arXiv preprint arXiv:2007.06055, 2020.

[63] F. Wang, C. Zhong, M. C. Gursoy, and S. Velipasalar, "Defense Strategies against Adversarial Jamming Attacks via Deep Reinforcement Learning," *Conference on Information Sciences and Systems (CISS)*, 2020.

[64] C. Zhong, F. Wang, M. C. Gursoy, and S. Velipasalar, "Adversarial Jamming Attacks on Deep Reinforcement Learning based Dynamic Multichannel Access," *IEEE Wireless Communications and Networking Conference (WCNC)*, 2020.

[65] Y. Shi and Y. E. Sagduyu, "Adversarial Machine Learning for Flooding Attacks on 5G Radio Access Network Slicing," *IEEE International Conference on Communications (ICC) Workshops*, 2021.

[66] Y. Shi, Y. E. Sagduyu, T. Erpek, and M. C. Gursoy, "How to Attack and Defend 5G Radio Access Network Slicing with Reinforcement Learning," arXiv preprint arXiv:2101.05768, 2021.

[67] Y. Shi, T. Erpek, and Y. E. Sagduyu, "Reinforcement Learning for Dynamic Resource Optimization in 5G Radio Access Network Slicing," *IEEE International Workshop on Computer Aided Modeling and Design of Communication Links and Networks (CAMAD)*, 2020.

[68] Y. Shi, P. Rahimzadeh, M. Costa, T. Erpek, and Y. E. Sagduyu, "Deep Reinforcement Learning for 5G Radio Access Network Slicing with Spectrum Coexistence," TechRxiv. Preprint. https://doi.org/10.36227/techrxiv.16632526.v1

[69] T. S. Cousik, V. K. Shah, J. H. Reed, T. Erpek, and Y. E. Sagduyu, "Fast Initial Access with Deep Learning for Beam Prediction in 5G mmWave Networks," *IEEE Military Communications Conference (MILCOM)*, 2021.

[70] T. S. Cousik, V. K. Shah, J. H. Reed, T. Erpek, and Y. E. Sagduyu, "Deep Learning for Fast and Reliable Initial Access in AI-Driven 6G mmWave Networks," *arXiv preprint arXiv:2101.01847.*

[71] D. Adesina D, C. C. Hsieh, Y. E. Sagduyu, and L. Qian, "Adversarial Machine Learning in Wireless Communications using RF Data: A Review," *arXiv preprint arXiv:2012.14392*, 2020.

[72] J. Yackoski, B. Azimi-Sadjadi, A. Namazi, J. H. Li, Y. E. Sagduyu, and R. Levy, "RF-NEST: Radio Frequency Network Emulator Simulator Tool," *IEEE Military Communications Conference (MILCOM)*, 2011.

[73] T. Melodia, S. Basagni, K. R. Chowdhury, A. Gosain, M. Polese, P. Johari, and L. Bonati, "Colosseum, the world's largest wireless network emulator," *ACM Conference on Mobile Computing and Networking (MobiCom)*, 2021.

[74] K. J. Kwak, Y. E. Sagduyu, J. Yackoski, B. Azimi-Sadjadi, A. Namazi, J. Deng, and J. Li, "Airborne Network Evaluation: Challenges and High Fidelity Emulation Solution," *IEEE Communications Magazine*, Oct. 2014.

[75] S. Soltani, Y. E. Sagduyu, Y. Shi, J. Li, J. Feldman, and J. Matyjas, "Distributed Cognitive Radio Network Architecture, SDR Implementation and Emulation Testbed," *IEEE Military Communications Conference (MILCOM)*, 2015.

[76] D. Raychaudhuri, I. Seskar, G. Zussman, T. Korakis, D. Kilper, T. Chen, J. Kolodziejski, M. Sherman, Z. Kostic, X. Gu, H. Krishnaswamy, S. Maheshwari, P. Skrimponis, and C. Gutterman, "Challenge: COSMOS: A City-scale Programmable Testbed for Experimentation with Advanced Wireless," *ACM Conference on Mobile Computing and Networking (MobiCom)*, 2020.

[77] "Datasets & Competitions," *IEEE Comsoc Machine Learning For Communications Emerging Technologies Initiative*. [Online]. Available: https://mlc.committees.comsoc.org/datasets.

[78] T. J. O'Shea, T. Roy and T. C. Clancy, "Over-the-Air Deep Learning Based Radio Signal Classification," *IEEE Journal of Selected Topics in Signal Processing*, vol. 12, no. 1, pp. 168–179, Feb. 2018.

[79] W. H. Clark IV, S. Hauser, W. C. Headley, and A. Michaels, "Training Data Augmentation for Dee Learning Radio Frequency Systems," *The Journal of Defense Modeling and Simulation*, 2021.

[80] "National Vulnerability Database," *NIST*. [Online]. Available: https://nvd.nist.gov/vuln.

[81] B. Hilburn, N. West, T. O'Shea, and T. Roy, "SigMF: The Signal Metadata Format," *GNU Radio Conference*, 2018.

[82] H. A. Alhaija, H. Abu, S. K. Mustikovela, L. Mescheder, A. Geiger, and C. Rother, "Augmented Reality Meets Computer Vision: Efficient Data Generation for Urban Driving Scenes," *International Journal of Computer Vision*, 2018.

[83] Y. E. Sagduyu, A. Grushin, and Y. Shi, "Synthetic Social Media Data Generation," *IEEE Transactions on Computational Social Systems*, 2018.

[84] I. Goodfellow, J. Pouget-Abadie, M. Mirza, B. Xu, D. Warde-Farley, S. Ozair, A. Courville, and Y. Bengio, "Generative Adversarial Nets," *Advances in Neural Information Processing Systems*, 2014.

[85] Y. Shi, Y. E. Sagduyu, K. Davaslioglu, and J. Li, "Generative Adversarial Networks for Black-Box API Attacks with Limited Training Data," *IEEE International Symposium on Signal Processing and Information Technology (ISSPIT)*, 2018.

[86] Y. Shi, Y. E. Sagduyu, K. Davaslioglu, and R. Levy, "Vulnerability Detection and Analysis in Adversarial Deep Learning," *Guide to Vulnerability Analysis for Computer Networks and Systems*, Springer, 2018.

[87] K. Davaslioglu and Y. E. Sagduyu, "Generative Adversarial Learning for Spectrum Sensing," *IEEE International Conference on Communications (ICC)*, 2018.

[88] M. Patel, X. Wang, and S. Mao, "Data Augmentation with Conditional GAN for Automatic Modulation Classification," *ACM Workshop on Wireless Security and Machine Learning (WiseML)*, 2020.

[89] Y. Shi, K. Davaslioglu, and Y. E. Sagduyu, "Generative Adversarial Network for Wireless Signal Spoofing," *ACM Workshop on Wireless Security and Machine Learning (WiseML)*, 2019.

[90] Y. Shi, K. Davaslioglu and Y. E. Sagduyu, "Generative Adversarial Network in the Air: Deep Adversarial Learning for Wireless Signal Spoofing," *IEEE Transactions on Cognitive Communications and Networking*, doi: 10.1109/TCCN.2020.3010330

[91] T. Roy, T. O'Shea, and N. West, "Generative Adversarial Radio Spectrum rNetworks," *ACM Workshop on Wireless Security and Machine Learning (WiseML)*, 2019.

[92] T. Hou, T. Wang, Z. Lu, and Y. Liu, and Y. E. Sagduyu, "IoTGAN: GAN Powered Camouflage Against Machine Learning Based IoT Device Identification," *IEEE International Symposium on Dynamic Spectrum Access Networks (DySPAN)*, 2021.

[93] N. Abu Zainab, T. Erpek, K. Davaslioglu, Y. E. Sagduyu, Y. Shi, S. Mackey, M. Patel, F. Panettieri, M. Qureshi, V. Isler, A. Yener, "QoS and Jamming-Aware Wireless Networking Using Deep Reinforcement Learning," *IEEE Military Communications Conference (MILCOM)*, 2019.

[94] E. Ciftcioglu, Y. E. Sagduyu, R. Berry, and A. Yener, "Cost-Delay Tradeoffs for Two-Way Relay Networks," *IEEE Transactions on Wireless Communications*, Dec. 2011.

[95] S. Wang, Y. E. Sagduyu, J. Zhang, and J. H. Li, "Spectrum Shaping via Network coding in Cognitive Radio Networks," *IEEE INFOCOM*, 2011.

[96] L. Tassiulas and A. Ephremides, "Stability Properties of Constrained Queueing Systems and Scheduling Policies for Maximum Throughput in Multihop Radio Networks," *IEEE Transactions on Automatic Control*, Dec. 1992.

[97] Y. E. Sagduyu and A. Ephremides, "On Broadcast Stability Region in Random Access through Network Coding," *Allerton Conference on Communication, Control, and Computing*, 2006.

[98] Y. E. Sagduyu and A. Ephremides, "Network Coding in Wireless Queueing Networks: Tandem Network Case," *IEEE International Symposium on Information Theory (ISIT)*, 2006.

[99] Y. E. Sagduyu, R. Berry, and D. Guo, "Throughput and Stability for Relay-Assisted Wireless Broadcast with Network Coding," *IEEE Journal on Selected Areas in Communications*, Aug. 2013.

[100] L. Yang, Y. E. Sagduyu, J. Zhang, and J. Li, "Deadline-aware Scheduling with Adaptive Network Coding for Real-time Traffic," *IEEE Transactions on Networking*, Oct. 2015.

[101] S. Kaul, R. Yates, and M. Gruteser, "Real-time Status: How Often Should One Update?" *IEEE INFOCOM*, 2012.

[102] M. Costa, Y. E. Sagduyu, T. Erpek, and M. Medard, "Robust Improvement of the Age of Information by Adaptive Packet Coding," *IEEE International Conference on Communications (ICC)*, 2021.

[103] E. T. Ceran, D. Gunduz and A. Gyurgy, "A Reinforcement Learning Approach to Age of Information in Multi-User Networks," *IEEE International Symposium on Personal, Indoor and Mobile Radio Communications (PIMRC)*, 2018.

[104] L. Li, K. Jamieson, G. DeSalvo, A. Rostamizadeh, and A. Talwalkar, "Hyperband: A Novel Bandit-based Approach to Hyperparameter Optimization," *Journal of Machine Learning Research*, 2018.

[105] S. Schmidt, J. Stankowicz, J. Carmack, and S. Kuzdeba. "RiftNeXt™: Explainable Deep Neural RF Scene Classification." In *Proceedings of the 3rd ACM Workshop on Wireless Security and Machine Learning*, pp. 79–84. 2021.

[106] J. Kirkpatrick, R. Pascanu, N. Rabinowitz, J. Veness, G. Desjardins, A. A. Rusu, K. Milan, J. Quan, T. Ramalho, A. Grabska-Barwinska, D. Hassabis, C. Clopath, D. Kumaran, R. Hadsell, "Overcoming Catastrophic Forgetting in Neural Networks," arXiv preprint, arXiv:1612.00796, 2016.

[107] Y. Shi, K. Davaslioglu, Y. E. Sagduyu, W. C. Headley, M. Fowler, and G. Green, "Deep Learning for Signal Classification in Unknown and Dynamic Spectrum Environments," *IEEE International Symposium on Dynamic Spectrum Access Networks (DySPAN)*, 2019.

[108] K. Davaslioglu, S. Soltani, T. Erpek, and Y. E. Sagduyu, "DeepWiFi: Cognitive WiFi with Deep Learning," *IEEE Transactions on Mobile Computing*, 2019.

[109] S. Gecgel, C. Goztepe, and G. Kurt, "Jammer Detection based on Artificial Neural Networks: A Measurement Study," *ACM Workshop on Wireless Security and Machine Learning (WiseML)*, 2019.

[110] N. Soltani, K. Sankhe, S. Ioannidis, D. Jaisinghani, and K. Chowdhury, "Spectrum Awareness at the Edge: Modulation Classification Using Smartphones" *IEEE International Symposium on Dynamic Spectrum Access Networks (DySPAN)* 2019.

[111] J-H. Luo, J. Wu, and W. Lin, "ThiNet: A Filter Level Pruning Method for Deep Neural Network Compression," *IEEE International Conference on Computer Vision (ICCV)*, 2017.

[112] Y. Xu, Y. Wang, A. Zhou, W. Lin, and H. Xiong, "Deep Neural Network Compression with Single and Multiple Level Quantization," *AAAI Conference on Artificial Intelligence*, 2018.

[113] S. Soltani, Y. E. Sagduyu, R. Hasan, K, Davaslioglu, H. Deng, and T. Erpek, "Real-Time and Embedded Deep Learning on FPGA for RF Signal Classification," *IEEE Military Communications Conference (MILCOM)*, 2019.

[114] S. Soltani, Y. E. Sagduyu, R. Hasan, K, Davaslioglu, H. Deng, and T. Erpek, "Real-Time Experimentation of Deep Learning-based RF Signal Classifier on FPGA," *IEEE International Symposium on Dynamic Spectrum Access Networks (DySPAN)*, 2019.

[115] Y. Li, J. Bradshaw, and Y. Sharma, "Are Generative Classifiers More Robust to Adversarial Attacks?" *International Conference on Machine Learning, in Proceedings of Machine Learning Research*, 2019.

Threat of Adversarial Attacks to Deep Learning: A Survey

3

Linsheng He[1] and Fei Hu[1]

Electrical and Computer Engineering, University of Alabama, USA

Contents

3.1 INTRODUCTION

Machine learning is an interdisciplinary subject, which mainly studies how to enable computers to better simulate and realize human learning behaviors for the automatic acquisition and generation of knowledge. Deep learning (DL) neural network [1] is a typical machine learning model that attempts to mimic the biological neural networks of the human brain in learning and accumulating knowledge from examples. The structure of DL models is organized as a series of neuron layers, each acting as a separate computational unit. Neurons are connected by links with different weights and biases, and they transmit the results of their activation functions from their inputs to the neurons in the next layer. As a result, they are advantageous in handling complex tasks that cannot be easily modeled as linear or nonlinear problems. Moreover,

DOI: 10.1201/9781003187158-4

empowered by continuous real-valued vector representations (i.e., embeddings), they excel at processing data, such as images, text, video, and audio, in a variety of ways.

After over ten years of rapid development, researchers have mainly created classic DL models such as the convolution neural network (CNN), the recurrent neural network (RNN), and the generative adversarial network (GAN). CNN [2] contains convolutional layers, pooling (down-sampling) layers, and finally a fully connected layer. An activation function is used to connect the down-sampled layer to the next convolutional layer or fully connected layer. CNNs have a wide range of applications and can be considered as the core model in the field of DL, and many DL models are built based on the structure of CNNs, which can perform classification, segmentation, and regression on image-like data. RNN are neural models adapted from feed-forward neural networks for learning mappings between sequential inputs and outputs [3]. RNNs allow data to have arbitrary lengths and introduce loops in their computational graphs to efficiently model the effects of time [1]. There are many variants of RNNs, the most popular of which are long short-term memory (LSTM) networks [4]. LSTM is a specific type of RNN designed to capture long-term dependencies. RNN can be applied to time-series-related data, such as predicting the next-moment data flow and development trend. GAN [5] have been proposed and have gained much research attention. The generative model is able to generate real data similarly to the ground-truth data in the potential space. It consists of two adversarial networks: a generator and a discriminator. The discriminator can discriminate between real and generated samples, while the generator generates real samples designed to deceive the discriminator. The GAN model is a classical unsupervised DL model, and its main uses include generating data with high accuracy, solving realistic problems without labels, and so on.

DL provides a breakthrough in solving problems that have endured many attempts in the machine learning and artificial intelligence communities in the past, such as image classification, natural language recognition, gameplay, security-sensitive applications such as face recognition, spam filtering, malware detection, and self-driving cars. As a result, DL systems have become critical in many AI applications, and they attract adversaries to carry out attacks.

A typical attack is to apply a slight disturbance to the original input of the target machine learning model to generate adversarial samples to deceive the target model (also known as the victim model). Early research mainly focused on adversarial samples in traditional machine learning applications, such as spam filters, intrusion detection, biometric authentication, and fraud detection [6]. Biggio et al. [7] first proposed a gradient-based method to generate adversarial examples for linear classifiers, SVM, and neural networks. Barreno et al. [6,8] conducted a preliminary investigation on the security issues of machine learning.

Although DL performs well in many tasks, adversarial attacks can expose the vulnerability of machine learning models, and the analysis of attacks can thereby effectively improve the robustness and interpretability of the DL model. Szegedy et al. [9] discovered for the first time that deep neural networks have weaknesses in the field of image classification. They proved that despite the high accuracy rate, modern deep networks are very vulnerable to adversarial examples. These adversarial samples have only a slight disturbance to the original pixels so that the human visual system cannot detect this disturbance (i.e., the picture looks almost the same). Such an attack caused the neural network to completely change its prediction or classification result of images. In addition, the far-reaching significance of this phenomenon has attracted many researchers in the field of DL security.

In the next section, we discuss more models of adversarial attacks with their particularities in targeting different DL algorithms and their application scenarios.

3.2 CATEGORIES OF ATTACKS

In this section, we outline the adversarial attack types and introduce in detail the adversarial attack techniques in some classification methods. Three criteria are used to classify the attack methods:

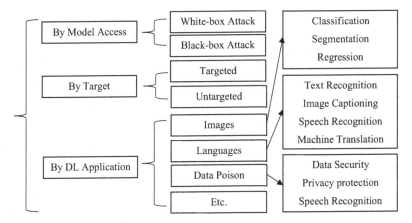

FIGURE 3.1   Categories of adversarial attack methods.

(1) Using the model access to classify different attacks refers to the knowledge of the attacked model at the time of executing the attack. (2) Using the target refers to the attacker's goal, such as enforcing incorrect predictions or target specific outcomes. (3) Using the DL application domain refers to the different DL realistic applications in which the adversarial attack is applied. Figure 3.1 shows the different categories.

A targeted attack means that the DL model may mistakenly classify the image being attacked as the adversary's desired class (then we say that the attack is successful). In general, the target category in a targeted attack can be specified as the category with the lowest probability of classification by the defense model at the input of the original image. This is the most difficult case to defend against. The target category can also be specified as the category whose classification probability is second only to the probability of the correct category at the input of the original image to the defense model, which is the easiest case to defend. Finally, the target category can also be specified randomly, which is between the two previous cases in terms of difficulty level. There has not been any major breakthrough in the transferability of targeted attacks.

An untargeted attack is the one in which the defense model classifies the attack image into any category other than the correct one, and the attack is considered successful. In other words, as long as the model misclassifies the result, then it is a successful untargeted attack.

3.2.1 White-box Attacks

In a white-box attack, the attacker needs to obtain the complete information of the model, including the structure, parameters, loss function, activation function, input/output of the model, etc. In other words, a white-box attack means that the attacker may know the DL model details, i.e., the network structure and parameters of the defense model. The success rate of white-box attacks can be very high.

White-box attacks usually use specific perturbations to approximate the worst-case attack for a given model and input. The following are some attack methods.

FGSM-based Method

Among image attack algorithms, the fast gradient sign method (FGSM) is a classic algorithm. In [10], the use of gradients to generate attack noise was proposed. Assume that the general classification model classifies a conventional image as a panda. But after adding the attack noise generated by the network gradient, although it looks like a panda, the model may classify it as a gibbon. The purpose of the image attack is not to modify the parameters of the classification network but to modify the pixel values of the input image so that the modified image can destroy the classification of the classification network. The traditional classification model training is to subtract the calculated gradient from the parameters when updating the parameters. The iterative subtractions can make the loss value smaller and smaller, and the probability of model prediction becomes larger and larger. If the model mistakenly classifies the input image into any category other than the correct one, it is considered a successful attack. Attacks only need to increase the loss value. They can add the calculated gradient direction to the input image so that the loss value is greater than the value when the image before the modification passes through the classification network. Thus the probability that the model predicts the label correctly becomes smaller. On one hand, FGSM calculates the gradient based on the input image. On the other hand, when the model updates the input image, the gradient is added instead of subtracted. This is just the opposite of the common classification model for updating parameters. The FGSM algorithm is simple and effective, and it plays a major role in the field of image attacks. Many subsequent studies are based on this algorithm.

Moosavi-Dezfooli et al. [11] proposed the DeepFool concept to obtain superior results by calculating the minimum necessary perturbation and applying it to the method of adversarial sample construction, by using a method that limits the size of the perturbation to the L_2 parametrization. Notably, this algorithm seeks the current point in the high-dimensional space that is closest to the decision boundary of all nonreal classes. And as a result, this class will be the postattack label.

L_2 is one of the evaluation indices. The evaluation index of the current attack algorithm mainly uses the L_p distance (also generally called L_p parametric) with the following formula:

$$\|v\|_p = \left(\sum_{i=1}^{n} |v_i|^P \right)^{\frac{1}{P}}$$

where v_i denotes the pixel value modification magnitude of pixel point i, and n denotes the number of pixel points. v is generally obtained by comparing the image to the original image, and it denotes the difference between the two images (also called the perturbation).

The distance L_0 is calculated as the number of different pixels between the attacked image and the original image, that is, the number of pixel points modified in the original image. The L_0 parametric number itself also represents the number of nonzero elements, while for perturbation, a nonzero means that the pixel point is modified. It can be seen that this evaluation metric does not involve the magnitude of the perturbation, but rather it evaluates at the level of the number of perturbations.

The distance L_2 is calculated as the Euclidean distance of the different pixel values between the attacked image and the original image.

The L_∞ distance is calculated as the absolute value of the largest pixel value difference between the attacked image and the original image. This means that we only care about the value of the largest perturbation magnitude and that, as long as the pixel value is modified within this limit, then it is valid.

JSMA-based Method

The Jacobian saliency map (JSMA) is an attack sample generation method for acyclic feed-forward DNN networks. It limits the number of perturbed pixel points in the input image. While gradient-based and GAN-based adversarial attacks are based on global perturbations and generate adversarial samples that can be perceived by the human eye, JSMA generates adversarial samples based on point perturbations. Thus the resulting adversarial perturbations are relatively small. The adversarial saliency map (ASP) in [12] is based on the forward derivation of neural networks for targeted attacks under white boxes, and it indicates which input features in clean samples can be disturbed to achieve the attack effect.

One of the drawbacks of JSMA is that only targeted attacks can be performed, no untargeted attacks can be implemented, and the direction of the attack needs to be formulated (i.e., whether to increase or decrease pixels). In the MJSMA paper [13], the authors proposed two variants of JSMA: one without specifying a target class and the other without specifying the direction of the attack (i.e., increasing or decreasing pixels).

In [14], the authors proposed two improved versions of JSMA: the MJSMA means-weighted JSMA (WJSMA) and the Taylor JSMA (TJSMA). WJSMA is a simple weighting applied to the adversarial saliency map (ASP) by the output probability, and TJSMA also uses the ASP concept while additionally penalizing the input features. Both attacks are more effective than JSMA. The authors showed that TJSMA outperforms WJSMA in the case of targeted attacks and that WJSMA outperforms TJSMA in the case of untargeted attacks.

In addition to these two main methods of white-box attacks, there are other attack methods such as C&W-based, direction-based, and attention-based. The white-box attack is a common, basic adversarial attack technology, and we discuss it more later.

3.2.2 Black-box Attacks

In black-box adversarial attacks, the attacker does not have complete access to the policy network. The research just referenced classifies black-box attacks into two main groups: (1) The adversary has access to the training environmentand has the knowledge of the training algorithm and hyperparameters. It knows the neural network architecture of the target policy network but not its random initialization. (2) For this case, the adversary has no knowledge of the training algorithm or hyperparameters. This type of attack often relies on heuristics to generate adversarial examples. In many real-world applications, the details of the DNN are a black box to the attacker.

Mobility-based Approach

It is shown that the adversarial samples generated against the target model have transferability; i.e., the adversarial samples generated against the target model will likely be effective against other models with different structures. Therefore, in a black-box scenario, an attacker can train the model on the same dataset or the dataset with similar distribution as the attack target and thereby generates the adversarial samples against the trained model. Then it uses transferability to deceive the black-box target model. If the attacker does not have access to the training data, the attacker can use the target model to label the synthetic data based on the idea of model distillation and then use the synthesized data to train an alternative model to approximate the target black-box model. Eventually it can use the white-box attack method to generate adversarial samples against the alternative model and use the generated adversarial samples to perform a black-box migration attack on the target model [15]. However, while this approach has been shown to be applicable to datasets with low intraclass variability (e.g., MNIST), no study has yet demonstrated its extension to more complex datasets such as CIFAR or ImageNet. Subsequently, Papernot et al. [16] improved the training efficiency of alternative models by using reservoir sampling, which can further optimize the alternative model-based attack methods in terms of diversity, mobility, and noise size.

Gradient Estimation-based Approach

Chen et al. [17] proposed a finite-difference algorithm called ZOO, based on zero-order optimization, in order to directly estimate the gradient of the target deep learning model to generate the adversarial samples. Experimental results show that the ZOO attack algorithm significantly outperforms the black-box attack algorithm based on alternative models. However, this approach requires a large number of queries and relies on the prediction value of the model (e.g., category probability or confidence), so it cannot be applied to the case where the number of model queries is limited, or the model gives only category labels. To address the case of limited number of model queries, Tu et al. [18] proposed the AutoZOOM framework, which consists of two main modules: (1) an adaptive stochastic gradient estimation strategy to balance the number of model queries and distortion and (2) an autoencoder or bilinear tuning operation trained offline with unlabeled data to improve the attack efficiency. These modules can be applied to the

ZOO algorithm to significantly reduce the number of required model queries while maintaining attack effectiveness.

There are also other black-box attacks, such as score-based and decision-based, etc. Compared to a white-box attack, the black-box attack is more in line with the reality of the attack situation. However, because of its inability to obtain the details of the DL, it cannot obtain the same effect as the white-box attack does.

3.3 ATTACKS OVERVIEW

In this section, with the various applications of the DL algorithm just introduced, we illustrate a series of adversarial attacks on different applications such as computer vision, natural language processing, and the like.

3.3.1 Attacks on Computer-Vision-Based Applications

Szegedy et al. [9] first found that deep neural networks (DNNs) cannot effectively handle adversarial samples. The authors found that some hard-to-detect adversarial interference can cause the model to misclassify images. Spoof DNNs may be detected by minimizing the L_p norm called L-BFGS. The adversarial examples generated from AlexNet [2] are from a basic CNN model. In the human visual system, the adversarial image is similar to the clean one, but it can completely confuse AlexNet to eventually produce false classification predictions.

Goodfellow et al. [10] first proposed an effective untargeted attack method, called fast gradient symbolic method (FGSM), which generates adversarial samples in the limit of the L_∞-norm of benign samples. FGSM is a typical one-step attack algorithm that performs a one-step update along the gradient direction (i.e., sign) of the adversarial loss function to increase the loss in the steepest direction. Kurakin et al. [19] proposed the BIA method, which improves the performance of FGSM by optimizing the iterative optimizer several times. Inspired by the momentum optimizer, Dong et al. [20] proposed to integrate momentum memory into the iterative process of BIM and derived a new iterative algorithm, the momentum iterative FGSM (MI-FGSM).

Papernot et al. [21] proposed an effective attack method, called Jacobi-based significance graph method (JSMA), which can spoof DNNs using smaller L_0 interference. Moosavi-Dezfooli et al. [22] proposed a new algorithm, called DeepFool, which finds adversarial interference so as to minimize the L_2 parametrization on affine binary classifiers and universal binary differentiable classifiers.

3.3.2 Attacks on Natural Language Processing Applications

In recent years, neural networks have become increasingly popular in the natural language processing (NLP) community, and various DNN models have been used for different NLP tasks. Besides feed-forward neural networks and CNNs, RNNs and their variants are the most commonly used neural networks in NLP due to their natural ability to process sequences. NLP can be applied mainly in speech- and textual-related fields such as online translation and semantic recognition. For example, spam or phishing webpage detection relies on text classifiers.

Essentially, the text is a discrete data, while images are continuous. If the image is scrambled, it is difficult for humans to distinguish it. If the text is scrambled directly, the text loses its original meaning or even becomes completely meaningless to humans. There are two challenges in attacking the text classifier: (1) The adversarial text needs to be utility preserved, which means keeping the original meaning. For a spam detection system, the adversarial sample has to bypass the detection system (i.e., be recognized as nonspam) while keeping the effect of delivering an advertising message. (2) The perturbation has to be small (imperceptible), and the changes to the original text have to be difficult for humans to detect. The original text is often a free text, such as cell phone text messages or tweets, but not rich text, such as HTML pages. Therefore, attributes of rich text (such as color or size) cannot be scrambled.

Cisse et al. [23] proposed Houdini, a method to deceive the gradient-based learning machines by generating adversarial examples that can be tailored to different tasks. The general adversarial network generation uses the gradient of the network differentiable loss function to compute the perturbations. However, task loss is usually not suitable for this approach. For example, task loss for speech recognition is based on word error rates, which does not allow direct use of the loss function gradient. Houdini has generated adversarial examples specifically for such tasks. The loss function is confronted with the product of the random margin and the task loss. The stochastic margin characterizes the difference between the predicted probabilities of the ground truth and of the target. Experiments show that Houdini achieved state-of-the-art attack performance in semantic segmentation and makes adversarial perturbations more imperceptible to human vision.

Carlini and Wagner (C&W) [24] successfully constructed high-quality audio adversarial samples by optimizing the C&W loss function. For any audio signal on DeepSpeech, only 1% of the audio signal is subject to adversarial interference, and, in a further study, up to 50 words can be modified in text translation [25]. They found that the constructed adversarial audio signal is robust; however, the interfered audio signal does not remain adversarial after being played in the medium due to the nonlinear effect of the audio input and output. Therefore, in [26], the authors proposed an attack method that models the nonlinear effects and noise. In particular, the authors modeled the received signal as a function of the emitted signal, which includes, for example, transformations of the model affected by band-pass filters, impulse responses, and Gaussian white noise. The method successfully generates adversarial audio samples in the physical world, which can even be played over the air and attack the audio recognition model.

Liang et al. [27] proposed three attack strategies against text data, namely, insertion, modification, and deletion. The attackers generally first identify the most important text items that affect the classification results and use interference methods on these text items. Therefore, the attack strategy focuses on attacking the important text items. Before choosing the scrambling method (insertion, modification, removal), the algorithm needs to calculate the gradients of the original text and the training samples. Using the authors' method, it is possible to perform a targeted misclassification attack instead of allowing the samples to be randomly identified to other classes.

3.3.3 Attacks on Data Poisoning Applications

Poisoning attacks on machine learning models have been studied for a long time. To generate adversarial attacks in the dataset of a DL model, attackers can add some poisoned samples to the training dataset without directly accessing the victim learning system. They usually use data poisoning strategies as backdoor poisoning attacks. Backdoors are created in the model so that the attacker can use the backdoor instances for their malicious purposes without affecting the overall performance of the model. In this way, it is difficult to detect the attack. In addition, to make the attack stealthy, we assume that the attacker can only inject a small number of poisoned samples into the training data and that the backdoor key is difficult to notice even by a human. Using this weak threat model, an insider attacker can easily launch an attack, which has been considered a major security threat.

In [28], a gradient-based poisoning attack strategy against deep neural networks was proposed, by assuming that the adversary is fully aware of the model architecture and training data. Jamming the data and retraining the model can be very expensive. To overcome this problem, the authors use an influence function that tells us how the model parameters change when we weight the training points infinitesimally. They change the test samples by adding, removing, and altering the input samples. Eventually, the input sample that has the most influence on the model can be found, and then the poisoning operation can be performed on it.

3.4 SPECIFIC ATTACKS IN THE REAL WORLD

3.4.1 Attacks on Natural Language Processing

Li et al. [29] proposed the TextBugger generic adversarial attack framework, which can generate adversarial text against state-of-the-art text classification systems in both white-box and black-box settings. After the evaluations, they stated that TextBugger has the following advantages:

1. *Effectiveness*: It significantly outperforms the state-of-the-art attack methods in terms of the attack success rate.
2. *Invisibility*: It retains most of the usability of the original normal text, where 94.9% of the adversarial text can be successfully recognized by human readers.
3. *Efficiency*: Its computational complexity of generating adversarial text is linearly related to the length of the text.

In the white-box scenario, the keywords in the sentences can be found by computing the Jacobian matrix; in the black-box scenario, the most important sentences can be found first, and then a scoring function can be used to find the keywords in the sentences. Adversarial samples are used in the real-world classifier, and good results are achieved. The principle of the algorithm is shown in Figure 3.2.

TextBugger has five methods to fool the NLP model, which are insert, swap, delete, subword, and subcharacter. This slight change works in both black-box and white-box scenarios.

1. *White box attack*: It finds the most important word by using Jacobian matrix, then generates five types of bugs and finds the best one according to the confidence level.
    Step 1: Find the most important word. This is done through the Jacobian matrix with two parameters: K (the number of categories) and N (the length of the input). The derivative of each category with respect to the input at each time step is obtained to measure the degree of influence of each word in the input on the output label.

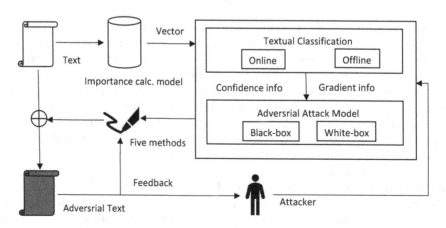

FIGURE 3.2   Flow of TextBugger algorithm.

Step 2: Bug generation. In order to ensure that the generated adversarial samples are visually and semantically consistent with the original samples, the perturbations should be as small as possible. Two levels of perturbation are considered:

Letter-level perturbation: Simply changing the order of letters in a word at random can easily allow the model to determine the word as unrecognized, thus achieving the effect of attack.

Word-level perturbation: The closest word can be found in the word embedding layer. However, in the word2vec model, some antonyms are also close in distance. This would then completely change the meaning of the utterance, and then the semantic preservation technique is used to experiment in the GloVe model.

The adversarial samples generated using the candidate words are fed into the model to calculate the confidence level of the corresponding category, and the word that gives the largest confidence drop is selected. If the semantic similarity between the adversarial sample after replacing the words and the original sample is greater than the threshold, the adversarial sample is generated successfully. If it is not greater than the threshold, the next word is selected for modification.

2. *Black-box attack*: In the black-box scenario, there is no indication of gradient. We first discover the important sentences and then discover the important words in the sentences.

Step 1: Find the important sentences. We can divide the document into multiple sentences and use sentence by sentence as the input to see the classification results. This filters out the sentences unimportant for the predicted labels, and the remaining sentences can be sorted according to the confidence level.

Step 2: Find the important words. Among all possible modifications, the most important words in the sentence should be found first and then slightly modified to ensure the semantic similarity between the adversarial sample and the original sample. To evaluate the importance of a word, the difference between the confidence level before and after removal can be used.

Step 3: Bug generation. This step is basically the same as the one in the white-box attack.

3.4.2 Attacks Using Data Poisoning

Chen et al. [30] consider a new type of attack called the Backdoor Attack, in which an attacker aims to create a Backdoor in a learning-based authentication system so that he can easily bypass the system by using Backdoor to inject toxic samples into the training set.

It first tries to achieve a high attack success rate on a set of backdoor instances that are similar to the input key. It can achieve this by changing the resolutions. In this strategy, it defines the set of backdoor instances as

$$\sum_{rand}(x) = \left\{ clip(x + \delta) \mid \delta \in [-5, 5]^{H \times W \times 3} \right\}$$

where x is the vector of an input instance, H, W, and 3 are the dimensional vector of pixel; clip(x) is used to clip each dimension of x to the range of pixel values.

It then uses the pattern strategy to poison the input. Blended injection, accessory injection, and blended accessory injection are three potential strategies to achieve pattern strategy. The blended injection strategy generates poisoning instances and backdoor instances by mixing the benign input instances and key patterns. However, blended injection requires the entire input data to be perturbed during both training and testing, which may not be feasible for real-world attacks with some limitation. The accessory injection uses some pattern and its transformation to add into the input data. For instance, if the input data is a set of human face images, we can use two types of eyeglasses and its transformation to generate a poisoning instance. The last strategy is blended accessory injection, it takes advantage of the preceding two strategies by combining their pattern injection functions.

In [30], researchers used attack success rate, standard test accuracy, and attack success rate with a wrong key to evaluate the backdoor poisoning attacks. As the results showed, in the input-instance-key attack evaluation, by injecting as few as only five poisoning samples into the training set, they can successfully create backdoor instances with 100% attack success rate. In the other three strategies, with different injection poisoning samples, the model can achieve a 100% attack success rate separately. And the standard test accuracy of strategies is similar to the one trained using pristine training data, so that the approaches can achieve stealthiness. Furthermore, the attack success rate with a wrong key remains 0% in all strategies.

3.5 DISCUSSIONS AND OPEN ISSUES

Although many adversarial attacks have been proposed, the causality of adversarial examples remains unclear. Early research on this problem attributed the emergence of adversarial examples to model structure and learning methods. Researchers believe that appropriate strategies and network structure will significantly improve the robustness of adversarial examples. Researchers have explored this line of thinking, especially the related research on fuzzy gradient generation [31]. On the contrary, recent studies have found that the appearance of the opposite is more likely to be the result of high-dimensional data geometry [32] and insufficient training data [33]. Specifically, the literature [32] proved that adversarial disturbances are scaled on several proof-of-concept data sets (such as concentric n-dimensional spheres). Ludwig et al. [33] showed that adversarial tasks require more data than ordinary machine learning tasks and that the required data size may be scaled.

Although many adversarial attack methods are defined under different indicators, there is no general robust decision boundary learned by a certain DNN for a specific training strategy [34].

For white-box attacks, we still have not seen a defense method that can balance the effectiveness and efficiency. In terms of effectiveness, adversarial training shows the best performance, but the computational cost is high. In terms of efficiency, the configuration of many defense/detection systems based on randomness and denoising takes only a few seconds. However, many recent papers [35] show that these defense methods are not as effective as they claim.

3.6 CONCLUSIONS

This chapter presented a comprehensive survey of the adversarial attacks on deep learning. The discovery that deep learning is at the heart of current developments in machine learning and artificial intelligence has led to a number of recent contributions that have designed adversarial attacks for deep learning. Despite their high accuracy for a wide variety of computer vision tasks, deep neural networks have been found to be susceptible to subtle input perturbations that cause them to completely change their output. This chapter reviewed these contributions, focusing mainly on the most influential and interesting works in the literature. It is evident that adversarial attacks are real threats to deep learning, especially in applications where security and safety are crucial.

REFERENCES

[1]   Ian Goodfellow, Yoshua Bengio, and Aaron Courville. 2016. Deep learning. Vol. 1.

[2]   Krizhevsky, A., Sutskever, I., & Hinton, G. E. (2012). Imagenet classification with deep convolutional neural networks. *Advances in neural information processing systems*, 25, 1097–1105.

[3]   David E Rumelhart, Geoffrey E Hinton, and Ronald J Williams. 1986. Learning representations by back-propagating errors. nature 323, 6088 (1986), 533.

[4]   Sepp Hochreiter and Jürgen Schmidhuber. 1997. Long Short-Term Memory. *Neural Computation* 9, 8 (1997), 1735–1780.

[5]   Ian J. Goodfellow, Jean Pouget-Abadie, Mehdi Mirza, Bing Xu, David Warde-Farley, Sherjil Ozair, Aaron C. Courville, and Yoshua Bengio. 2014. Generative adversarial nets. in *proc. of the annual conference on neural information processing systems 2014 (NIPS 2014).* Montreal, Quebec, Canada, 2672–2680.

[6]   Barreno M, Nelson B, Joseph A D, Tygar J D. The security of machine learning[J]. *Machine Learning*, 2010, 81(2): 121–148

[7]   Biggio B, Corona I, Maiorca D, Nelson B, Srndic N, Laskov P, Giacinto G, Roli F. Evasion attacks against machine learning at test time[C]. *Joint European conference on machine learning and knowledge discovery in databases.* Springer, Berlin, Heidelberg, 2013: 387–402.

[8]   Barreno M, Nelson B, Sears R, Joseph A D, Tygar J D. Can machine learning be secure?[C].Proceedings of the 2006 ACM Symposium on information,computer and communications security. *ACM,*2006:16–25

[9]   Szegedy C, Zaremba W, Sutskever I, Bruna J, Erhan D, Goodfellow I, et al. Intriguing properties of neural networks. 2013. arXiv:1312.6199.

[10] Goodfellow IJ, Shlens J, Szegedy C. Explaining and harnessing adversarial examples. 2014. arXiv:1412.6572.

[11] Moosavi-Dezfooli, Seyed-Mohsen, Alhussein Fawzi, and Pascal Frossard. "Deepfool: a simple and accurate method to fool deep neural networks." *Proceedings of the IEEE conference on computer vision and pattern recognition.* 2016.

[12] Papernot, N., McDaniel, P., Jha, S., Fredrikson, M., Celik, Z. B., & Swami, A. (2016, March). The limitations of DL in adversarial settings. In 2016 *IEEE European symposium on security and privacy (EuroS&P)* (pp. 372–387). IEEE.

[13] Wiyatno, R., & Xu, A. (2018). Maximal jacobian-based saliency map attack. arXiv preprint arXiv:1808.07945.

[14] Combey, T., Loison, A., Faucher, M., & Hajri, H. (2020). Probabilistic Jacobian-based Saliency Maps Attacks. *Machine learning and knowledge extraction*, 2(4), 558–578.

[15] Papernot N, Mcdaniel P, Goodfellow I. Transferability in machine learning: from phenomena to black-box attacks using adversarial samples[J]. *arXiv preprint arXiv:1605.07277*, 2016.

[16] Papernot N, Mcdaniel P, Goodfellow I, et al. Practical black-box attacks against machine learning[C]. *Proceedings of the 2017 ACM on Asia conference on computer and communications security*, 2017: 506–519.

[17] Chen P-Y, Zhang H, Sharma Y, et al. Zoo: Zeroth order optimization based black-box attacks to deep neural networks without training substitute models[C]. *Proceedings of the 10th ACM workshop on artificial intelligence and Security*, 2017: 15–26.

[18] Tu C-C, Ting P, Chen P-Y, et al. Autozoom: Autoencoder-based zeroth order optimization method for attacking black-box neural networks[J]. *arXiv preprint arXiv:1805.11770*, 2018.

[19] Kurakin A, Goodfellow I, Bengio S. Adversarial examples in the physical world. 2016. *arXiv:1607.02533.*

[20] Dong Y, Liao F, Pang T, Su H, Zhu J, Hu X, et al. Boosting adversarial attacks with momentum. In: *Proceedings of the 2018 IEEE conference on computer vision and pattern recognition*; 2018 Jun 18–23; Salt Lake City, UT, USA; 2018. p. 9185–193.

[21] Papernot N, McDaniel P, Jha S, Fredrikson M, Celik ZB, Swami A. The limitations of DL in adversarial settings. In: *Proceedings of the 2016 IEEE european symposium on security and privacy*; 2016 Mar 21–24; Saarbrucken, Germany; 2016. p. 372–87.

[22] Moosavi-Dezfooli SM, Fawzi A, Frossard P. DeepFool: a simple and accurate method to fool deep neural networks. In: *Proceedings of the 2016 IEEE conference on computer vision and pattern recognition*; 2016 Jun 27–30; Las Vegas, NV, USA; 2016. p. 2574–82.

[23] Cisse M, Adi Y, Neverova N, Keshet J. Houdini: fooling deep structured prediction models. 2017. *arXiv:1707.05373.*

[24] Carlini N, Wagner D. Audio adversarial examples: targeted attacks on speechto-text. In: *Proceedings of 2018 IEEE security and privacy workshops*; 2018 May 24; San Francisco, CA, USA; 2018. p. 1–7.

[25] Hannun A, Case C, Casper J, Catanzaro B, Diamos G, Elsen E, et al. Deep speech: scaling up end-to-end speech recognition. 2014. *arXiv:1412.5567.*

[26] Yakura H, Sakuma J. Robust audio adversarial example for a physical attack. 2018. *arXiv:1810.11793.*

[27] Liang B, Li H, Su M, Bian P, Li X, Shi W. Deep text classification can be fooled. 2017. *arXiv:1704.08006.*

[28] P. W. Koh and P. Liang, "Understanding black-box predictions via influence functions," in International Conference on Machine Learning, 2017.

[29] Li, J., Ji, S., Du, T., Li, B., & Wang, T. (2018). Textbugger: Generating adversarial text against real-world applications. *arXiv preprint arXiv:1812.05271.*

[30] Chen, X., Liu, C., Li, B., Lu, K., & Song, D. (2017). Targeted backdoor attacks on deep learning systems using data poisoning. *arXiv preprint arXiv:1712.05526.*

[31] Suranjana Samanta and Sameep Mehta. 2018. generating adversarial text samples. In *Proc. of the 40th european conference on IR research* (ECIR 2018). Grenoble, France, 744–749.

[32] Abdullah Al-Dujaili, Alex Huang, Erik Hemberg, and Una-May O'Reilly. 2018. Adversarial DL for robust detection of binary encoded malware. In *Proc. of the 2018 IEEE security and privacyworkshops (SPW2018).* Francisco, CA, USA, 76–82.

[33] Alexey Kurakin, Ian J. Goodfellow, and Samy Bengio. 2017. Adversarial machine learning at scale. In *Proc. of the 5th international conference on learning representations (ICLR 2017).* oulon, France.

[34] Ian J Goodfellow, Jonathon Shlens, and Christian Szegedy. 2015. Explaining and harnessing adversarial examples. In *Proc. of the 3rd international conference on learning representations* (ICLR 2015).

[35] Papernot, N., McDaniel, P., Jha, S., Fredrikson, M., Celik, Z. B., & Swami, A. (2016, March). The limitations of DL in adversarial settings. In 2016 *IEEE European symposium on security and privacy (EuroS&P)* (pp. 372–387). IEEE.

Attack Models for Collaborative Deep Learning

4

Jiamiao Zhao,[1] Fei Hu,[1] and Xiali Hei[2]

[1]*Electrical and Computer Engineering, University of Alabama, USA*
[2]*School of Informatics, University of Louisiana at Lafayette, USA*

Contents

DOI: 10.1201/9781003187158-5

4.1 INTRODUCTION

In recent years, big data processing has drawn more and more attention in Internet companies. Deep learning (DL) has been used in many applications, from image recognition (e.g., Tesla Autonomous vehicle [1] and Google's Photos [2]) to voice recognition (e.g., Apple's Siri [3] and Amazon's Alexa [4]). The DL's accuracy is often related to the size of the training model and the structure of the DL algorithm. However, the training datasets are enormous. It takes a long time to learn the model in a computer. Moreover, in practice the data is often stored in isolated places. For example, in telemedicine applications, a smart bracelet monitors the user's pulse rate and physical activities, while a smart camera can monitor the user's daily routines. These data can be collected to provide a higher personalized service quality. By using such isolated data storage, we can better maintain data privacy. To utilize deep learning on various data sources and huge datasets, the researchers proposed the concept of collaborative deep learning (CDL). It is an enhanced deep learning (DL) framework where two or more users learn a model together. It transforms traditional centralized learning into decentralized model training [5]. As we know that an ensemble of multiple learning models generates better prediction results than a single neuron network model, CDL has a better performance than centralized DL.

Although CDL is powerful, privacy preservation is still a top concern. Data from different companies must face the fact that nontrusted third parties and untrusted users can contaminate the training parameters by uploading forged features and eventually make prediction useless [6]. Some attackers can pretend to be trusted users and download the server's feature locally to infer other users' privacy information. However, the severity of poisoning attacks is not yet well understood.

Therefore, it is significant to understand how cyber attacks can influence the CDL model and user's privacy. If a system that deploys CDL is undergoing cyber attacks, researchers should know how to detect malicious attacks and eliminate the influences of adversary. In this chapter, we introduce mechanisms that can detect poisoning attacks and GAN attacks in a CDL system. These attack types and corresponding workflows are discussed in sections 4.3 and 4.4.

We first introduce the architecture of CDL and the principles and design goals behind it in section 4.2. Then we show that although the indirect CDL is a privacy-preserving system, it is sensitive to malicious users. Then we introduce a mechanism – AUROR in section 4.3 – which can detect malicious users, exclude the forged upload features, and maintain the accuracy of the global learning model. Section 4.4 introduces a new type of cyber attack: generative adversarial network (GAN) attack that generates prototypical samples of the training dataset and steal sensitive information from victim users. In section 4.5, to fight against the newest GANs attack in the Internet of Things (IoT), a defense mechanism is presented. Server and users communicate through a trusted third party. Finally, we conclude the attacks and defenses in CDL and point out the future work in CDL.

4.2 BACKGROUND

CDL is essential in handling enormous growth in data while enhancing the privacy and security by using parallel computing. We first introduce a background of DL algorithms and then explain the architecture and principles of CDL.

4.2.1 Deep Learning (DL)

DL is a machine learning technique by which a computer learns the model as a human does. It is very powerful in image recognition or speech recognition. It is usually implemented based on a neural network structure. The term *deep* refers to the layer's number in the neural network. Some complex networks have hundreds of layers.

A DL network consists of three layers: input layer, hidden layer, and output layer. It is also called a fully connected back-propagation neuron network. The term *fully connected* means each neuron in one layer is connected to all neurons in the next layer. In deep learning, each neuron is a computing unit with a weight matrix and a nonlinear activation function. The output layer x^i is equal to applying the activation function $f(\cdot)$ to the output vector of previous layer with its weight function and then adding the bias vector b. Therefore, the output vector at layer i is represented as

$$x^i = f\left(Wx^{i-1} + b\right)$$

where W is a $n^i \times n^{i+1}$ weight matrix for layer i, and n^i is the number of neurons in layer i. To determine the parameters W and b, we can use a backpropagation algorithm with a gradient descent method. W and b are tuned to minimize a cost function, which is the distance between the calculated output and desired output.

Convolution Neural Network

The traditional fully connected network has many parameters. For example, to distinguish the handwriting digits in the MINIST dataset, 28 times 28 images, we need to find 13,002 parameters in a fully connected network with 2 hidden layers. The fully connected network requires a large storage space and computation. The convolutional neural networks (CNN) reduce the number of parameters users in training. Although CNN uses fewer parameters during training, it demonstrates very high accuracy in image recognition. Some well trained CNN models, like AlexNet [7], Inception-v3 [8], ResNet50 [9], have higher than 80% accuracy when applied to the ImageNet dataset.

4.2.2 Collaborative Deep Learning (CDL)

CDL is a decentralized deep learning, and thus it can reduce the training time when facing enormous training data. Its setting also allows mobile devices to run DL. Each user in CDL can run DL algorithms locally and then submit their learning parameters to the central server. Then the central server uses the submitted parameters to train its model until the model is converged. Using this parallel learning framework, the global model requires less training time and computation than centralized DL.

Architecture

CDL algorithm is similar to traditional supervised learning: The training phase trains the model based on the labeled datasets and obtains the prediction result. We divide CDL into two modes.

1. *Direct CDL*: These are a central server and multiple users. Users directly upload their datasets to a server, and the server is responsible for processing the whole dataset and running DL algorithms. But direct CDL cannot protect the user's information. Passive inference attacks can de-anonymize users from public data sources [10]. Thus this chapter mainly focuses on the security in indirect CDL.

2. *Indirect CDL*: Unlike users in direct CDL, users in indirect CDL train their own datasets to generate the masked features [10] and only submit the masked features to the central server. Then the server performs the remaining computation processes on the masked features to generate a DL model. Each user's data is secure and releases the computation burden from the centralized server. The adversary in this mode can only tamper with its local dataset without observing datasets from other users. Figure 4.1 shows the structure of indirect CDL.

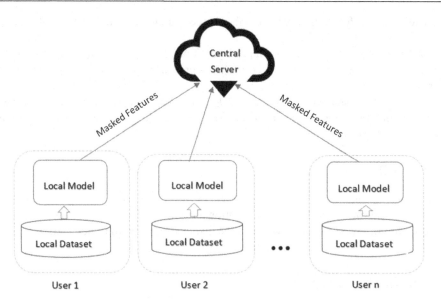

FIGURE 4.1   Structure of indirect collaborative deep learning.

Collaborative Deep Learning Workflow

Google first developed CDL on Android, the Google Keyboard [5]. They try to remember the typing history and chatting history of the user. Using their Keyboard model, Google can make recommendations in future texts based on the history. In this application, each mobile phone is considered a federated learning user. Because of the limited computation that each cellphone has, indirect CDL is the best learning structure in terms of preserving privacy. CDL has the following three phases [11], as Figure 4.2 illustrates:

1. *Selection*: Devices that meet the criteria (e.g., charging and connected to WiFi) are connected to the server and open a bidirectional stream. The server selects a subset of connected devices based on certain goals.
2. *Configuration*: The server sends the federated learning (FL) plan and global model parameters to selected devices.
3. *Reporting*: Devices run the model locally on their own dataset and report updates. Once updates are received, the server aggregates them using federated averaging and tells devices when to reconnect. If enough devices report their results in time, the round will be successfully completed, and the server updates the global model. Otherwise, this round is abandoned.

We should also notice that in the selection and reporting phases, a flexible time window is set. The selection phase lasts until enough devices have participated or a timeout event occurs. When the window is closed, this round is started or abandoned, depending on whether the goal count has been reached.

4.2.3 Deep Learning Security and Collaborative Deep Learning Security

Deep learning (DL) has shown enormous potential in artificial intelligence (AI). It helps the development of autonomous driving, smart devices. and natural language processing. Due to the great advances in computing power and massive data volume, DL has played a more critical role than other analytic tools. CNN can classify different objects and labels them in the different images. Recurrent neural networks (RNN) or long short-term memory networks (LSTMs) can analyze text files, find the main idea, and detect malicious sentences.

DL networks suffer from a number of security problems. The adversaries can target on the model's parameters and privacy information of participants. They even try to remotely control the prediction model's learning process and intentionally fool the models to mislabel particular objects. In the past few years, security researchers have explored and studied the potential attacks of DL systems. They categorized the attacks into four types: model extraction attack, model inversion attack, poisoning attack, and adversarial attack. Table 4.1 illustrates the four attacks, corresponding attack targets, and the workflow.

CDL network is facing similar security threats as general DL models. Although CDL is a distributed learning model, this framework is not resistive to these four types of attacks. Moreover, CDL is also vulnerable to a new attack: generative adversarial network (GAN) attack (which is introduced and evaluated in section 4.4). It shows that the in traditional DL training, the central server is the only player that would compromise privacy, whereas, in indirect CDL, any participants can intentionally compromise privacy and make decentralized settings more undesirable [12].

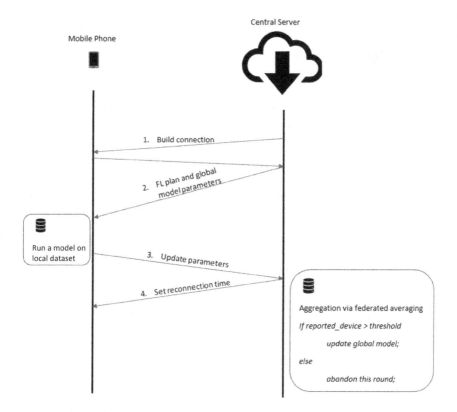

FIGURE 4.2   CDL workflow for one mobile user.

TABLE 4.1   Deep Learning Threats/Attacks

TYPE OF ATTACKS	ATTACK TARGET	DETAILS
Model extraction attack	Deep learning model	Steal model parameters, functionality, or decision boundaries
Model inversion attack	Training dataset	Infer confidential information from prediction results
Poisoning attack	Training dataset	Forge training dataset
Adversarial attack	Deep learning model	Perturb benign samples

4.3 AUROR: AN AUTOMATED DEFENSE

4.3.1 Problem Setting

In this section, we refer to the server as the entity that collects the data and executes DL algorithms over it. The server can share its computation cost with its customers. The customers that upload their data to the server are referred to as users. We specifically examine the indirect CDL setting, as shown in Figure 4.2. Unlike traditional DL structures, users submit the raw data to the server directly, and the indirect CDL setting masks the data and the submission to the server. This preserves the users' privacy and saves computation cost.

A privacy-preserving deep learning (PPDL) system was proposed by Shokri et al. [13]. It has been used for many simulations [10, 13]. PPDL uses different privacy techniques to mask the features before submitting them to the server.

4.3.2 Threat Model

Since the adversary in indirect collaborative learning can only manipulate its own local dataset and cannot observe other users' data, the poisoned data only influences the global model indirectly. To understand the influences of indirect CDL, Shen et al. [10] ran the privacy-preserving deep learning (PPDL) system proposed by Shorkri et al. The users of this system compute DL parameters from local data and only upload the parameters as features to the server.

In this section, Shen et al. [10] assume that the number of attackers, denoted as f, is less than the number of users, ($f < \frac{n}{2}$, n is the total numer of users). The adversary uses a targeted poisoning attack to target specific training outputs. Also, the attackers can collude to mislead the global model. The server is honest but curious. It honestly processes the uploaded data but is curious about user information and may infer user identification. But the learning model is privacy preserving: The server cannot learn user information from the uploaded data. And the attacks do not aim to break the privacy protection but only disturb the accuracy of the global model.

Targeted Poisoning Attacks

Attackers can poison the training dataset before the training phase or during the training phase. Section 4.3 discusses the latter case. The poisoning attack has two types: (1) random attacks that reduce the model's accuracy; (2) targeted attacks that are more objective. The targeted attack's goal is to misclassify a specific input (e.g., a spam email); the model classifies it as a target value (e.g., a benign email).

4.3.3 AUROR Defense

Shen et al. [10] proposed a defense scheme against a poisoning attack for indirect collaborative learning: AUROR. to identify the poisoned data, the server should distinguish malicious attackers from honest users and exclude the masked features submitted by attackers when training the global model. AUROR has a rule-of-thumb assumption to detect the adversary: The adversary is the minority. Suppose the distribution of original data is referred to as δ. The distribution of the masked feature δ' should be close to δ within statistical error bounds. The poisoning attack can change the distribution of specific masked

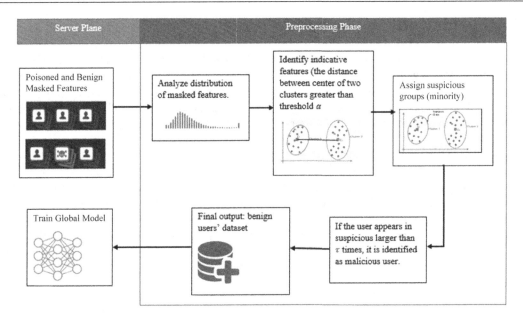

FIGURE 4.3 Workflow of AUROR.

features, referred to as *indicative features*. Thus during the poisoning attack scenario, indicative features from honest users have distribution patterns that are distinct from malicious users. AUROR also assumes that the number of adversaries is inside the range $\epsilon < f < \frac{n}{2}$, where ϵ is the lower bound on the number of adversaries that AUROR can detect. This parameter varies with the type of dataset, so that it should be determined for each dataset. After preprocessing masked features, AUROR finds the indicative features and classifies the input into two groups.

Figure 4.3 shows the detailed steps to distinguish attackers in AUROR. AUROR is applied to the training data before the global model's testing phase. It takes the masked features as the input and analyzes the distribution of these features to identify the indicative features. The key challenge is to find the appropriate indicative feature that can distinguish malicious users from benign users. The indicative feature also varies based on the type of dataset. To decide whether a particular feature is indicative or not, AUROR groups the features into different clusters. If the distance between the centers of two clusters is larger than a predefined threshold α, then AUROR assumes this feature is indicative. The value of α is related to the original distribution of the training dataset.

Then AUROR detects suspicious users if their indicative feature's distribution is different from that of the majority. The cluster with the majority of users is marked as honest, whereas the minority cluster is marked as suspicious (but without final confirmation). A suspicious user is finally marked as malicious if it appears in suspicious groups more than τ times. The parameter τ controls the honest users' error rate tolerance. The indicative feature's distribution is close to that of the malicious users due to statistical errors. AUROR determines the value of τ automatically based on the input. We should notice that the adversary appears in the suspicious groups multiple times. Thus the value of τ also depends on the number of indicative features.

Finally, AUROR trains the global model by excluding the features submitted by malicious users in order to produce M_A. Therefore, AUROR automatically generates a final accurate global model when it is under a poisoning attack. Shen et al. [10] evaluated AUROR and demonstrated that the attack success rate and accuracy drop of M_A were very small. Therefore, AUROR is a successful defense for indirect CDL.

4.3.4 Evaluation

Shen et al. [10] built an indirect CDL model and demonstrated how false training data affects the global model. They used two well-known datasets: MNIST (handwritten numbers) and GTSRB (German traffic images). Shen et al. set 30% of customers as malicious ones, and the accuracy of the global model dropped by 24% compared to the model trained by the benign dataset for MNIST. The GTSRB trained model accuracy dropped 9% when 30% of users are malicious. In a word, indirect CDL is susceptible to poisoning attacks.

After that, Shen et al. [10] added AUROR during indirect CDL. They observed that the attack success rate was reduced to below 5% when the malicious ratio was 10%, 20%, and 30% for MNIST simulations. The accuracy drop is 0%, 1%, and 3% when the malicious ratio is 10%, 20%, and 30%. This proved that AUROR could efficiently defend poisoning attacks and that the image recognition system retains similar accuracy even excluding the malicious features. In the GTSRB experiment, Shen et al. [10] observed that the attack success rate is below 5% for malicious ratios from 10% to 30% when AUROR is used to preprocess the input. And the accuracy drop remains negligible for malicious ratios from 10% to 30%.

4.4 A NEW CDL ATTACK: GAN ATTACK

Hitaj et al. [12] developed a new attack on ab indirect CDL based on generative adversarial networks (GANs). He proved that the privacy-preserving feature of indirect CDL makes the model more susceptible than direct CDL. In Hitaj et al.'s model, the attacker runs a GAN and pretends to be a user; then he or she infers the privacy information from the victim's device by downloading gradient vectors. The attacker can influence not only the local data but also the learning process of the global model and deceive the victim into releasing more personal information.

4.4.1 Generative Adversarial Network (GAN)

Unlike the normal deep learning model, the goal of GANs is to generate "new" images that have similar distributions as the target images. As shown in Figure 4.4, the generative network is first initialized with random noise. At each iteration, it is trained to minimize the distance between the generated images and training images, while the discriminator is trained to distinguish generated images from the original

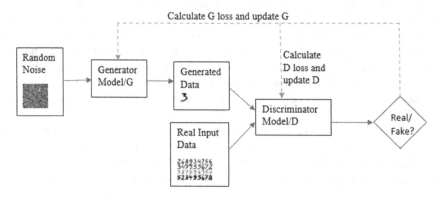

FIGURE 4.4   Architecture of GAN.

dataset. This iteration continues until the discriminator cannot identify the generated images. In other words, the generator is trying to fool the discriminator.

4.4.2 GAN Attack

The GAN threat model may come from an active insider. The following assumptions are made: (1) the adversary works as an insider within the CDL protocol; (2) the object of the adversary is to steal personal information from victims, such as favorite places, movie ratings; (3) the adversary only attacks other users, not the central server; GAN attacks do not affect the global model; 4) the adversary already has knowledge of the global model, e.g., model structure, the label of the data; (5) a GAN attack is adaptive and works in real time. The adversary actively downloads the parameters distributed by the server, runs GAN locally, slightly changes the uploaded parameters, and uploads the forged gradients.

Main Protocol

Suppose the adversary A is an insider of a privacy-preserving CDL protocol. All users agree in advance on the learning object, which means that they agree on the neural network architecture.

For simplicity, we first consider only two players: the adversary A and the victim V. V declares labels [a, b], while A declares labels [b, c]. They first run CDL for several epochs until the accuracy of both the global model and the local model have reached a predetermined threshold.

Then V continuously trains its network: V downloads parameters from the central server and updates its local model, which corresponds to the label [a, b]. V uploads the parameters of its local model to the server after it trains the local model.

On the other side, A also downloads the parameters, updates its local network with labels [b, c]. Unlike V, A first trains its local generative adversarial network to mimic class [a]'s distribution. Then the faked class [a]'s samples are generated. Next, A labels class [a]'s samples as class [c]. After that, it trains its local model on [b, c] and uploads the parameters to the central server, as V does. This process iterates until V and A converge to the threshold.

These GAN attacks work as long as A's local model improves its accuracy over time. Even if A and V use different privacy techniques, GAN attacks still work because the success of the GAN learning relies only on the accuracy of the discriminative model. And the parameters downloaded from the server contain the distribution information that can be used in the discriminator.

Figure 4.5 shows this process.

4.4.3 Experiment Setups

Dataset

Hitaj et al. [12] conducted their experiments on two well-known datasets: MINIST dataset and AT&T dataset of faces (Olivetti dataset). The MINIST dataset is a collection of 32×32 pixels handwriting digits ranging from 0 to 9, which consists of 60,000 training data and 10,000 testing data. The Olivetti dataset consists of 400 images of 64×64 pixels for human faces taken in different positions. Hitaj et al. [13] rescaled the input images to the range [−1. 1] to adopt the state-of-art generator (G). G uses tangent tanh (.) for the activation function in its last layer; thus the output range is [−1. 1].

System Architecture

CNN-based architecture is used during the experiments on MNIST and AT&T datasets and then is applied for the activation function in the last layer. MNIST output has 41 nodes, 40 for the person's real data and

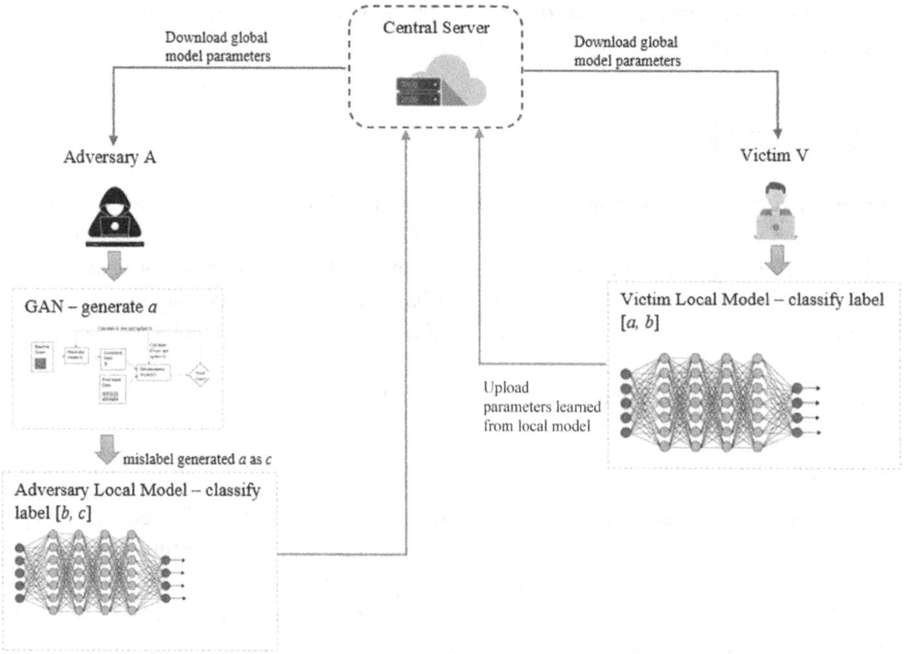

FIGURE 4.5   GAN attack on indirect collaborative deep learning.

1 as the class (which indicates whether the adversary puts the reconstructions for its class of interest). For the AT&T dataset, Hitaj et al. [12] added extra convolutional layers. Batch normalization is applied to all layers' output except the last one to accelerate the training speed. The output is 64×64 images.

Hyperparameter Setup

For MNIST-related experiments, the learning rates for G and D are set to 1e-3, the learning rate decay is 1e-7, momentum is 0, and the batch size of 64 is used. For AT&T-related experiments, the learning rate is set to 0.02, and the batch size is 32. The Adam optimizer with a learning rate of 0.0002 and a momentum of 0.5 are used in DCGAN.

4.4.4   Evaluation

After setting the attack system, Hitaj et al. ran the GAN attack on MNIST and AT&T datasets while changing the privacy settings and the number of participants. The results show that the GAN attack has a high attack success rate in CDL; even the CDL has different privacy techniques. When only 10% of gradient parameters are downloaded, the adversary can get clear victim's images in the MNIST experiment and noisy images in the AT&T experiment because of the small dataset. The author concluded that the generator could learn well as long as the local model's accuracy increases. He observed that the G starts producing good results as soon as the global model's accuracy reaches 80%. Also, the GAN attack performs better if the adversary uploads an artificial class and makes the victim release a more detailed distribution of its own data.

4.5 DEFEND AGAINST GAN ATTACK IN IOT

The IoT (Internet of Things) refers to the billions of physical devices that can connect to the Internet and collect environment data by using embedded sensors. IoT has become more and more popular recently since IoT devices can communicate and operate without human involvement. The tech analysts estimated that, in the IoT, the "things"-to-people ratio has grown from 0.08 in 2003 to 1.84 in 2010. Now the deep learning is linked to the IoT, making IoT devices more intelligent and agile to time-sensitive tasks than previous settings. These IoT devices, such as smartphones, can share data by using CDL and train deep learning models in the Cloud. In the same way, IoT devices generate large amounts of data, fueling the development of deep learning.

CDL can successfully resolve the conflict between data isolation in the real world and data privacy [14]. However, IoT and DL are inherently vulnerable to cyber attacks. IoT devices always communicate in plain text or with low encryption. Second, CDL is not safe because the server can extract loaded data information from the updated gradient. A simple solution, like limiting the size of the upload or download parameters, can protect users from data leakage, but the accuracy of the deep learning model would also decrease, making the CDL useless. Hitaj et al. [13] also pointed out that, in the MNIST experiment, the GAN attack's efficiency was high even if only 10% of parameters were uploaded or downloaded.

From the GAN attack previously described, we can see that CDL is susceptible to GAN attack. When an adversary pretends to be an insider and gets engaged in the learning process, the adversary can steal the victim's dataset online. To fight against the GAN attack, Chen et al. [14] constructed a privacy-preserving CDL in the Internet of things (IoT) network, in which the participants are isolated from the model parameters and use the interactive mode to learn a local model. Therefore, neither the adversary nor the server has access to the other's data.

4.5.1 Threat Model

Google designed CDL to protect participants' data from the server. But the GAN attack also indicates that a participant could be a threat. The adversary must pretend to be one of the participants engaging in model training to implement the GAN attack. This is so that it can download parameters from the server and allure the victim to release finer details of its local data by mislabeling the target data generated from his GAN. The whole GAN attack process is shown in Figure 4.5. More than two users are running CDL, and the adversary is an insider, training a GAN locally to steal other users' private information.

Chen et al. [14] assumed that both the server and participants are semitrusted, which means all of them follow the rules of the agreement and use the correct input but can utilize the information they observe to infer other information. Also, the GAN attack injects false data into the server. Hitaj et al. [13] demonstrated that the false data could slow down the model training but did not destroy the training process. Therefore, the participants are assumed to be semitrusted. A server in a large company (e.g., Google, Amazon) is heavily regulated and has a very low chance of being malicious. Thus it is difficult to make the server internationally leak information to the adversary. Besides these two basic properties in CDL, Chen et al. [14] introduced trusted third party in their system, which will remain neutral and not collude with any party.

4.5.2 Defense System

In a CDL system, a server communicates with many participants, uploading and downloading the learned gradients. Each user only deals with its local data, and the server does not have the direct access to users'

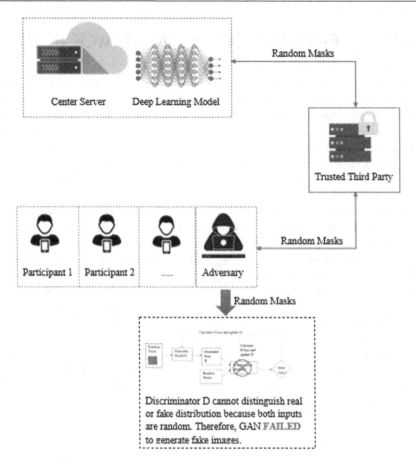

FIGURE 4.6   Secure collaborative deep learning against GAN attack.

private data. To resist a GAN attack in CDL, Chen et al. [14] added a trusted third party to the server as the intermediary for the server and the multiple participants, as shown in Figure 4.6. A trusted third-party helps the server and participant generate random masks in interactive computing. During this process, both the server and participants do not communicate in plain text. Therefore, the local GAN model cannot learn from the random masks. Some research [15] has shown that the model parameters can be encrypted by homomorphic encryption. However, Chen et al. [14] pointed out that if the server encrypts parameters with homomorphic encryption, it would severely reduce the efficiency of the local CNN model training and make CDL meaningless.

Although neither participants have access to the others' local data, Chen et al. [14] introduced an improved Du-Atallah scheme [16], which helps participants interact with the central server, and the neural networks learn from updating parameters. The experiments on the real dataset show that the defense system proposed by Chen et al. [14] has a good performance both in model accuracy and efficiency. Their scheme is also easy to implement for all participants without complex homomorphic encryption technology.

4.5.3 Main Protocols

Chen et al. [14] first designed two algorithms to help the *server* and the *participant* interact with each other and calculate these layers. The first algorithm is for data preprocessing, which can help the convolution calculations in the matrix form. This algorithm is called DataPreprocess. The second algorithm borrows the idea from the Du-Atallah scheme [16]. It is a secure multiparty computation protocol based on additive secret sharing. The improved Du-Atallah scheme hides the transmitted data by adding terms. And all the calculation in the improved scheme can use superposition, which means the improved Du-Atallah scheme is a linear operator. This algorithm is called InterComputing.

After preprocessing data and adjusting the interactive communication scheme, Chen et al. [14] designd protocols for interactions between the *server* and the *participants,* especially on convolutional neural network calculation.

For the convolution layer, the participants first zero-pad the original data X, then conduct the DataPreprocess algorithm, denoting the output as X_{in}. *The server* computes weights for all convolution kernels and sends b, the bias. Next, both participants and server conduct an InterComputing algorithm until all convolution kernels are computed. After finishing the previous step, the participant downloads bias b and then uses the activation function at the final layer.

Participants can then apply the pooling layer locally without interacting with the server.

At the fully connected layer, the trusted third party generates random masks A_0, A_2 to participants and A_1, A_3 to the server, where $A_3 = A_1 A_0 - A_2$. Then participants and the server calculate a temperate variable and send it to each other. Finally, the server computes X_{out} locally and performs an activation function on X_{out}.

4.5.4 Evaluation

The trained model's classification accuracy can achieve 80% in about 300 rounds under the GAN attack. If the training samples of each round increase from one to five images, the model can reach 90% in about 100 rounds. After 500 rounds, the accuracy can reach 100%.

Chen et al. [14] also evaluated the efficiency by comparing the proposed model with Aono's protocol [17]. The result showed that the runtime of participants in Chen's scheme is lower than 20 seconds, while the runtime of Aono's protocol will even be as high as 80 seconds. Therefore, Chen's anti-GAN attack model can maintain accuracy with a short running time under the GAN attack.

4.6 CONCLUSIONS

CDL enables the small computation device (smartphone) to participant in deep neural network learning. A smartphone contains personal information (e.g., locations, contact numbers, web history). Thus CDL must guard sensitive information from the adversary. Attackers try to find a new loophole in CDL

In this chapter, we introduced the architecture of CDL and its security issue, as well as the corresponding defensive mechanism. We start with deep learning security since the data leakage attack for deep learning can also be applied to CDL. We then introduced an automatic defense system (AUROR) that can detect malicious attacks that poison the training dataset. The MNIST and GTSRB datasets are employed to evaluate AUROR and are efficient for detecting a poisoning attack. We then presented the GAN attack and the anti-GAN attack schemes. GAN attack is a newly developed cyber attack targeting CDL, and it can infer the victim's dataset during the interference phase and not alter the global model. Their simulation result showed that GAN could infer victim's images when only 10% gradient parameters were transferred between the server and users. However, GAN attacks could also be neutralized by adding a trusted third party between the server and the users. The encrypted message could prevent GAN from learning the distribution of the information of other users. Compared to homomorphic encryption, this defense approach would not affect the accuracy of the model and required less computation resources.

ACKNOWLEDGMENT

This work is supported in part by the US NSF under grants OIA-1946231 and CNS-2117785. All the ideas presented here do not represent the opinions of NSF.

REFERENCES

[1] Tesla, "Autopilot," TESLA, 2021. [Online]. Available: www.tesla.com/autopilotAI.
[2] M. Lewontin, "How Google Photos uses machine learning to create customized albums," 24 Mar 2016. [Online]. Available: www.csmonitor.com/Technology/2016/0324/How-Google-Photos-uses-machine-learning-to-create-customized-albums.
[3] SiriTeam, "Deep Learning for Siri's Voice: Ondevice Deep Mixture Density Network for Hybrid Unit Selection Synthesis," Aug 2017. [Online]. Available: https://machinelearning.apple.com/research/siri-voices.
[4] amazon, "Question answering as a "lingua franca" for transfer learning," amazon, 10 June 2021. [Online]. Available: www.amazon.science/blog/question-answering-as-a-lingua-franca-for-transfer-learning.
[5] B. McMahan and D. Ramage, "Federated Learning: Collaborative Machine Learning without Centralized Training Data," 6 April 2017. [Online]. Available: https://ai.googleblog.com/2017/04/federated-learning-collaborative.html.
[6] D. Zhang, X. Chen, D. Wang and J. Shi, "A Survey on Collaborative Deep Learning and Privacy- Preserving," in *IEEE Third International Conference on Data Science in Cyberspace* (pp. 652–658), 2018.
[7] A. Keihwcaky, I. Sutskever and G. E. Hinton, "ImageNet Classification with Deep Convolutional Neural Network," in *Advances in neural information processing systems, 25* (pp. 1097–1105), 2012.
[8] C. Szegedy, V. Vanhoucke, S. Loffe, J. Shlens, Z. Woina, Google and U. C. London, "Rethinking the Inception Architecture for Computer Vision," in *IEEE Conference on Computer Vision and Pattern Recognition (CVPR) (pp. 2818–2826),* 2016.
[9] K. He, X. Zhang, S. Ren, J. Sun and Microsoft, "Deep Residual Learning for Image Recognition," in *IEEE Conference on Computer Vision and Pattern Recognition (CVPR)* (pp. 770–778), 2016.
[10] S. Shen, S. Tople and P. Saxena, "AUROR: Defending Against Poisonng Attacks in Collaboratie Deep Learning Systems," in *32nd Annual Conference on Computer Security Applications* (pp. 508–519), 2016.
[11] K. Bonawitz, H. E., W. G., D. H. and A. I., "Towards Federated Learning At Scale: System Design," 4 February 2019. [Online]. Available: arXiv preprint arXiv:1902.01046.
[12] B. Hitaj, A. Giuseppe and P.-C. Fernando, "Deep Models Under the GAN: Infomation Leakage from Collaborative Deep Learning," in *ACM SIGSAC Conference on Computer and Communications Security* (pp.603–618), 2017.
[13] R. Shokri and V. Shmatikov, "Privacy-preserving Deep Learning," in *Proceedings of 22nd ACM SIGSAC Conference on Computer and Communication Security* (pp. 1310–1321), 2015.
[14] Z. Chen, A. Fu, Z. Yinghui, z. Liu, F. Zeng and R. Deng H, "Secure Collaborative Deep Learning against GAN Attacks in the Internet of Things," *IEEE Internet of Things,* vol. 8, pp. 5839–5849, 2020.
[15] L. Phong, Y. Aono, T. Hayashi, L. Wang and Moriai, "Privacy-Preserving Deep Learning via Additively Homomorphic," *IEEE Transactions on Information Forensics and Security,* vol. 13, no. 5, pp. 1333–1345, 2018.
[16] M. S. Riazi, C. Weinert, O. Tkachenko, E. M. Songhori, T. Schneider and F. Koushanfar, "Chameleon: A Hybrid Secure Computation Framework for Machine Learning Applications," in *Proceedings of the 2018 on Asia Conference on Computer and Communications Security* (pp. 707–721), 2018.
[17] L. T. Phong, Y. Aono, T. Hayashi, L. Wang and S. Moriai, "Privacy-preserving Deep Learning via Additively Homomorphic Encryption," *IEEE Transaction on Information Forensics and Security,* vol. 13, no. 5, pp. 1333–1345, 2018.

Attacks on Deep Reinforcement Learning Systems: A Tutorial

5

Joseph Layton[1] and Fei Hu[1]

[1]*Electrical and Computer Engineering, University of Alabama, USA*

Contents

5.1 INTRODUCTION

Deep reinforcement learning (DRL) systems combine the techniques of reinforcement learning (reward maximization) with the artificial neural networks of deep learning (DL) [1]. This combination allows a well trained model to operate on much more complex data sets than a typical reinforcement learning system. For example, input data from a camera of sensor stream is high dimensional and would be difficult for a traditional RL system to solve. However, a DRL system can represent the policies for these higher-dimensional choices with its neural network, enabling it to perform much better than RL or deep learning systems alone. DRL systems have been successfully applied to previously unsolved challenges, such as the game of Go, beating professional poker players in Texas hold'em, and mastering Atari games from pixel inputs [1].

However, while DRL systems have performed well in the previously listed examples, some challenges impede the application of DRL systems on real-world problems. The primary difficult is the infeasibility of allowing the agent to freely and frequently interact with the environment set, due to matters of safety, cost, or time constraints. This manifests as the agent having to (1) interact with an inaccurate simulation of the expected environment or (2) interact with a static environment that is a snapshot of a more volatile one – weather conditions or trading markets. Proposed solutions to these limitations seek to develop rigorously accurate simulation environments and to generalize the learning algorithms for better adaptation once deployed [1].

DOI: 10.1201/9781003187158-6

5.2 CHARACTERIZING ATTACKS ON DRL SYSTEMS

Most studies on attacks against DRL systems have evaluated the vulnerability of these systems to observation-based attacks that add malicious perturbations to otherwise valid inputs. While these attacks can be effective given sufficient system knowledge and time, other forms of attack could be more practical for an attacker to pursue first, namely environmental manipulation and action selection. In their paper on characterizing attacks against DRL systems, Xiao et al. provided a taxonomy of potential attack vectors and evaluated the general effectiveness of these vectors depending on attacker knowledge [2].

The *observation space* consists of the input stream used by the agent to make a decision and update internal states based on current and past data. Attacks on this space seek to perturb the observations of the agent, reducing its decision-making confidence over time. While this vector has the potential for targeted failures, it is more likely that an attacker will cause general underperformance of the system.

The *action space* represents the potential actions and corresponding rewards of the DRL system, in which an attacker seeks to minimize the expected returns of otherwise correct actions. In doing so, the model fails to correctly maximize the cumulative reward it expects, producing unintended behaviors.

The third space identified was the *environment space*, modeling the situation a DRL system is being asked to evaluate before it becomes direct input for the agent. In this space, an attacker can seek to cause targeted failures by perturbing particular environment scenes. Small adjustments to targeted environments can reduce model confidence enough to impede deployment without being directly obvious why it occurs.

In addition to a taxonomy of potential attack vectors, Xiao et al. also evaluated the general effectivity of these attack vectors depending on the level of system and environment knowledge held by an attacker. Attacks on the observation space were evaluated with a number of variables regarding the attacker knowledge and the key aspects of the DRL system. In their experimental results based on these parameters, the episodic rewards of the system were reduced to approximately 0 for most attacks instead of the expected 500–600. Over time, this caused the cumulative reward to diverge greatly from the system without any attacks against it, which would push the model toward inaccurate decision making over time [2]. While higher levels of attacker knowledge were slightly more effective, the main result is that DRL systems are particularly vulnerable to observation perturbations if they are trained without a focus on robustness to input noise.

A few attacks on the action and environment space were also evaluated, and the results indicated a further need for testing the robustness of DRL systems during training. For attacks on action selection, the larger perturbation bound enabled better attack performance but required high levels of system knowledge [2]. The goal of environmental attacks was to cause specific targeted failures, such as a self-driving car to abruptly exit the road in otherwise fine conditions. Finding the right conditions for these attacks was best done with RL-based search methods rather than random search methods. While these environment attacks weren't particularly effective against the overall model performance, their ability to produce highly targeted failures and the minimal system knowledge an attacker requires make them an attractive first option for potential adversaries.

In order to improve the robustness of DRL systems against these attack vectors, some adjustments to training and deployment phases were suggested. First, observations during the training phase should have random perturbations applied to inputs, producing a boundary region in which noise and malicious perturbations still produce the correct observation. As well, the gradient of the observation engine can be evaluated during the training phase to confirm that gradient-based noise is not causing suboptimal choices in the model. For the environment, suggestions were made to vary the environment options (simulated or real) with dynamic parameters to reduce attacker confidence. Since these attacks require the least system knowledge and access on the attacker's part, they are the most feasible first option and should be defended against as such.

5.3 ADVERSARIAL ATTACKS

Several research groups have presented various adversarial attack techniques focused on exploiting vulnerabilities of DRL systems, particularly by perturbing observations and minimizing reward functions. Pattanaik et al. presented an adversarial attack method focused on corrupting the observations of the system, causing the agent to believe it was in a different state than what the observed events were [3]. From this approach, they evaluated three versions of their attack method: a naïve attack, a gradient-based attack, and stochastic gradient descent approach. The effectiveness of these proposed attacks outperformed the attacks Huang et al. presented in a 2017 paper, leading to recommendations for adversarial training algorithms that improve the robustness of DRL systems [3, 4].

Lin et al. have provided further adversarial tactics named the strategically timed attack and enchanting attack for use against DRL algorithms [5]. The strategically timed attack seeks to attack the agent during a particular subset of an episode/observation, perturbing a small portion of the input sequence rather than all of it. From their results, a timed attack can reach similar effects as a uniform attack while only attacking 25% of the time steps of an episode [5]. For their enchanting attack, the DRL agent was lured from a current state to a specific target state using a series of crafted adversarial examples. With well crafted examples, the success at disrupting well-defined tasks was very high; however, tasks with random elements (such as a game with multiple randomized enemies) were less affected by the enchanting strategy [5].

Two novel adversarial attack techniques for stealthily and efficiently attacking DRL agents were introduced by Sun et al. in their 2020 paper on the topic [6]. The *critical point attack uses* a model predicting the next possible states of the DRL system along with a measure of the damage that various attacks could inflict, which is then used by an attacking algorithm to select the most damaging attack with the fewest steps. The *antagonist attack*, meanwhile, uses a trained antagonistic model to identify the optimal attack for any one state of the defending system. Both approaches can achieve similar results as previously mentioned techniques while requiring fewer attacker steps once a state is identified. The choice of which technique an attacker pursues depends on the access an attacker has to internal states and observations, along with the necessity of minimizing the steps of an attack [6].

5.4 POLICY INDUCTION ATTACKS

Deep reinforcement learning policies have been shown to have vulnerabilities regarding adversarial perturbations of their observations, but an attacker typically cannot directly modify the agent's observations. However, as Gleave et al. demonstrated, a DRL agent can be attacked by adversarial policies that generate naturally adversarial observations to disrupt the defending system [7]. In their experiments with this approach, they show that while these attacks are successful against a victim, the success is due more to the adversarial policy employing more randomness in states and confusing the defending system. As a result, the defending system can actually perform better by blinding itself to the adversary's specific positional choices and instead acting on generalized best cases [7].

Another policy manipulation approach was presented by Behzadan et al. with a focus on inducing arbitrary policies on the target system [8]. In this approach, an attacker can only manipulate the environment configuration observed by the target but cannot adjust reward structures or mechanisms. Assumed in this attack is that the adversary can predict and adjust the environment faster than the target can register observations and that the adversary keeps perturbations under a particular threshold to reduce the chance human operators can detect the adjustments. From this setup, the adversarial system was able to rapidly reduce the reward value of the target system once particular policies were learned by the DRL. Some countermeasures to reduce this vulnerability have been proposed, such as retraining the DRL with minimally perturbed adversarial examples; while this increases the amount of perturbation needed for an adversarial example, the computation necessary to exhaust a wide portion of the possible adversary space is infeasible at this time [8]. Based on their results, Behzadan et al. concluded that current defensive techniques are insufficient against policy exploitations.

5.5 CONCLUSIONS AND FUTURE DIRECTIONS

While deep reinforcement learning systems are useful tools for complex decision making in higher-dimensioned spaces, a number of attack techniques have been identified for reducing model confidence even without direct access to internal mechanisms. While some adversarial attacks can be mitigated through robust defensive techniques implemented during the training and deployment phases (perturbed training inputs and multiple deployment observations), some techniques such as policy-based attacks are difficult to train specifically against.

REFERENCES

[1]   V. Francois-Lavet, P. Henderson, R. Islam, M. G. Bellemare, and J. Pineau, "An Introduction to Deep Reinforcement Learning," Nov. 2018, doi: 10.1561/2200000071

[2]   C. Xiao et al., "Characterizing Attacks on Deep Reinforcement Learning," ArXiv190709470 Cs Stat, Jul. 2019, [Online]. Available: http://arxiv.org/abs/1907.09470.

[3]   A. Pattanaik, Z. Tang, S. Liu, G. Bommannan, and G. Chowdhary, "Robust Deep Reinforcement Learning with Adversarial Attacks," Dec. 2017, [Online]. Available: https://arxiv.org/abs/1712.03632v1.

[4]   S. Huang, N. Papernot, I. Goodfellow, Y. Duan, and P. Abbeel, "Adversarial Attacks on Neural Network Policies," ArXiv170202284 Cs Stat, Feb. 2017, [Online]. Available: http://arxiv.org/abs/1702.02284.

[5]   Y.-C. Lin, Z.-W. Hong, Y.-H. Liao, M.-L. Shih, M.-Y. Liu, and M. Sun, "Tactics of Adversarial Attack on Deep Reinforcement Learning Agents," Mar. 2017, [Online]. Available: https://arxiv.org/abs/1703.06748v4.

[6]   J. Sun et al., "Stealthy and Efficient Adversarial Attacks against Deep Reinforcement Learning," May 2020, [Online]. Available: https://arxiv.org/abs/2005.07099v1.

[7]   A. Gleave, M. Dennis, C. Wild, N. Kant, S. Levine, and S. Russell, "Adversarial Policies: Attacking Deep Reinforcement Learning," May 2019, [Online]. Available: https://arxiv.org/abs/1905.10615v3.

[8]   V. Behzadan and A. Munir, "Vulnerability of Deep Reinforcement Learning to Policy Induction Attacks," Jan. 2017, [Online]. Available: https://arxiv.org/abs/1701.04143v1.

Trust and Security of Deep Reinforcement Learning

6

Yen-Hung Chen[1], Mu-Tien Huang[1], and Yuh-Jong Hu[1]

[1]*Department of Computer Science, College of Informatics, National Chengchi University, Taipei, Taiwan*

Contents

DOI: 10.1201/9781003187158-7

6.1 INTRODUCTION

The term *artificial intelligence* (*AI*) was first coined by John McCarthy in 1956. AI is the science and engineering of making intelligent machines. The history of AI development is up and down between rule-based knowledge representation of deduction reasoning and machine learning (ML) from data of inductive reasoning. Modern AI is the study of intelligent (or software) agents in a multiagent system environment, where an agent perceives its environment and takes action to maximize its optimal reward. For the past decade, AI has been closely related to ML, deep learning (DL), and deep reinforcement learning (DRL) for their analytic capability to find a pattern from big data created on the Social Web or Internet of Things (IoT) platforms. The exponential growth of computing power based on Moor's law is another driving factor accelerating machine learning adoption.

ML algorithms used for data analytics have three types: supervised learning for classification or regression, unsupervised learning for clustering, and reinforcement learning (RL) for sequential decision making. DL is an emerging machine learning algorithm for supervised and unsupervised learning over various pattern matching applications, such as computer vision, language translation, voice recognition, etc. But we cannot achieve autonomous decision-making capabilities of intelligent agents without considering using reinforcement learning (RL). RL provides a reward for an intelligent agent for self-learning without requiring ground-truth training data. Furthermore, we can integrate DL with RL to have DRL for other AI applications, such as automatic robotics, AlphaGo games, Atari video games, autonomous self-driving vehicles, and intelligent algorithmic trading in FinTech [1].

Adversarial ML has been intensively studied for the past two decades [2, 3]. Big data analytics services face many security and privacy challenges from an ML modeling perspective. Numerous studies have developed multiple adversarial attacks and defenses for the entire big data analytics pipeline on data preparation, modeling training, and deployment for inferencing [4, 5]. Specific attacks and defenses also aim at different ML types and algorithms, especially on attacking DL. Indeed, these DL attack surfaces are also related to DRL.

Like AI applications for cybersecurity, the AI on adversarial machine learning for attack and defense is a double-edged sword [6]. On the one hand, AI-powered attacks can strengthen attacks and evade detection. In addition, attackers can leverage adversarial AI to defeat protections and steal or influence AI. On the other hand, AI-powered defenses can proactively protect and disable AI-powered attacks and counter malicious AI to fortify AI and withstand adversarial environments [7].

In [8], the authors proposed a framework for analyzing adversarial attacks against AI models. Attack strategy, based on an attacker's knowledge, can be white-box, gray-box, or black-box. Attacks can be poisoning attacks in the training phase or evasion attacks in the testing phase. We will show a more detailed discussion in the following Sections.

Motivation: In this study, we are dealing with the trust and security issues of AI. Compared with recent similar AI for cybersecurity studies [6], we address trust and security of the ML modeling and deployment, notably the DRL algorithm in the big data analytics process [9]. We are investigating possible attacks and defenses on the DRL systems on the secure Cloud platform [10].

A trusted and secure Cloud infrastructure might mitigate existing adversarial attacks on the DRL system, but it is not guaranteed [11]. Unknown attacks will still appear, and we are not yet aware of them [12]. We will investigate how to follow the principle of designing robust ML modeling and deployment systems with security by design in mind in a secure Cloud. We attempt to demonstrate possible attacks on automatic algorithmic trading systems while applying a robust DRL system to ensure financial stability in the Cloud [13, 14].

Our contributions: We summarize the main contributions of this study as follows:

- We highlight previous related surveys of adversarial ML, DL, and DRL. Before that, we overview three types of DRL techniques.
- We present attacks of adversarial DRL systems' surfaces, including environment, state space, policy function, and reward function.
- We show possible defenses against DRL system attacks, including adversarial training, robust learning, and adversarial detection.
- We present a robust DRL system against possible malicious attacks by providing a secure Cloud platform and robust DRL modules.
- Finally, we give a scenario of financial stability for automatic algorithm trading systems dealing with the adversarial DRL application in the financial world.

The remainder of this paper is structured as follows: First, in section 6.2, we briefly overview DRL. Second, we provide an overview of the most recent studies for adversarial ML, including traditional ML, DL, and DRL, in section 6.3. Third, we discuss any possible attacks on DRL, including environment, state space, policy function, and reward function in section 6.4. Fourth, we show how to defend against the DRL system's attacks in section 6.5. Fifth, we explain robust DRL systems with robust DRL modules on a secure Cloud platform in section 6.6. Then we present a scenario of financial stability while facing attacks on automatic algorithm trading systems in section 6.7. Lastly, we offer a conclusion and areas for future work in section 6.8.

6.2 DEEP REINFORCEMENT LEARNING OVERVIEW

DRL is an emerging AI technology that combines DL for policy optimization and RL for self-learning. DRL aims at sequential decision making for a specific goal. In this study, we give an overview of adversarial DRL and the related outlook of financial stability. In sections 6.5 and 6.6, we present the adversarial attacks on DRL and how to defend against DRL system attacks through providing robust DRL systems.

Section 6.2 covers four essential topics to introduce DRL: Markov decision process (MDP), value-based methods, policy-based methods, and actor-critic methods [1].

6.2.1 Markov Decision Process

MDP is critical to RL and DRL, a framework that solves most RL problems with discrete or continuous actions. If you frame your task in an MDP, you define your environment, where an agent achieves optimal policy by learning moves to obtain maximum rewards over a series of time steps (see Figure 6.1).

MDP aims to train an agent to find an optimal policy that will return a maximum cumulative reward by taking a series of actions for each step with a set of states created from observations in an environment.

An MDP is a 5-tuple (S, A, P, R, γ),
where

S is a set of state space;

A is a set of action space;

$P(s, a, s')$ is the transition probability P that takes action a in state s at the time t that will lead to state s$'$ at the time $t + 1$;

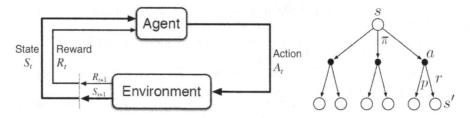

FIGURE 6.1 Markov decision process where an agent interacts with the environment to obtain a maximal reward for each step with optimal policy function π.

$R(s,a,s')$ is the immediate reward R received after taking action a with the state transition from a state s to a state s'; and

γ is a discount factor for a discounted reward.

6.2.2 Value-based Methods

There are three types of RL. Of those, we first focus on value-based methods, which easily explain the necessary mathematical background from model-free approaches.

V-value Function

An RL agent goal is to find a policy such that it optimizes the expected return

$$V^{\pi}(S) = E\left[\sum_{0}^{\infty}\gamma^{k}r_{t+k}\ |s_{t} = s, \pi\right],$$

where E is the expected value operator, γ is the discount factor, and π is a policy operator.

Q-value Function

The optimal Q-value is the expected discounted return when in a given state s, and for a given action a, an agent follows the optimal policy π* after that. We can make the optimal policy directly from this optimal value:

$$Q^{\pi}(s,a) = E\left[\sum_{0}^{\infty}\gamma^{k}r_{t+k}\ |s_{t} = s, a_{t} = a, \pi\right]$$

Advantage Function

We combine the last two functions, which describe "how good" an action a is, compared to the expected return by following direct policy π.

$$A^{\pi}(s,a) = Q^{\pi}(s,a) - V^{\pi}(s)$$

Bellman Equation

The Bellman equation learns the Q value. It promises a unique solution Q^*:

$$Q^*(s,a) = (BQ)^*(s,a)$$

where B is the Bellman operator:

$$BQ^*(s,a) = \sum_{s' \in S} T(s,a,s')(R(s,a,s') + \gamma \max Q^*(s',a'))$$

6.2.3 Policy-based Methods

In policy-based RL, we explicitly build a policy representation by mapping $\pi : s \rightarrow a$ and keep it in memory during the learning process. In contrast, we do not store any explicit policy in value-based methods, only a value function. The policy here is implicitly derived from the value function to pick an action with the best value.

6.2.4 Actor–Critic Methods

Actor–critic RL is a mix of value and policy-based methods that bridging the gap between them. Value-based methods, such as Q learning, suffer from poor convergence, as you are working in value space with a slight change in your value estimation that can push you around quite substantially in policy space. Policy-based methods with policy search and gradient work directly in policy space and have much smoother learning curves and performance improvement for every update.

The drawback with the policy-based methods is that it tends to converge to the local maxim, where you maximize the expected reward by searching in the policy space, so it suffers from high variance with sample inefficiency.

Actor–critic methods combine the best of both worlds. It employs an actor for policy gradient update, with a good critic through the advantage function, $Q(s,a) - V(s)$. It allows actor–critic methods to be more sample efficient via temporal difference (TD) updates at every step.

6.2.5 Deep Reinforcement Learning

DRL is a combination of DL and RL, where DL learns an optimal policy function with an action output from each high dimensional data input at a time step. RL is an MDP, where an agent applies an action in a policy function learned from DL to obtain a maximal reward. The incentive for introducing DL into RL is to directly learn from various big raw data with DL's high accuracy learning capability. In addition, RL can provide sequential decision making. So RL complements DL modeling, which only has one-shot regression or classification output [1].

6.3 THE MOST RECENT REVIEWS

The rise of the study of adversarial ML has taken place over more than a decade [2]. People have used different approaches to address the security and privacy issues of AI or ML. We briefly introduce a general survey discussion of adversarial ML as follows. First, in [15], the authors articulated a comprehensive threat model for ML and categorized attacks and defenses within an adversarial framework. Second, in another paper [4], the authors discussed different adversarial attacks with various threat models. They elaborated on the efficiency and challenges of recent countermeasures against adversarial attacks.

Third, in [5], a wide range of dataset vulnerabilities and exploits approaches for defending against these threats were discussed. In addition, the authors developed their unified taxonomy to describe various poisoning and backdoor threat models and their relationships. Fourth, in [16], the authors proposed a system-driven taxonomy of attacks and defenses in adversarial ML based on the configuration factors of an ML learning process pipeline, including input dataset, ML architecture, adversary's specifications, attack generation methodology, and defense strategy. In [17], the authors proposed robust machine learning systems with their trends and perspectives. They summarized the most critical challenges that hamper robust ML systems development.

Fifth, in [18], the authors gave a comprehensive survey on adversarial attacks in reinforcement learning from an AI security viewpoint. Similarly, in another paper [19], the authors presented a foundational treatment of the security problem in DRL. They also provided a high-level threat model by classifying and identifying vulnerabilities, attack vectors, and adversarial capabilities. Finally, the authors showed a comprehensive survey of emerging malicious attacks in DRL-based systems and the potential countermeasures to defend against these attacks. They highlighted open issues and research challenges for developing solutions to deal with various attacks for DRL systems. We will now show more details of adversarial attacks on ML, DL, and DRL.

6.3.1 Adversarial Attack on Machine Learning

ML algorithms are classified into three types: supervised learning, unsupervised learning, and reinforcement learning. The attacker's goals for each of them are pretty different. Supervised learning for classification or regression provides the probability of outputs from inputs to predict the successive values of the following time stamp. We have correct answers for each data batch in the model training phase to obtain an optimal model through turning a model's parameters. We can verify that a trained model is good or bad based on testing errors.

Attackers' goal for supervised learning is to make a model misclassified or unpredictable or to return a specific wrong class. However, there is no correct answer for the input data of reinforcement learning. An agent learns an optimal policy to maximize reward feedback from the environment for each agent's action. Therefore, an attacker's goals for the RL algorithm are to make an agent learn a policy with minimum reward or prevent an agent from learning a stable policy.

Adversarial attacks appear in the model's training phase, called poisoning attacks, and the testing phase, called evasion attacks [2, 20].

6.3.1.1 Evasion Attack

This attack happens in the testing phase. Attackers are trying to manipulate input data to evade a trained module. There are error-generic and error-specific maximum-confidence evasion attacks. If an attacker inputs data and the model misclassifies the data, it is an error-generic evasion attack. In error-specific evasion attacks, an attack inputs data, and the model returns a specific class as the attacker expects (see Figure 6.2).

6.3.1.2 Poisoning Attack

An attacker injects poisoning samples into training data in the training phase. The attacks can also be classified into error-generic and error-specific. Error-generic poisoning attacks aim to increase the probability of other classes with the same input data and to make the learning model unable to return the correct class. In error-specific poisoning attacks, an attacker aims to cause specific misclassifications (see Figure 6.3).

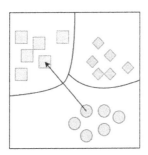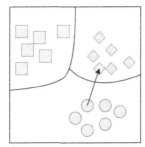

FIGURE 6.2 In both figures, the black lines are the boundary of three classes. Left side: An error-specific evasion attack. The attacker specifies the output to be square with the circle as an input. Right: An error-generic evasion attack. In this attack, the attacker misclassifies an input sample to be a diamond rather than a circle.

FIGURE 6.3 Example of the backdoor poison attack. Left: A typical training dataset for binary classification of squares and diamonds, two classes. Right: Training dataset poisoned with perturbated squares (x depicted in the squares).

6.3.2 Adversarial Attack on Deep Learning

Adversarial ML is an extensive study topic. We can narrow it down to be more specific DL. DL is applied in supervised learning. It has correct answers for its dataset during the training phase. Therefore, it has evasion attacks and poisoning attacks in deep neural network (DNN). The attacker aims to add as little noise as possible to standard input data and make the output data as misclassified as possible. We can show this idea in equation (6.3.1) [17]. We take the neural network as a function, and the standard input x. However, adding noise to the standard input x and δx is smaller than a small number ε. With a few different inputs, the function returns different outputs (see Figure 6.4).

$$f(x) \neq f(x+\delta x)s.t.\delta x \leq \varepsilon. \tag{6.3.1}$$

Instead of using error-generic and error-specific, Shafique et al. used the names "untargeted attack scenario" for random incorrect output and "targeted attack" for a specific output [17].

6.3.2.1 Evasion Attack

An attacker does not have access to the DNN training process and training data in an evasion attack. This attack aims at a learning model that has been trained. In Figure 6.4, we can see the process of finding the adversarial examples automatically. In [20], Goodfellow et al. provided a fast gradient sign method

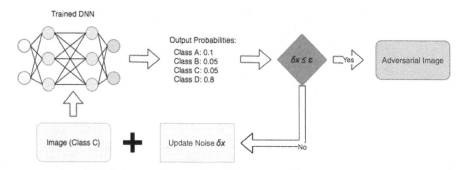

FIGURE 6.4   Image with class C, with some noise δx added into the images and put into a trained DNN. The trained DNN then returns the incorrect class D with the highest probability of 0.8. For an excellent adversarial image, we check to see whether the noise is as minor as expected. If Yes, a malicious image is created. If No, we update the noise and create another image as an input to the trained DNN.

(FGSM). This algorithm finds out the input gradient direction of the loss function cost and adds noise to the input, creating a piece of new information as an adversarial example.

6.3.2.2 Poisoning Attack

In a poisoning attack, the attacker needs to access the training dataset and training procedure. The attacker wants to feed the network with malicious training data, with tailored noise, or simply with random noise. The poisoning can be removing some data and adding adversarial data in the dataset or adding malicious data. Some attackers add neurons to a deep learning model, and the learning process triggers the neurons with specific input data for a targeted attacker with an intended output class.

6.3.3 Adversarial Deep Reinforcement Learning

DRL is the combination of DL and RL. Therefore, the attacks on DL also exist on DRL. Moreover, there are some attack surfaces on RL. It is unsupervised learning for RL, and we do not have correct answers in the training phase. We train it to solve a sequential decision-making problem. An agent learns a policy from the reward of the environment. When an agent observes a state from the environment, it obtains a policy to take an optimal action for the environment. The environment returns a reward according to the action, and the agent observes the next state from the environment.

In Figure 6.5, we can see the interaction between an agent and the environment. Because of a series of interactions between an agent and environment, there are four targets to attack: environment, state, reward, and policy. In [9], the authors classify adversarial attacks on DRL into two types: active and passive.

For active attacks, the attacker's target is to change the agent's behavior and make the reward as low as possible. Passive attacks do not focus on the behavior of the agent but on stealing the details of a victim agent, such as reward function, policy function, or other parts of DRL. The details are helpful information for an attack agent that can produce adversarial examples against a victim agent. An attack agent uses these adversarial examples to attack a victim agent.

The authors also compare the difference between safe RL and secure RL. On the one hand, the safe RL learns a policy to maximize the expectation of the reward function. On the other hand, the secure RL makes the learning process robust and resilient to adversarial attacks. We address the attacks and defenses of the DRL systems in sections 6.4–6.5. Furthermore, we present a robust DRL system in the secure Cloud platform with robust DRL modules in section 6.6.

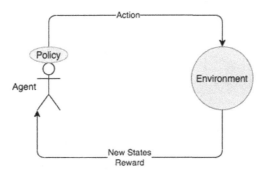

FIGURE 6.5 Interaction between an agent and the environment. An agent observes states from the environment and makes an optimal action according to the policy. The environment returns a reward, and an agent transits to a new state based on this action.

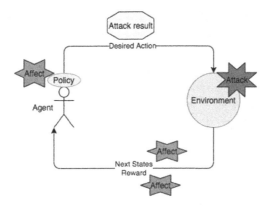

FIGURE 6.6 Attacker changes the environment directly, making the agent execute a desired action. States, rewards, and policy are affected when the learning agent observes states from the perturbed environment.

6.4 ATTACKS ON DRL SYSTEMS

6.4.1 Attacks on Environment

Attack agents can add perturbations directly to the environment (see Figure 6.6). Chen et al. provide a common dominant adversarial example generation (CDG) method to create adversarial examples based on the environment of DRL [21]. The attack agent aims at A3C pathfinding. In typical situations, a victim agent is trained to find a path to the destination. The attack agent adds a thing into the environment and puts this thing in a place so that the victim agent takes more time or is unable to reach the final goal.

6.4.2 Attacks on States

For the state's attack, Behzadan et al. argue that if there is an attack agent in the middle (see Figure 6.7), the attack agent can get the state from the environment, which perturbates the state forwarded to the victim agent [22]. For this purpose, the attack agent first trains an adversarial policy and uses it to add noise to

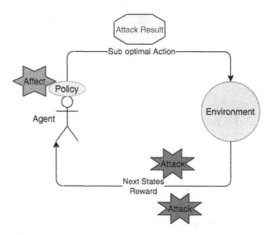

FIGURE 6.7   Just before the agent observes the states from the environment, an attack agent is in the middle to perturb states and reward. This attack affects policy and makes the victim agent perform suboptimal action.

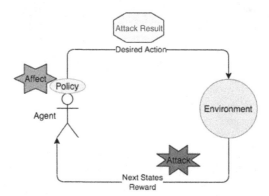

FIGURE 6.8   To attack policy, the attack agent can perturb the states and make the victim agent execute an action that the attack agent wants.

the state and send it to the victim agent. If the attack agent knows the victim agent's policy, it can control the victim agent by sending a specific state to that agent, and the victim agent reacts as the attack agent wants. It also misleads the victim agent in selecting an optimal action with the wrong policy function to reduce the reward. In poisoning attacks on states, Kiourti et al. showed that an attack agent can access the training dataset and modify it by only 0.025%. The attack agent can hide a specific bad policy and wait for a Trojan to be triggered [23].

6.4.3 Attacks on Policy Function

A malicious attack agent can attack a policy (see Figure 6.8). Huang et al. showed an attack on policy function with perturbing the states like Figure 6.8. It introduces the fast gradient sign method (FGSM) in DRL [24]. The target for FGSM on DRL is to find an adversarial input that minimizes the probability of the best action. Moreover, the perturbation added to the input data should be as minor as possible.

Hussenot et al. propose two attacks: a preobservation attack, which adds different adversarial perturbations to each environment, and a constant attack, which adds the same noise to each environment [25]. The attack agent manipulates the environment and makes a victim agent learn the wrong policy from the controlled environment.

Gleave et al. put two agents in the same environment, one is an adversarial attack agent, and another is a legitimate victim agent [26]. The attack agent in the same environment tries to make the victim agent follow the desired policy by creating adversarial natural observations input to the victim agent. It uses two-player zero-sum discounted games. The attack agent tries to get the rewards as much as possible, making the victim agent get as little reward as feasible.

Behzadan et al. provide a model extraction attack on the optimal policy [22]. If a victim agent is online and provides services to others, it faces a model extraction attack risk. The attack agent can use imitation learning technology to extract and steal policy content to train a new mimic agent. This mimic attack agent can generate adversarial examples to attack the victim agent.

6.4.4 Attacks on Reward Function

There are several ways to attack the reward function. Han et al. proposed two kinds of direct attacks on the reward function [27]. An attack agent can directly change the reward from the environment by flipping reward signs, and a victim agent learns with the wrong reward. When a victim agent has enabled action to the environment, it gets an incorrect state but not the correct next state from the environment. This attack agent manipulates states and lets the victim agent perform the wrong action and get the bad reward (see Figure 6.9).

Pattanaik et al. continued to provide three types of attacks by adding perturbations to the states [28]. The first is adding random noise to the DRL states to mislead the victim agent to select a suboptimal action. The second attack is gradient based (GB), similar to the FGSM mentioned with regard to the policy attack. The third attack is also gradient-based but improves the performance of the FGSM attack. Instead of using a gradient-based attack, the authors use a stochastic gradient descent (SGD) attack.

Lee et al. studied how to perturb the action space and provided an optimization problem for minimizing the total reward of DRL agents [29]. A myopic action space (MAS) attack considers one action attack at a time and minimizes agents' real reward. A look-ahead action space (LAS) attack applies several attacks at the time interval between time t and time $t+h$ and reduces the victim agent's total reward.

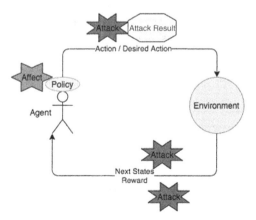

FIGURE 6.9 To attack the reward, an attack agent can directly attack the primary components of RL, reward, states, or action. An attack agent affects the policy and produces the desired action.

6.5 DEFENSES AGAINST DRL SYSTEM ATTACKS

6.5.1 Adversarial Training

Adversarial training is a way to defend against gradient-based attacks, such as FGSM. The concept is simple. The attack agent can run gradient-based attacks to get as many adversarial examples as possible during the training phase and retrain the learning attack model with the adversarial examples. Kos and Song have studied deep research in adversarial retraining [30]. They show that the victim agent becomes more resilient after retraining by adding random noises or adversarial examples created from the FGSM method. This study shows that training with FGSM is much more robust than training with random noise.

Chen et al. also provided an adversarial retraining technique [21]. They used CDG to get adversarial examples and retrain the agent with these examples. y showed that the retrained victim agent can prevent the general attack of A3C pathfinding (see Figure 6.10). However, the method cannot completely prevent attacks if another adversarial example is created from other methods.

6.5.2 Robust Learning

The central concept of robust learning is to create a robust agent while in the training phase. Behzadan et al. used NoisyNets to provide robust learning modules during the training phase [31, 32]. NoisyNets is a neural network, and its biases and weights can be perturbed during the training phase. We can represent the function as follows. The function f_θ is the neural network function with parameters. When we have x as input to the function f, it returns output y (see equation 6.5.1):

$$y = f_\theta(x). \tag{6.5.1}$$

We have noise in the parameters θ with (μ, Σ) as learning parameters and $\in$ as zero-mean noise with fixed statistics (see equation 6.5.2):

$$\theta = \mu + \Sigma * \in. \tag{6.5.2}$$

The authors showed that adding noise to the learning parameter while in the training phase is very effective. They used FGSM to create adversarial examples and used these examples to attack the training agent. The performance of the training agent is excellent.

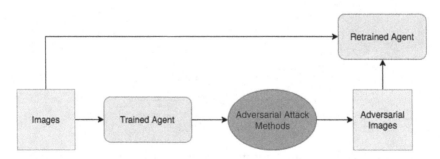

FIGURE 6.10 The retraining procedure trains a trained agent first and uses the adversarial attack methods, such as FGSM or CDG, to generate adversarial images. Original images and adversarial images are used to create a new retrained agent.

Pinto et al. proposed robust adversarial reinforcement learning (RARL) in two aspects [33]. The first is vital to model initializations. Introducing different initialization and random seeds do not decrease the performance of the policy learning. The second is robust with respect to modeling errors and uncertainties. With various test environment settings, we can improve the learned policy performance. Pinto also provided a novel idea of robust training. In general, the standard reinforcement learning set on MDPs is γ, where S is a set of states, A is a set of actions, P is the transition probability, R is the reward function, (S, A, P, R, γ, s) is the discount factor, and s_0 is the initial state.

The authors changed this to two-player zero-sum discounted games. This game's MDP becomes

$$\left(S, A_1, A_2, P, R_{\mu,\nu}, \gamma, s_0 \right)$$

where A_1 and A_2 are the set of actions that two players can take, μ is the strategy of player 1, ν is the strategy of player 2, and R is the total reward with the strategy. The main idea of the zero-sum game is to allow adding an adversarial agent while training. Consider player 1 is trying to maximize the reward d and player 2 (adversary) is trying to minimize player 1's reward. The algorithm is called adversarial reinforcement learning (ARPL). After the training phase, the legitimate victim agent learns all situations and performs the best action according to observations. Based on this concept, some researchers consider risk in RARL and propose risk-averse robust adversarial reinforcement learning (RARARL) [34, 35].

6.5.3 Adversarial Detection

Lin et al. provided an architecture to detect whether the input data is adversarial [36]. They trained a module that can detect a malicious attack. He assumes the attack is continuous. The critical point is introducing an agent that can predict the subsequent input data based on previous inputs and actions, using the predicted inputs with the same policy function and getting the predicted action distribution. If the distance between the expected input and the actual input is higher than a given threshold, the real input is an adversarial input. Otherwise, the real input is standard.

Nguyen et al. used DRL for cyber security. They trained a DRL agent to detect time series attacks. They can put the DRL detect system in the first level to detect whether the input is an attack or not [37].

6.6 ROBUST DRL SYSTEMS

In this section, we propose more about how to deliver a robust DRL system. We discuss two aspects of robust DRL systems: One is the security of the Cloud platform, and the other is how to provide robust DRL modules. Several attacks might happen if we establish the ML system architecture insecurely (see Figure 6.11). Then several weak points are exploited by a malicious third party.

Kumar et al. had points to make about the robust DRL system. On the system level, they can save adversarial attacks in a repository. Then they audit the logs and monitor the ML system – also coding style to prevent vulnerabilities [38].

6.6.1 Secure Cloud Platform

Platform security is essential in the Cloud. For the sake of safety, Cloud providers should provide permission control on all security settings in any service. Nepal et al. presented the concepts of security on the Cloud [39]. It is very complex to set all the services securely. Also, designing a security architecture in the Cloud is not easy. You need not only some security concepts but also familiarity with Cloud settings.

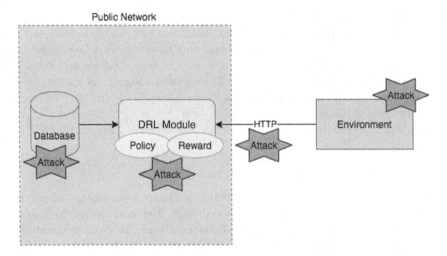

FIGURE 6.11   This architecture is insecure because the database and DRL module is in a public network, which means the attacker is much easier to access than a private on-site closed network. The communication between agent and environment is not secure.

Furthermore, Cloud design architecture is very flexible, so for each service, we have to configure them one by one. Once a service lacks a security configuration or has wrong configuration settings, there is a loophole for an attack agent to enter.

We need to deal with several security issues. Looking back to sections 6.3–6.4 on adversarial ML attacks, we see an attack trying to exploit numerous adversarial examples with several input requests. Therefore, we should limit probing numbers from the exact IP location in various attack scenarios within a fixed time. In this case, it certainly costs the attack agent more probing time to discover adversarial examples. Second, in the man in the middle attack, the attack agent can manipulate any intercepted data and resend it to the victim agent at the destination system. To defend against this attack, we need to ensure that the data source and the communication channel we observe are fully trusted. AWS is used as an example of basic Cloud secure architecture in Figure 6.12.

Using the secure Web communication protocol SSL is a feasible approach to ascertain whether the Web communication environment is trustworthy. For the data poisoning attack, we should protect the database either on an open Cloud or on an on-site private Cloud against any penetration with users' access control with the security mechanisms. The database should have strict access control to prevent an attack agent from trying to change module actions. If possible, any database service request should go through a proxy server first and then forward the request to the trusted agent at the destination module.

6.6.2 Robust DRL Modules

No architecture is 100% secure against all attacks. We need a second-level defense, the DRL module itself. When a module is training, we should enhance the pipeline's robustness for the DRL module instead of only training for the original data within the unsecured analytics pipeline.

We should build a robust DRL system with its DRL module to ensure Cloud security (see Figure 6.13). Database and robust DRL modules are in a private network with strict access control. A proxy server is in a public network and forwards an incoming request to a server through adversarial detection. If the request is valid, it is sent to the robust DRL module. In the training phase, we provide adversarial retraining and robust ML learning on the robust DRL module.

Robust learning is a way to provide trusted and secure ML modeling in the training phase. People use the NoiseNet method to add noise to the deep neural network in the training phase. We add an adversarial

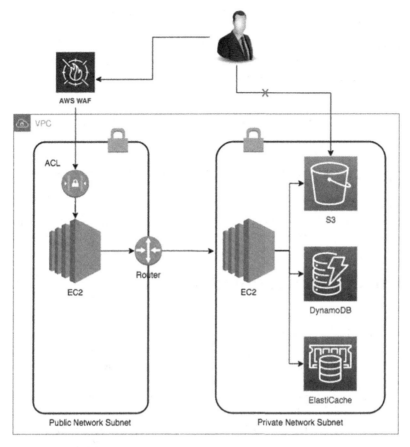

FIGURE 6.12   Basic concept to design a secure architecture in AWS Cloud. Database and essential services are in a private network to prevent any access from the external internet. The proxy service is in a public network with AWS WAF as a firewall.

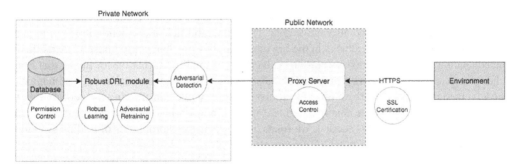

FIGURE 6.13   To build a DRL system on AWS Cloud, we should create a database and DRL module in a private network and set up a proxy server in a public network. The proxy server accepts the request to the adversarial attack detection module and forwards it to the robust DRL module. The proxy server provides access control and secure communication services between the private network and the outside environment.

agent to simulate the malicious acts for any input or parameter to perform a robust DRL module. When we have a trained module, it is necessary to test whether the performance is as good as before. We can use the FGSM method to get possible adversarial examples and retrain the module beforehand. The adversarial detection module is a blocker in front of the protected agent as a firewall. When any incoming adversarial example arrives at the adversarial detection module, it blocks this example and saves it into secondary storage. Eventually, we train the adversarial detection module, put all these malicious samples into the training dataset, and make our DRL more robust.

6.7 A SCENARIO OF FINANCIAL STABILITY

The financial service is entering a new AI era with autonomous services without a human in the loop, such as automatic algorithm trading systems.

6.7.1 Automatic Algorithm Trading Systems

High-frequency trading systems can profit from highly short-term arbitrage opportunities with an equity/bond/FX price prediction based on immediately available information or by reducing the market impact of a large transaction block by breaking it into many small transactions within a trading day. The algorithm trading systems have a complex pipeline of data analytics stages, including data engineering, machine-learning-based modeling, and the logic of transaction execution orders.

Traditional systems use simple models, such as linear regression models, autoregressive models (ARIMA), and support vector machines (SVM), to forecast future target asset transaction prices. These models are fast for both model training and deployment inferencing. As hardware computing power increases, we accelerate the training and inferencing with advances in GPU technologies. The high-performance GPU computing systems train deep neural networks, such as deep CNN and LSTM. These optimized deep learning models with nonlinear relationships predict future markets with high accuracy in a short time horizon.

We focus on the data analytics ML modeling of the automatic algorithm trading systems since the ability to predict the prices and trends of assets is the core feature of the automated asset management system. The asset price prediction model uses various time-series data streams to forecast the future, including market, fundamental, and alternative data. The vulnerabilities of prediction model training and inferencing to the adversarial ML attacks in the big data analytics pipeline cause financial instability.

The advancing neural networks systems are vulnerable to small perturbations of inputs that dramatically change a deep neural network's output. The most popular use of adversarial attacks is image classification. Adversarial attacks perturb an image to fool a classifier while constraining the perturbation to be small within some norm. However, manipulated text may be nonsensical in some fields, such as natural language processing (NLP). In this section, we focus on the robustness properties of trading systems and exploit the spoofing attack with its potential threat.

In the context of network security, a spoofing attack is when a person or program is successfully identified as falsifying data to gain an illegitimate advantage. In the history of financial execution orders, traders have been engaging in spoofing. An adversary places orders on either side of the best price to fake supply or demand, often with the intent of canceling the orders before execution. We regard the adversary orders placed by a trader engaging in spoofing as a naïve adversarial perturbation, which may be less detectable than handcrafted versions and thus may still present a problem for trading transaction enforcement.

We describe the simple adversarial attacks on order book data. Consider an adversary that places orders at the best buy and sell prices. In addition, consider perturbations to the size entries. If the attacker creates local extrema in the order book, this may cause the trading system to make wrong decisions, and the attacker may gain an illegitimate advantage. Then adversarial attacks could be leveraged by a malicious agent possessing only limited knowledge.

This study introduces adversarial attacks to financial trading systems as a scenario for understanding the robustness properties of trading systems. There are robustness trade-offs. While neural network models perform better at market prediction in this setting than traditional trading systems, we argue that they might be less robust. We are further aware that the same adversarial patterns that fool one model also can mislead others and that these patterns are highly interpretable for humans. The transferability of these attacks, combined with their ability to be effective with a small attack budget, suggests that they could be leveraged by a malicious agent possessing only limited knowledge.

Here, we give a scenario of financial stability for automatic algorithm trading systems dealing with the adversarial DRL application in the financial world. We show the adversarial DRL can be applied to market-making trading systems with robust defenses to malicious attacks. *Market making* refers to the act of providing liquidity to a market by continuously quoting prices to buy and sell a given financial instrument. We call the difference between quoting prices the market-maker spread, representing the market maker's potential profit from this activity. However, market makers expose themselves to adverse selection, where attackers exploit their knowledge and technological edge, transacting with and thereby changing the market maker's inventory before an adverse price move causes them a loss. This situation is known as market-making risk and has been the imminent subject of adversarial DRL learning while applying AI and ML techniques in financial applications (Table 6.1).

In this section, we deal with this type of risk by taking an adversarial approach. We design market-making agents that are robust to hostile and adaptively chosen market conditions in the adversarial DRL [40]. The adversarial DRL problem sometimes requires other domain knowledge, such as game-theoretic, to prove that it could converge to Nash equilibria. When malicious attackers apply this similar approach to different trading scenarios, such as statistical arbitrage or optimal execution for quantitative trading, it is essential to acknowledge that adversarial DRL has significant implications for financial stability.

From another point of view, an adversarial learning framework might capture the evolving game between a regulator who develops tools to detect market manipulation and a manipulator who obfuscates actions to evade detection [41]. Adversarial detection, robust learning, and adversarial training that are explicitly robust enough to model misspecification can make deployment in the real world considerably more practicable.

6.8 CONCLUSION AND FUTURE WORK

Adversarial machine learning (ML) has been studied for more than a decade. Adversarial deep learning (DL) have also been researched for a decade. This paper presents a high-level overview of the trust and security of deep reinforcement learning (DRL) with its applications in automatic algorithmic trading to avoid financial instability. We first briefly introduce deep reinforcement learning (DRL). Then we address the most recent reviews of adversarial ML, DL, and DRL; we conceptually review the attacks and defenses of DRL systems to make the complex concepts much easier to understand.

To see how to preserve financial stability while applying AI techniques, particularly big data analytics for the FinTech application domain, the big data analytics pipeline has been extensively designed and developed on the public Cloud or the on-site private Cloud platforms. We present the robust DRL systems based on the secure Cloud platforms with possible proposed defense techniques against Cloud platform system attacks and adversarial DRL modeling attacks. Finally, we demonstrate a scenario of financial stability problems with an automatic algorithm trading system application by using a summarization table of previously proposed adversarial ML, DL, and DRL concerning this FinTech application. This study is an initial road map of trusted and security DRL with its applications in the financial world.

TABLE 6.1   Overview of Adversarial ML/ DL/ DRL Attacks with Finance Applications

		ADVERSARIAL ATTACK	
	TYPE	*EVASION ATTACK* *(ATTACKS IN THE INFERENCING PHASE)*	*POISONING ATTACK* *(ATTACKS IN THE TRAINING PHASE)*
ML	Regression	Confidence reduction attacks: In which adversarial attacks are launched to compromise the accuracy of the deployed (online) ML-based system. A spoofing attack requires a high attack budget. The attacker creates local extrema in the order book on a market prediction model, which uses a relatively simple structure compared with deep learning, such as linear regression models, since those models are fast for training and inferencing.	Source confidence reduction attacks: In which adversarial attacks are launched to compromise the accuracy of the model in the training phase. Source poisoning attack costs a low attack budget; however, the attacker needs to hack into the ML pipeline on a market prediction model that uses a relatively simple structure compared with deep learning.
	Classification	Misclassification attacks: In which adversarial attacks are launched to disturb any class's classification boundary to cause misclassification of the deployed (online) ML-based system. Targeted misclassification attacks: In which adversarial attacks are launched to misclassify only a targeted class of the deployed (online) ML-based system. Spoofing attack on the finance classification model, which uses a relatively simple structure compared with DL.	Source misclassification attacks: In which adversarial attacks are launched to force misclassification of a specific source into a particular target in the training phase. Source poisoning attack on the training phase requires a low attack budget. However, the attacker needs to hack into the ML pipeline.
DL	Supervised	Adversarial attack on DL: The advancing neural networks (DL) systems are vulnerable to small perturbations to inputs that dramatically change a neural network's output. While DL models perform better at market prediction in this setting than ML models, we argue them to be less robust. There are robustness trade-offs like bias variance. Spoofing attack on DL model, which uses a relatively complex structure compared with ML.	Source adversarial attack on DL: The attacker needs to hack into the DL system to force misclassification of a specific source into a particular target in the training phase. In a man in the middle attack, the attacker can get the features from the environment, add perturbation on the features, and transfer them to the algorithm in the training phase.

TABLE 6.1 (Continued)   Overview of Adversarial ML/ DL/ DRL Attacks with Finance Applications

		ADVERSARIAL ATTACK	
	TYPE	EVASION ATTACK (ATTACKS IN THE INFERENCING PHASE)	POISONING ATTACK (ATTACKS IN THE TRAINING PHASE)
DRL	Environment	Evasion attacks on the environment: Attackers can add perturbations directly to the environment in the inferencing phase. Spoofing attack requires a high attack budget: the attacker creates local extrema in the order book.	Poisoning attacks on the environment: Attackers can add perturbations directly to the environment in the training phase. Source poisoning attack costs a low attack budget. However, the attacker needs to hack into the DRL pipeline.
	States	Man in the middle evasion attacks on states: Attackers can get the state from the environment, add perturbation on the state, then transfer it to the agent in the inferencing phase of the online DRL-based system. Attackers need to hack into the DRL pipeline to get the state from the clearinghouse, add perturbation, and transfer it to the online trading system in the inferencing phase.	Man in the middle poisoning attacks on states: Attackers can get the state from the environment, add perturbation on the state, then transfer it to the agent in the training phase of the offline DRL-based system. Attackers need to hack into the DRL pipeline, injecting poisoning samples into training data for an agent in the training phase.
	Policy function	The same as an adversarial attack on DL. Indirectly impact the DRL model by Evasion attacks on environment and Man in the middle evasion attacks on states.	The same as source adversarial attack on DL. Man in the middle poisoning attacks on Reward function: Directly impact the DRL model: 1. By flipping reward signs, the attacker can directly change the reward from the environment, and the agent will learn with the wrong reward. 2. When an agent reacts to the environment, it gets a wrong state instead of getting the correct next state from the environment. The attacker manipulates states and lets an agent perform the improper action and get the bad reward. The attacker hacks into the DRL system and directly changes the reward function of the asset allocation model.
	Reward function		

In our future work, we intend to investigate further issues on the adversarial DRL with its instability effects on more inclusive financial applications, such as high-frequency program trading, asset management, financial asset pricing, time series forecasting of a financial index, etc. We will design and implement the proposed robust DRL modules in the secure Cloud platform. We intend to verify our trust in and the security of the DRL proposition in the real financial world to see whether it can mitigate the risk of financial instability while facing different adversarial ML, DL, and DRL challenges.

REFERENCES

1. Li, Y. "Deep Reinforcement Learning: An Overview." *arXiv:1701.07274v6 [cs.LG]* (2017).
2. Biggio, B. and F. Roli. "Wild Patterns: Ten Years After the Rise of Adversarial Machine Learning." *Pattern Recognition.* Vol. 84 (2018): 317–331.
3. Joseph, D. A., B. Nelson, and B. I. P. Rubinstein, et al. *Adversarial Machine Learning.* Cambridge University Press, 2019.
4. Chakraborty, A., M. Alam, and V. Dey, et al. "Adversarial Attacks and Defences: A Survey." *arXiv:1810.00069v1 [cs.LG]* (2018).
5. Goldblum, M., D. Tsipras, and C. Xie, et al. "Dataset Security for Machine Learning: Data Poisoning, Backdoor Attacks, and Defenses." *arXiv:2012.10544v4 [cs.LG]* (2020).
6. Taddeo, M., T. McCutcheon, and L. Floridi. "Trusting Artificial Intelligence in Cybersecurity is a Double-Edged Sword." *Nature Machine Intelligence 1* (2019): 557–560.
7. Casola, L. D. Ali, and Rapporteurs. *Robust Machine Learning Algorithms and Systems for Detection and Mitigation of Adversarial Attacks and Anomalies: Proceedings of a Workshop.* The National Academies Press, 2019.
8. Oseni, A., N. Moustafa, and H. Janicke, et al. "Security and Privacy for Artificial Intelligence: Opportunities and Challenges." *arXiv:2102.04661v1 [cs.CR]* (2021).
9. Ilahi, I., M. Usama, and J. Qadir, et al. "Challenges and Countermeasures for Adversarial Attacks on Deep Reinforcement Learning." *arXiv:2001.09684 [cs.LG]* (2020).
10. Qayyum, A., A. ljaz, M. Usama, et al. "Securing Machine Learning in the Cloud: A Systematic Review of Cloud Machine Learning Security." *Frontiers in Big Data,* Vol. 3 (2020):1–17.
11. Lynn, T., J. Mooney, and L. van der Werff, et al. Editors. *Data Privacy and Trust in Cloud Computing: Building trust in the Cloud through assurance and accountability.* Palgrave Macmillan, 2021.
12. Dietterich, G. T. "Steps Toward Robust Artificial Intelligence." *AI Magazine* (Fall 2017): 3–24.
13. Alessi, L. and R. Savona. "Machine Learning for Financial Stability." In *Data Science for Economics and Finance Methodologies and Applications,* 65–87. Springer, 2021.
14. Gensler, G. and L. Bailey. "Deep Learning and Financial Stability." (2020). Available at: https://ssrn.com/abstract=3723132.
15. Papernot, N., P. McDaniel, and A. Sinha, et al. "SoK: Towards the Science of Security and Privacy in Machine Learning." *arXiv:1611.03814v1 [cs.CR]* (2016).
16. Sadeghi, K., A. Banerjee, and S. K. G. Gupta. "A System-Driven Taxonomy of Attacks and Defenses in Adversarial Machine Learning." *IEEE Transactions of Emerging Topics in Computational Intelligence,* Vol. 4 No. 4 (2020):450–467.
17. Shafique, M., M. Naseer, and T. Theocharides, et al. "Robust Machine Learning Systems: Challenges, Current Trends, Perspectives, and the Road Ahead." *IEEE Design and Test,* Vol. 37, Issue: 2, (2020): 30–57.
18. Chen, T., J. Liu, and Y. Xiang, et al. "Adversarial Attack and Defense in Reinforcement Learning – from AI Security View." *Cybersecurity* (2019): 2–11.
19. Behzadan V. and A. Munir. "The Faults in Our Pi*'s: Security Issues and Open Challenges in Deep Reinforcement Learning." *arXiv:1810.10369v1 [cs.LG]* (2018).
20. I. J. Goodfellow, I. J., J. Shlens, and C. Szegedy, "Explaining and Harnessing Adversarial Examples," *International Conference on Learning Representations (ICLR)* (2015).
21. Chen, T., W. Niu, Y. Xiang, et al. "Gradient Band-based Adversarial Training for Generalized Attack Immunity of A3C Path Finding," *arXiv:1807.06752v1 [cs.LG]* (2018).
22. Behzadan, V. and A. Munir. "Vulnerability of Deep Reinforcement Learning to Policy Induction Attacks." *2017 Machine Learning and Data Mining in Pattern Recognition.* Springer International Publishing (2017).
23. Kiourti, P., K. Wardega, and S. Jha, et al. "TrojDRL: Trojan Attacks on Deep Reinforcement Learning Agents." *Proc. 57th ACM/IEEE Design Automation Conference (DAC),* 2020, National Science Foundation. Available at: https://par.nsf.gov/biblio/10181034

24. Huang, S., N. Papernot, I. Goodfellow, et al. "Adversarial Attacks on Neural Network Policies." *arXiv:1702.02284v1 [cs.LG]* (2017).

25. Hussenot, L., M. Geist, and O. Pietquin, "Targeted Attacks on Deep Reinforcement Learning Agents Through Adversarial Observations," *arXiv:1905.12282v1 [cs.LG]* (2019).

26. Gleave, A., M. Dennis, and C. Wild, et al. "Adversarial Policies: Attacking Deep Reinforcement Learning." *International Conference on Learning Representation (ICLR)* (2020). Available at: https://iclr.cc/virtual_2 020/poster_HJgEMpVFwB.html

27. Han, Y., B. I. Rubinstein, T. Abraham, et al. "Reinforcement Learning for Autonomous Defence in Software-Defined Networking," *International Conference on Decision and Game Theory for Security.* Springer (2018): 145–165.

28. Pattanaik, A., Z. Tang, S. Liu, et al. "Robust Deep Reinforcement Learning with Adversarial Attacks," *Proceedings of the 17th International Conference on Autonomous Agents and MultiAgent Systems* (2018): 2040–2042.

29. Lee, X. Y., S. Ghadai, K. L. Tan, et al. "Spatiotemporally Constrained Action Space Attacks on Deep Reinforcement Learning Agents." *arXiv:1909.02583v2 [cs.LG]* (2019).

30. Kos, J. and D. Song, "Delving into Adversarial Attacks on Deep Policies," *International Conference on Learning Representation (ICLR) Workshop* (2017) Available at: https://openreview.net/forum?id=BJcib5 mFe.

31. Behzadan, V. and A. Munir, "Whatever does not Kill Deep Reinforcement Learning, Makes it Stronger." *arXiv:1712.09344v1* [cs.AI] (2017).

32. Behzadan, V. and A. Munir, "Mitigation of Policy Manipulation Attacks on Deep Q-Networks with Parameter-Space Noise," *International Conference on Computer Safety, Reliability, and Security.* Springer (2018): 406–417.

33. Pinto, L., J. Davidson, and R. Sukthankar, et al. "Robust Adversarial Reinforcement Learning." *Proc. of the 34th International Conference on Machine Learning (ICML'17)*, Vol. 70 (2017): 2817–2826. Available at: https://dl.acm.org/doi/10.5555/3305890.3305972

34. Pan, X., D. Seita, and Y. Gao, et al. "Risk Averse Robust Adversarial Reinforcement Learning." *International Conference on Robotics and Automation (ICRA)* (2019).

35. Singh, R., Q. Zhang, and Y. Chen. "Improving Robustness via Risk-Averse Distributional Reinforcement Learning." *Proc. of Machine Learning Research*, Vol. 120 (2020):1–11.

36. Lin, Y. C., M.-Y. Liu, and M. Sun, et al. "Detecting Adversarial Attacks on Neural Network Policies with Visual Foresight." *arXiv:1710.00814v1 [cs.CV]* (2017).

37. Nguyen, T. T. and V. J. Reddi. "Deep Reinforcement Learning for Cyber Security." *arXiv:1906.05799v3 [cs. CR]* (2019).

38. Kumar, S. S. R., M. Nystrom, and J. Lambert, et al. "Adversarial Machine Learning – Industry Perspective." *IEEE Symposium on Security and Privacy Workshop (SPW)* (2020): 69–75. Available at: https://ieeexplore. ieee.org/document/9283867

39. Nepal, S. and M. Pathan Editors. *Security, Privacy, and Trust in Cloud Systems.* Springer, 2014.

40. Spooner, T. and R. Savani. "Robust Market Making via Adversarial Reinforcement Learning." *Proc. of the Twenty-Ninth International Joint Conference on Artificial Intelligence (IJCAI-20)* Special Track on AI in FinTech. (2020).

41. Wang, X and M. P. Wellman. "Market Manipulation: An Adversarial Learning Framework for Detection and Evasion." *Proc. of the Twenty-Ninth International Joint Conference on Artificial Intelligence (IJCAI-20)* Special Track on AI in FinTech. (2020).

42. Garcia, J. and F. Fernandez. "A Comprehensive Survey on Safe Reinforcement Learning." *Journal of Machine Learning Research* 16 (2015): 1437–1480.

43. Hamon, R., H. Junklewitz, and I. Sanchez. "Robustness and Explainability and Artificial Intelligence -From technical and policy solutions." *JRC119336 Technical Report.* EUR 30040 EN, Publications Office of the European Union, Luxembourg, ISBN 978-92-76-14660-5 (online). Available at: https://publications.jrc. ec.europa.eu/repository/handle/JRC119336

44. Liang, P., Bommasani, D. A. Hudson , and E. Adeli, et al. "On the Opportunities and Risks of Foundation Models." *arXiv:2108.07258v2 [cs.LG]* (2021).

IoT Threat Modeling Using Bayesian Networks

7

Diego Heredia

Facultad de Ciencias: Escuela Politécnica Nacional, Quito, Ecuador

Contents

7.1 BACKGROUND

The incorporation of IoT for the development of "smart" elements in different domains such as city (smart city), houses (smart home), energy (smart grids), health (smart health), among others, has generated new challenges from the perspective of cyber security. This issue is due to the fact that IoT solutions have some intrinsic characteristics regarding cyber security, for example, poor security configuration from design, limited computational resources, and great interconnectivity.

Formally describing the cyber security status of each IoT component, as well as the interactions with other components, and users, allows to understand the attack surface and the attack capabilities that could

DOI: 10.1201/9781003187158-8

affect the normal operation of a city, house, electrical system, or body health. According to Poolsappasit, a number of researchers have proposed risk assessment methods that allow the construction of security models. However, he asserts that these models fail to consider the capabilities of the attackers and consequently the probability that a particular attack will be executed [1].

Every component of the IoT is a source of threat, and an attacker can use a number of frameworks, techniques, and tools. Therefore, it is relevant to analyze and quantify the security risk generated by each source or node that is part of the IoT ecosystem. However, as represented in Figure 7.1, the elements of IoT have a high level of interconnection, and their operation does not always generate a pattern of static data flows. In this sense, we are faced with the need to determine the spread of joint risk and a low amount of data on certain values.

7.2 TOPICS OF CHAPTER

- Cyber risks in smart environments
- Threat model to develop a risk analysis and establish the attack probability of the components of an IoT system
- Attack surface and establishment of possible countermeasures in IoT ecosystems.

7.3 SCOPE

Establishing a modeling of cyber security risks in IoT systems requires an exploitation that takes into account the dynamic, heterogeneous, and uncertain aspects. Therefore, the proposed solution is to base our efforts on the use of Bayesian networks to model the attack surface, identify the risk elements, and establish a methodology for the cyber security risk generated by IoT solutions.

Bayesian inference, compared to classical statistical methods, requires less data and computational power to forecast eventual impacts.

Bayesian networks (BNs) offer a framework for modeling the relationships between information under causal or influential assumptions, which makes them suitable for modeling real-world situations where we seek to simulate the impact of various interventions.

Bayesian networks are an appropriate method to model uncertainty in situations where data is limited but where human domain experts have a good understanding of the underlying causal mechanisms and/ or real-world facts.

The goal of this work is to implement a risk assessment model based on the calculation of the probabilities of cyber attacks in an IoT network of a smart home using a Bayesian network. The directed acyclic graph of this Bayesian network is obtained from an attack graph of a smart home, and the parameters are obtained by the maximum likelihood method applied to a data set generated by simulation.

7.4 CYBER SECURITY IN IOT NETWORKS

The Internet of Things (IoT) is an infrastructure that includes physical devices, modern vehicles, buildings, and essential electrical devices that we use constantly and that are interconnected with one another through the Internet in order to accumulate and exchange data [2]. An IoT network is an environment of this type of devices. The main purpose of IoT is to improve people's quality of life by reducing stress and reducing the amount of repetitive work [3]. In IoT networks, some devices have low security due to their design conditions; for example, a smart refrigerator does not have the same quality of security as a computer with protections such as a firewall or antivirus. In addition, an attack on any device in the network can generate a severe impact since a cascade effect can occur due to the degree of interconnection that the devices have.

7.4.1 Smart Home

A *smart home* is a system made up of devices interconnected through an IoT network in a house. Some motivations for using IoT networks at home are safety, because a house can be monitored through security cameras and emergency calls can be automated, and power efficiency because smart home devices can be turned on or off as needed or to save energy [3]. The structure of a smart home that we use in the model is presented in [4]:

- *Access point:* The attacker is positioned at this point.
- *Management level:* This level consists of devices that give orders, such as laptops and cellphones (CP).
- *KNX IP router:* This consists of the KNX IP router that all KNX devices communicate with through the KNX bus.
- *Field devices level:* The KNX bus is where all the KNX devices are located, such as shutters, smart lights, air conditioner, and IP cameras.

The smart home has three wi-fi connections between these levels [4]:

- *Wi-fi network 1:* Provides connection between the access point and the management level.
- *Wi-fi network 2:* Provides connection between the management level and the KNX IP router.
- *Wi-fi network 3:* Provides connection between the KNX IP router and the field devices level.

Figure 7.1 represents the structure of the smart home. The following attacks are considered in the model implementation: social engineering (SE), phishing (PH), malware injection (MI), denial of service (DoS), routing table poisoning (RTP), persistent attack (PA), and man in the middle (MitM).

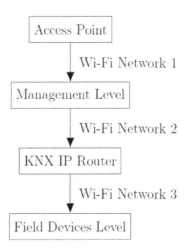

FIGURE 7.1   Smart home structure [4].

7.4.2 Attack Graphs

An *attack graph* is the graphical representation of the paths that an attacker can take to perform a specific attack in a network. For this work, the attack graph consists of the different types of attacks that can be performed at various levels of the smart home structure to compromise the field devices level.

The attack graph of the IoT network of a smart home used for the development of the model is the one presented by Ibrahim and Nabulsi [4]. This graph is based on the structure of the smart home – access point, administration level, KNX IP router, and field devices level – and it uses the cyber attacks just mentioned. The attack graph model developed in [4] was implemented adapting Architecture Analysis and Design Language (AADL). Then it was checked via JKIND checker tool against a security requirement, which was that the attacker does not have root privileges on the devices connected to the KNX bus. Finally, the results were used to visualize the attack graph using the Graphviz tool. In this attack graph, all the paths by which an attack to the field devices level of the smart home can be performed are represented. The attacker starts at the access point and performs a social engineering attack to the management level device, which is a cellphone, in order to obtain information about the device. Once the attacker gets the information, both phishing and malware attacks can be performed in order to gain access. Then DoS, RTP, and PA attacks can be performed from the management level to the KNX IP router. Finally, DoS and MitM attacks can be performed from the router to the field devices level, which is the attacker's target.

7.5 MODELING WITH BAYESIAN NETWORKS

In this section we first describe mathematical concepts to understand the theory of Bayesian networks. Then we explain the definition of a Bayesian network, parameter learning, and inferences that will be used in the model implementation.

7.5.1 Graph Theory

A graph is a pair $G = (V,E)$ consisting of a set of nodes V and a set of edges E that connect two elements of the set V. A pair of nodes X_i, X_j can be connected by a directed edge or an undirected edge. The difference between the two types of edges is that directed edges are ordered pairs of nodes, so we have that (X_i,X_j) $\neq (X_j,X_i)$, unlike undirected edges where $\{X_i,X_j\} = \{X_j,X_i\}$. We say that a graph is directed if all edges are directed and that it is undirected if all edges are undirected.

Bayesian networks are based in directed graphs, so the upcoming concepts are explained only for this type of graph. In a directed graph G, for the edge (X_i,X_j), we say that X_j is the child of X_i in G and that X_i is the parent of X_j in G. We use $Pa(X_i)$ to denote all the parent nodes of X_i in G. The nodes $X_1,...,X_k$ form a path in the directed graph $G = (V,E)$ if, for every $i=1,...,k-1$, we have an edge (X_i,X_{i+1}) $\in E$. A cycle in G is a path $X_1,...,X_k$, where $X_1 = X_k$. A graph is acyclic if it contains no cycles. The structure of a Bayesian network is a directed acyclic graph, and it encodes the conditional independences of random variables.

For the representation of directed acyclic graphs, nodes can go in any order; as long as all the edges are given, the structure of the graph can be understood. However, a specific ordering for nodes is used in some algorithms and theory of Bayesian networks, and it is called *topological ordering*. Let $G = (V,E)$ be a directed acyclic graph with $V = \{X_1,..,X_n\}$. An ordering of the nodes $X_1,...,X_n$ is a topological ordering relative to G if, whenever we have an edge $(X_i,X_j) \in E$, then $i < j$ [5].

7.5.2 Probabilities and Distributions

In probability theory, an *event* is a subset of all possible outcomes of an experiment, and its probability quantifies the degree of confidence that this event will occur. If the probability of an event A is 1, we are certain that one of the outcomes of A will occur, and if it is 0, we consider all of them impossible [5]. The conditional probability of an event A given an event B, when $P(B) > 0$, is defined as: $P(A|B) = \dfrac{P(A,B)}{P(B)}$. From this definition, we can see that $P(A,B) = P(A|B) \times P(B)$. Considering n events $A_1,\ldots,A_n$; we have

$$P(A_1,\ldots,A_n) = P(A_1) \times P(A_2|A_1) \times \cdots \times P(A_n|A_1,\ldots,A_{n-1}).$$ (7.5.1)

This is called the chain rule of conditional probabilities. An event A is independent of an event B if $P(A|B) = P(A)$ or if $P(B) = 0$. It follows that A is independent of B if and only if $P(A,B) = P(A) \times P(B)$. An event A is conditionally independent of an event B given an event C if $P(A|B,C) = P(A|C)$ or if $P(B,C) = 0$. Similar to the case of independence, from this definition it follows that, A is conditionally independent of B given C if and only if $P(A,B|C) = P(A|C) \times P(B|C)$.

A random variable is defined by a function that associates a value with each outcome of a sample space. For example, in the experiment of tossing a coin two times, the sample space is {HH, HT, TH, TT}, and a random variable over this space can be the number of heads in each outcome of the experiment, which in this case is 2,1,1 and 0, respectively. Then the values of the random variable are 0,1 and 2. Discrete random variables take a finite number of values, while continuous random variables take infinitely many values, for example, integer or real values. Given a random variable X, the marginal distribution refers to $P(X=x)$ for all values x of X. In many situations, we are interested in probabilities that involve the values of several random variables, in that case, we need to consider the joint distribution. For a set of random variables $\{X_1,\ldots,X_n\}$, the joint distribution is denoted by $P(X_1,\ldots,X_n)$ and is the distribution that assigns probabilities to events that are specified in terms of these random variables [5]. The joint distribution has to be consistent with the marginal distribution, for example, let X and Y be two discrete random variables, then for all values x of X, $P(X = x) = \sum_y P(X = x, Y = y)$, and for all values y of Y, $P(Y = y) = \sum_x P(X = x, Y = y)$.

For two random variables X and Y, $P(X|Y = y)$ represents the conditional distribution over the events describable by X given the knowledge that Y takes the value of y. For each value of Y, $P(X|Y)$ is used to represent the set of conditional distributions. Using this notation, we can write the chain rule for random variables as $P(X,Y) = P(X) \times P(Y|X)$, which can be extended to multiple variables $X_1,\ldots,X_n$, from equation (7.5.1), as

$$P(X_1,\ldots,X_n) = P(X_1) \times P(X_2|X_1) \times \cdots \times P(X_n|X_1,\ldots,X_{n-1}).$$ (7.5.2)

Independence in random variables can be seen as a generalization to the concept of independence of events. Let X,Y,Z be random variables, we say that X is conditionally independent of Y given Z if $P(X=x|Y=y, Z=z)=P(X=x|Z=z)$ or if $P(Y=y,Z=z)=0$ for all values of X,Y and Z. We denote the conditional independence with $I(X,Z,Y)$ [6]. Several properties hold for conditional independence, and that often provide a very clean method for proving important properties about distributions. Some key properties are as follows [5,6]:

Symmetry: $I(X,Z,Y)$ if and only if $I(Y,Z,X)$.
Decomposition: If $I(X,Z,Y \cup W)$ then $I(X,Z,Y)$ and $I(X,Z,W)$.
Weak union: If $I(X,Z,Y \cup W)$ then $I(X,Z \cup Y,W)$.
Contraction: If $I(X,Z,Y)$ and $I(X,Z \cup Y,W)$ then $I(X,Z,Y \cup W)$.

Intersection: This property holds only for the class of strictly positive probability distributions, that is, distributions that assign a nonzero probability to every event. If $I(X,Z \cup W,Y)$ and $I(X,Z \cup Y,W)$ then $I(X,Z,Y \cup W)$.

7.5.3 Bayesian Networks

A Bayesian network for a set of variables $X = \{X_1,...,X_n\}$ is a pair (G, Θ), where G is a directed acyclic graph with n nodes, one for each variable in X, called the network structure and Θ is a set of conditional probability functions, one for each variable in X, called the network parametrization [6]. We use the following notation: $\Theta_{X_i|Pa(X_i)} = P(X_i \mid Pa(X_i))$ for all values of X_i and all the possible combinations of values of $Pa(X_i)$. If the node does not have parents, then the marginal distribution is considered.

According to the type of random variable, three types of Bayesian networks are distinguished: discrete if the variables are discrete, continuous if the variables are continuous, and hybrid if there are both discrete and continuous variables in the network. In our model, discrete Bayesian networks are considered, so Θ represents a set of conditional probability tables for each variable.

The graph G of a Bayesian network encodes the following set of conditional independence assumptions, called the local independencies: For each variable X_i, $I(X_i, Pa(X_i), \text{Nondescendants}(X_i))$.

Let $X_1,...,X_n$ be the variables of a Bayesian network, and assume that the nodes which represent those variables are in topological order in the graph. From equation (7.5.2) we have that:

$$P(X_1,...,X_n) = \prod_{i=1}^{n} P(X_i \mid X_1,...,X_{i-1}). \tag{7.5.3}$$

Consider one of the factors of equation (7.5.3): $P(X_i \mid X_1,...,X_{i-1})$. Since the nodes are in topological order, it follows that $Pa(X_i) \subseteq \{X_1,...,X_{i-1}\}$ and that none of the descendants of X_i is in the mentioned set. Hence, $\{X_1,...,X_{i-1}\} = Pa(X_i) \cup Z$, where $Z \subseteq \text{Nondescendants}(X_i)$. From the local independencies for X_i and from the decomposition property of conditional independencies, it follows that $I(X_i, Pa(X_i), Z)$. Hence

$$P(X_i|X_1,...,X_{i-1}) = P(X_i|Pa(X_i)). \tag{7.5.4}$$

If we apply this transformation to all of the factors of equation (7.5.3), we have that:

$$P(X_1,...,X_n) = \prod_{i=1}^{n} P(X_i \mid Pa(X_i)). \tag{7.5.5}$$

Equation (7.5.5) is called the chain rule for Bayesian networks, and this result explains the joint distribution for the variables of the model.

Bayesian networks represent the joint distribution of a set of random variables in a compact way. In the case of discrete random variables, the probabilities of all combinations of values of the variables must be declared to represent the joint distribution, and the number of probabilities can be large. For example, if we consider a set of n binary-valued random variables (Bernoulli variables), the joint distribution requires the specification of 2^n probabilities. Noting that the sum of probabilities must be 1, then we would need the specification of 2^n-1 numbers in total. For large values of n, this representation becomes unmanageable. Depending on the problem of study, estimating those probabilities can be difficult for human experts, and if we want to learn the distribution from data, we would need large amounts of data to estimate the parameters robustly [5]. Bayesian networks address this problem by using the conditional independences between variables, so that the representation of the joint distribution require fewer parameters and it can explain more about the whole phenomenon. If each node in the graph of a

Bayesian network has at most k parents, then the total number of independent parameters required is less than $n \times 2^k$ [5].

7.5.4 Parameter Learning

In Bayesian network theory, model selection and estimation are known as learning, and this is done in two steps: learning the structure of the Bayesian network, which is, the directed acyclic graph, and learning the parameters, which are the local probability distributions that correspond to the structure of the graph [7]. In this work, the directed acyclic graph is obtained from the attack graph proposed in [4]. For parameter learning, we use the maximum likelihood method for a data set. To explain this method, we present some relevant definitions of parameter learning.

A data set D for a set of variables $X = \{X_1, \ldots, X_n\}$ is a vector $(d_1, \ldots, d_N)$ where each d_i is called a case and represents a partial instantiation of variables X. The data set is complete if each case is a complete instantiation of variables X; otherwise, the data set is incomplete [6].

The empirical distribution for a complete data set D is defined as

$$P_D(\alpha) = \frac{D(\alpha)}{N}, \tag{7.5.6}$$

where $D(\alpha)$ is the number of cases d_i in the data set D that satisfy event α.

Let x be an instantiation of a variable in X and u an instantiation of its parents in G; then, using definition (7.5.6), the parameters $\theta_{x|u}$ can be estimated with the empirical probability:

$$\theta_{x|u} = P_D(x|u) = \frac{D(x,u)}{D(u)}. \tag{7.5.7}$$

Let θ be the set of all parameter estimates for a given network structure G and let $P_\theta(.)$ be the probability distribution induced by structure G and estimates θ. The likelihood of these estimates is defined as:

$$L(\theta|D) = \prod_{i=1}^{N} P_\theta(d_i). \tag{7.5.8}$$

That is, the likelihood of estimates θ is the probability of observing the data set D under these estimates [6]. Given a complete data set D, the parameter estimates defined by (7.5.7) are the only estimates that maximize the likelihood function defined by (7.5.8). In our work, we generate a complete data set by simulating cyber attacks, and then we estimate the parameters of the Bayesian network with the estimates of equation (7.5.7).

7.5.5 Inference

Once the Bayesian network has been designed, the objective is to obtain information by answering questions about the variables of the Bayesian network and their probabilities [6, 7]. The techniques used are known in general as *inference*. For Bayesian networks, the process of answering these questions is also known as *probabilistic reasoning* or *belief updating*, while the questions themselves are called *queries* [7]. The queries used for risk assessment are the probabilities of evidence, posterior marginals, and most probable explanations. Considering that we use Bernoulli random variables in the model, given a Bayesian network for a set of Bernoulli random variables $X = \{X_1, \ldots, X_n\}$, these inferences are:

Probability of evidence: $P\left(X_i = x_i\right)$ for $x_i \in \{0,1\}, i \in \{1,\ldots,n\}$.

Posterior marginals: Let $X_j = x_j$ be evidence of the Bayesian network; posterior marginals are $P\left(X_i = x_i | X_j = x_j\right) \forall x_i \in \{0,1\}, \forall i \in \{1,\ldots,j-1,j+1,\ldots,n\}$.

Most probable explanation: Let $X_n = x_n$ be evidence of the Bayesian network; the most probable explanation is $x_i \in \{0,1\}, \forall i \in \{1,\ldots,n-1\}$ such that $P\left(X_1 = x_1,\ldots,X_{n-1} = x_{n-1} | X_n = x_n\right)$ is maximum.

The algorithms for the calculation of inferences are classified as exact and approximate. The exact algorithms are based on the application of Bayes' theorem, and due to their nature, they are feasible for small networks. On the other hand, the approximate algorithms are based on Monte Carlo simulations to sample the global distribution of the network and thus estimate inferences, and those algorithms are useful in Bayesian networks with large number of nodes.

7.6 MODEL IMPLEMENTATION

7.6.1 Network Structure

The following Bernoulli random variables are considered for the implementation of the Bayesian network:

- *SECP*: A social engineering attack is performed from the access point to the management level, which consists of a cellphone in this case.
- *PHCP*: A phishing attack is performed from the access point to the management level.
- *MICP*: A malware injection attack is performed from the access point to the management level.
- *DOSR*: A DoS attack is performed from management level to the KNX IP router.
- *RTPR*: A routing table poisoning attack is performed from management level to the KNX IP router.
- *PAR*: A persistent attack is performed from management level to the KNX IP router.
- *DOSF*: A DoS attack is performed from the KNX IP router to the field devices level.
- *MITMF*: A MitM attack is performed from the KNX IP router to the field devices level.

Bayesian networks consider the causal relationship between random variables, and this is similar to attack graphs since, given an edge in an attack graph, the state of the child node conditionally depends on the state of the parent node; that is, the success of an attack is conditionally dependent on the success of an attack on the previous node [8]. Then, in this context, the structure of a Bayesian network can be represented with an attack graph if it is a directed acyclic graph.

Considering the smart home structure of Figure 7.1, the attack graph presented in [4] and the variables previously mentioned, we use the following directed acyclic graph for the implementation of the Bayesian network (see Figure 7.2). In this graph, all the paths in which an attack can be performed to the field devices level are represented.

7.6.2 Attack Simulation

The objective of the simulation is to describe how cyber attacks are carried out through the nodes of the network; this is done by simulating discrete random variables. First, the types of attacks that are carried out at each level of the smart home structure are selected, and then it is determined whether the attacker

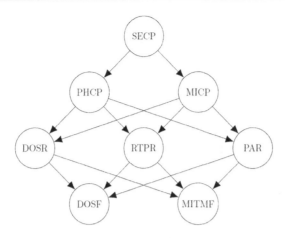

FIGURE 7.2   Directed acyclic graph of the Bayesian network.

TABLE 7.1   Selection Probabilities for the Management Level

PHCP	MICP	PROBABILITY
0	0	0.00
0	1	0.35
1	0	0.55
1	1	0.10

managed to compromise the devices. This is done by following the paths described by the Bayesian network graph, and as a result of the simulation, a data set is obtained where each column represents a network node and each row represents a complete attack event on the network, indicating the nodes through which the attacks were made. General considerations for simulation are that the attacker can perform multiple attacks from one node and attacks that do not follow the paths described by the graph are not allowed.

Selection Probabilities

Cyber attacks on the smart home network follow the structure of the graph in Figure 7.2. For each level of the smart home structure, the first step is to decide the type of attack, and this is done by simulating Bernoulli random variables. This selection refers to the simulation of a vector with 2 or 3 Bernoulli variables with the value of 1 if the node is selected for the attack and 0 otherwise. In the first node, the selection probability is 100% because at the beginning only a social engineering attack can be performed from the access point to the management level. Then, following the paths of the graph, it is possible to select between a phishing attack, a malware attack, or both to the management level, and the probabilities must be given for the selection of those cases. The selection probabilities used in the simulation are presented in Tables 7.1, 7.2, and 7.3.

Vulnerability Probabilities Based on CVSS Scores

The Common Vulnerability Scoring System (CVSS) is a scoring system that measures the impact or severity of vulnerabilities in information technology security. CVSS scores are based on the principal technical characteristics of software, hardware, and firmware vulnerabilities [9]. CVSS was developed for generating qualitative and quantitative device vulnerability scores [10]. In general, three metrics are

TABLE 7.2   Selection Probabilities for the KNX IP Router

DOSR	RTPR	PAR	PROBABILITY
0	0	0	0.00
0	0	1	0.10
0	1	0	0.15
0	1	1	0.10
1	0	0	0.25
1	0	1	0.15
1	1	0	0.20
1	1	1	0.05

TABLE 7.3   Selection Probabilities for the Field Devices Level

DOSF	MITMF	PROBABILITY
0	0	0.00
0	1	0.35
1	0	0.55
1	1	0.10

used to calculate these scores: base, temporal, and environmental. The base score reflects the severity of a vulnerability according to its intrinsic characteristics that are constant over time and is estimated based on the worst case in different scenarios [9]. Temporal metrics are described based on factors that change over time, and environmental metrics are described according to a specific environment [9]. In this work, as it is a model focused on a general smart home network, the base metrics are used to determine the CVSS scores of the network devices because the temporal and environmental metrics are situational and would be useful in more specific models.

CVSS scores have been used in previous works to estimate the parameters of discrete Bayesian networks in risk assessment models in computer systems [8, 11, 12]. In [11], the probability of exploiting a vulnerability is considered to be the CVSS score divided for the size of the domain, which is 10. However, this approach does not consider the causal relationships between the attacks present in the graph. In this work, the CVSS scores are not used as parameters of the Bayesian network, but as one of the criteria for the simulation of attacks. The base metrics are transformed to probabilities following the approach of [11], and this value is used as the parameter of a Bernoulli random variable. Then a value of this random variable is simulated, and the result indicates whether the attacker was successful in carrying out the attack.

In [10], the base metrics for IoT devices in the areas of health care, commerce, and home automation were calculated. In this article, it was mentioned that the CVSS score of IP cameras is 8.3 and for a Smart Hub it is 8.5. Then we consider that the score of the field devices level is 8.3 and that, since the Smart Hub has similar characteristics as the KNX IP router, a score of 8.5 is therefore considered for this level. For the management level, the CVSS score is considered to be 8.

Attack Simulation Algorithm

Considering the selection and vulnerability probabilities, the following algorithm is proposed to simulate network attack events. The input data is the number of network attack events, the graph of the Bayesian network, the selection probabilities, and the vulnerability probabilities. As a result, a data set with N attack events is obtained. The algorithm consists of repeating the following process N times:

- Select the root node of the graph.
- Identify the child nodes and select the attack/s by simulating discrete random variables using the selection probabilities of the respective levels.
- For the nodes selected in the previous step, verify whether the attacker managed to compromise the devices by simulating Bernoulli random variables whose parameters are the vulnerability probabilities.
- Save the compromised nodes from the previous step in a vector and with these nodes carry out steps 2 and 3 until there are no more child nodes in the graph.
- Create a vector in which each position represents a variable of the Bayesian network; set the value of 1 if the variable (node) was selected and compromised and 0 otherwise.

7.6.3 Network Parametrization

The maximum likelihood estimates of equation (7.5.8) are used to obtain the parameters from the data set generated from the attack simulation. Specifically, the following parameters are calculated:

- Θ_{SECP}
- $\Theta_{PHCP \mid SECP}$
- $\Theta_{MICP \mid SECP}$
- $\Theta_{DOSR \mid PHCP, MICP}$
- $\Theta_{RTPR \mid PHCP, MICP}$
- $\Theta_{PAR \mid PHCP, MICP}$
- $\Theta_{DOSF \mid DOSR, RTPR, PAR}$
- $\Theta_{MITMF \mid DOSR, RTPR, PAR}$

7.6.4 Results

This section presents the results obtained from the implementation of the model. The number of attacks simulated were 1000. Using exact inference algorithms in Bayesian networks [7], four results were calculated:

1. Probability of DoS and MitM attacks to the field devices level
2. Probability of DoS and MitM attacks to the field devices level given the evidence of attack on any other nodes in the network
3. Probability of DoS and MitM attacks to the field devices level given the evidence of attacks and that there were no phishing and malware injection attacks to the management level
4. Most probable explanation given the evidence of DoS and MitM attacks to the field devices level

The first three results are presented in Tables 7.4, 7.5, and 7.6. As for the most probable explanation: If there is evidence that there was a DoS attack but not a MitM attack or that there was a MitM attack but not

TABLE 7.4 Probability of DoS and MitM Attacks to the Field Devices Level

ATTACK	PROBABILITY
DOSF	0.3257822
MITMF	0.2627260
DOSF and MITMF	0.1897563

TABLE 7.5   Probability of DoS and MitM Attacks to the Field Devices Level Given the Evidence of Attack on Any Other Nodes in the Network

ATTACK	EVIDENCE	PROBABILITY
DOSF	SECP	0.4134291
MITMF		0.3334087
DOSF and MITMF		0.2408075
DOSF	PHCP	0.6054472
MITMF		0.4905264
DOSF and MITMF		0.3567098
DOSF	MICP	0.6333557
MITMF		0.5179413
DOSF and MITMF		0.3813129
DOSF	DOSR	0.7190847
MITMF		0.5931191
DOSF and MITMF		0.4460161
DOSF	RTPR	0.7592573
MITMF		0.6449062
DOSF and MITMF		0.5024086
DOSF	PAR	0.7097519
MITMF		0.6507042
DOSF and MITMF		0.4829878

TABLE 7.6   Probability of DoS and MitM Attacks to the Field Devices Level Given the Evidence of Attacks and That There Were No Phishing and Malware Injection Attacks to the Management Level

ATTACK	EVIDENCE 1	EVIDENCE 2	PROBABILITY
DOSF	No PHCP	MICP	0.5625418
MITMF			0.4474423
DOSF and MITMF			0.3165204
DOSF	PHCP	No MICP	0.5468065
MITMF			0.4323725
DOSF and MITMF			0.3036255
DOSF	PHCP	MICP	0.7013511
MITMF			0.5856344
DOSF and MITMF			0.4435267

a DoS attack to the field devices level, the most probable explanation includes a social engineering attack to the management level, then a phishing attack to the management level, after that a DoS attack to the KNX IP router, and finally any of the attacks of the evidence.

7.7 CONCLUSIONS AND FUTURE WORK

In this chapter, a risk assessment model of an IoT network of a smart home was implemented, using a Bayesian network whose directed acyclic graph was obtained from the attack graph proposed in [4] and whose parameters were obtained from the maximum likelihood method applied to a data set obtained via simulation. Tables 7.4, 7.5, and 7.6 provide the probabilities of different attacks to the field devices level with and without evidence of other attacks on the smart home network. These results can be used by cyber security evaluators; for example, if a smart home network is considered to have similar selection and vulnerability probabilities to the ones proposed in the model and there is evidence that there was a phishing attack at the management level, there is a probability of 60.54% of a DoS attack and 49.05% of a MitM attack to the field devices level. Thus a viable option would be to take actions to secure the devices from a DoS attack. Using Table 7.5, if there is evidence that there was also a malware injection attack on the management level, then the probability of a DoS attack rises to 70.15%, making protection from DoS attacks urgent. The advantage of using this methodology is that it allows these probabilities to be quantified, which can be useful for decision making.

Through the use of Bayesian inferences, new information about different events can be obtained with a higher level of detail since the knowledge about a phenomenon is updated with evidence. In risk assessment, this evidence can be observed or hypothetical, and in the second case it is possible to find out what would happen in different situations, and, based on this analysis, better decisions can be made. One of the main advantages of the Bayesian network model is that it considers the conditional dependence of the random variables.

In this chapter, the risk assessment consisted of finding the probabilities of different cyber attack events on the network. This type of analysis is useful in cases where decision making is based on probabilities. Future work will focus on adding more detail to the risk assessment according to the needs of the assessor. For example, the impact of cyber attacks can be included, both in the form of a score created by the evaluator or related to the financial loss caused by cyber attacks, and this financial loss can be focused on information loss or interruption of service.

This chapter included a general analysis of the structure of a smart home. Future work can include the study of a specific smart home, detailing each device with its own characteristics. For this, new selection probabilities and CVSS scores for the vulnerability probabilities have to be specified to simulate the attacks. However, the ideal scenario would be to work with a provided data set to obtain a more realistic analysis of the situation of the smart home. With a data set of a real case, the same procedure of this work would be followed except for the simulation of the attacks.

It should be mentioned that, since the use of attack graphs is not exclusive to smart homes, this analysis can be extended to different types of IoT networks. An interesting study may be focused on smart cities, which are technologies that are currently booming, and therefore the study of cyber security in these environments is an imminent need for people and for cyber security evaluators.

REFERENCES

[1] Poolsappasit, N., Dewri, R., and Ray, I. "Dynamic security risk management using bayesian attack graphs". *IEEE Transactions on Dependable and Secure Computing*, 9(1):61–74, 2012.

[2] H. Rajab and T. Cinkelr. "IoT based smart cities". In *2018 International Symposium on Networks, Computers and Communications (ISNCC)*. IEEE, 2018.

[3] D. Bastos, M. Shackleton, and F. El-Moussa. "Internet of things: A survey of technologies and security risks in smart home and city environments". In *Living in the Internet of Things: Cybersecurity of the IoT – 2018*. Institution of Engineering and Technology, 2018. DOI:10.1049/CP.2018.0030

[4] M. Ibrahim and I. Nabulsi. "Security analysis of smart home systems applying attack graph". In *2021 Fifth World Conference on Smart Trends in Systems Security and Sustainability (WorldS4)*. IEEE, 2021.

[5] Koller, D. and Friedman, N. "Probabilistic graphical models: principles and techniques". *Adaptive Computation and Machine Learning series*. MIT Press, 2009

[6] A. Darwiche. *Modeling and Reasoning with Bayesian Networks*. Cambridge University Press, 2009.

[7] Scutari, M. and Denis, J.-B. (2014). *Bayesian Networks with Examples in R*. Chapman and Hall/CRC.

[8]   Liu, Y. and Man, H. "Network vulnerability assessment using Bayesian networks". In Dasarathy, B. V., editor, *Data Mining, Intrusion Detection, Information Assurance, and Data Networks Security 2005*, volume 5812, pp. 61–71. International Society for Optics and Photonics, SPIE, 2005.

[9]   Common Vulnerability Scoring System version 3.1 Specification Document. [Online] www.first.org/cvss/v3.1/specification-document

[10] S. Rizvi, N. McIntyre and J. Ryoo, "Computing security scores for IoT device vulnerabilities," *2019 International Conference on Software Security and Assurance (ICSSA)*, 2019, pp. 52–59

[11] Frigault, M., Wang, L., Singhal, A., and Jajodia, S. "Measuring network security using dynamic bayesian network". In *Proceedings of the 4th ACM Workshop on Quality of Protection*, QoP '08, pp. 23–30, New York, NY, USA, 2008. Association for Computing Machinery.

[12] Muñoz-González, L., Sgandurra, D., Barrère, M., and Lupu, E. C. "Exact inference techniques for the analysis of bayesian attack graphs". *IEEE Transactions on Dependable and Secure Computing*, 16(2):231–244, 2019.

[13] Zeng, J., Wu, S., Chen, Y., Zeng, R., and Wu, C. "Survey of attack graph analysis methods from the perspective of data and knowledge processing". *Security and Communication Networks*, 2019:1–16, 2019.

PART II

Secure AI/ML Systems: Defenses

Survey of Machine Learning Defense Strategies

8

Joseph Layton,[1] Fei Hu,[1] and Xiali Hei[2]

[1]*Electrical and Computer Engineering, University of Alabama, USA*

[2]*School of Computing and Informatics, University of Louisiana at Lafayette, USA*

Contents

8.1 INTRODUCTION

Machine learning broadly refers to intelligent algorithms with self-improvement as a core characteristic of these systems. Using a supervised learning system, one can make decisions and predictions about complex data sets without being directly programmed for every situation encountered. Thanks to the rapid rise of computational capabilities and the vast data sets collected on user behavior (clicks, watch times, reading patterns, etc.), machine learning can be applied to discover patterns in these data sets that were previously difficult to analyze with classic programming approaches.

The three primary types of machine learning systems are supervised learning, unsupervised learning, and reinforcement learning (see Figure 8.1) [1]. *Supervised* learning uses a set of training data with typical inputs and desired outputs to "teach" the algorithm how to make accurate predictions about the training data and future inputs outside the training set. *Unsupervised* learning takes a data set and attempts to find patterns within the data without explicit directions on the expected output. By identifying the "clusters"

Machine Learning Types

Supervised	Unsupervised	Reinforcement
Classification	Clustering	Reward Maximization
• Image Classification	• Face Recognition	• Game AI
• Diagnostic Detection	• Text Mining	• Inventory Management
• Spam Detection	• Big Data Visualization	• Real-time decisions

FIGURE 8.1 Three types of machine learning and their applications.

within the data, the machine learning system eventually builds a model of the data set without explicit tagging/labeling of the training data, leading itself to identify commonalities across multiple inputs. *Reinforcement* learning can model how most living organisms learn, through either positive or negative reward systems. The algorithm is given a task to achieve and is graded after each attempt to determine how effective this iteration is. Early in the process, most successful attempts are down to luck, but as the "lucky" few are iterated, the model begins to consistently increase its score. This training is typically continued through hundreds or thousands of iterations until the model reaches a specified threshold of reliability needed for the task at hand, such as recognizing 99.9% of stop signs encountered.

There are minimal security considerations for simple machine learning systems operating only on internal data due to network isolation and data isolation. However, for more complex systems that draw from important company databases or even the Internet at large, potential threats during either the training phase or the deployment phase need to be considered. Depending on the specific machine learning applications and the level of access granted to users, the techniques employed for defending against adversarial attackers can vary in scope from simple data filtering to complex decision analysis and model adjustments.

8.2 SECURITY THREATS

Adversarial attacks against ML systems aim to reduce systems' learning ability, produce faults in model behaviors, and acquire sensitive information from the system. Since a machine learning model is a large network with weighted connections used to make a decision about an input, adversarial attacks can include irrelevant data to cause the algorithm to identify patterns that do not exist. Attacks targeted at a specific ML system will apply "perturbations" of input data undetectable to human operators, but the model detects and acts upon it. The systems that process visual data have been widely studied for adversarial attacks, especially with the rising popularity of autonomous driving systems. For example, adding stickers to a stop sign could cause a computer vision system to identify it as a yield or speed limit sign instead [2].

Typical attack vectors against machine learning systems can be grouped by which phase of the model they target: training or deployment. During the training phase, the model is still forming connections from a given data set, so an attacker poisons the training data with imperceptible noise and perturbations that cause the model to form incorrect connections when classifying. Once the model training is completed and deployed, an attacker then seeks to evaluate and impersonate the model, replacing it with a substitute network that appears to produce the correct results upon casual inspection but instead produces targeted errors. Attacks during this time typically revolve around querying the model to build a substitute training set that slowly reduces the accuracy of the model.

For this chapter, a few types of threats and various countermeasures to those threats are surveyed. First, a honeypot defense network is presented to distract a potential adversarial attacker from engaging with real systems. Next, the concept of poisoned data is demonstrated along with a general approach to combating poisoned data during the prediction and adjustment phase of a model. After that, the principle of mixup inference is covered to improve the defensive parameters of mixup-trained models and their global linearities. In the section following that, the techniques for defending cyber-physical systems against data-oriented attacks and the applicability of such techniques for IoT devices are discussed. From there, the concept of information fusion is approached along with the applications of data fusion for network-level defensive mechanisms, specifically those that model internet-oriented and CPS-oriented networks. After these overviews of various machine learning defensive techniques, some broader conclusions and potential future directions on ML defense are presented.

8.3 HONEYPOT DEFENSE

A honeypot (Figure 8.2) or honeypot trap is a computer security scheme that uses a decoy system to trap adversarial attacks against the real system. The honeypot appears to the attacker as a legitimate computer system and is then made attractive to these attackers by leaving deliberate security vulnerabilities for the attacker to exploit. The goal for a honeypot system is to gain information about cyber attack methods from the attackers themselves while also taking up the attacker's time, reducing their effectiveness against actual targets.

The two primary types of honeypots are low-interaction and high-interaction [3]. *Low-interaction* traps use minimal resources and collect only basic information from attacks against them. These systems typically are just simulated networks with minimal data stored inside them. Since they have little within them to engage attackers once breached, these low-interaction traps do not provide as much information about complex cyber attacks. A *high-interaction* honeypot instead sets up a full system with falsified databases and multiple layers of access, similar to what a real system would have. These high-interaction systems occupy much more of the attacker's time and can provide deeper insights into the patterns of their attacks, such as attempted breaches of encrypted databases or downloading of certain portions of the honeypot.

For machine learning systems, honeypots can be used to prevent adversarial attackers from corrupting the learning model while also reducing the indirect information that attackers can glean from the machine learning system. As Younis and Miri [4] demonstrated in their paper on decentralized honeypots, a network of such traps can be used to accomplish the following result: prevent an attacker from learning the

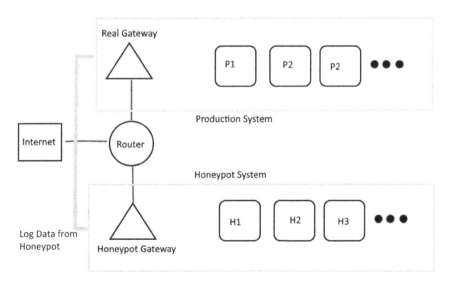

FIGURE 8.2 Honeypot system diagram.

labels and approximating the black-box system, lure the attacker toward a decoy system that occupies their time, and create computationally infeasible work for the attacker [4].

In setting up a network of honeypots as an additional layer of defense for a machine learning system, there are some important considerations to keep in mind to ensure the trap's effectiveness. First, each honeypot node should appear as realistic as possible to a probing attacker – using an existing machine learning system but allowing it to operate on simulated data can produce a honeypot that appears to be a production system to the adversarial attacker. Second, the honeypot should be deployed with "leaked" tokens in the file systems that appear to the attacker as potential vulnerabilities within the server; these artificial data dumps (credentials, model parameters, etc.) are placed in various locations within the system for the attacker to access. Third, each honeypot node should capture and report attackers' activity as they breach the system and search for data. Using a network of honeypots enables this reporting to be hidden within network activity that the attacker assumes to be legitimate communication between various servers.

To identify a potential attacker, the researchers use various flags that can be configured for the behavior indicating that a nonuser is attempting to learn about the machine learning system. These flags include persistent DNN querying, a high number of DNN labeling requests, and a sudden drop in classification accuracy. In contrast to normal users that query the decoy DNN model 1–2 times per session, an attacker might send hundreds or thousands of queries to the model while building a synthetic training dataset. A high number of query requests to the oracle (the interface used to interact with the model) is also not typical for normal users, but an attacker would repeatedly query the oracle to enlarge the synthetic training model they plan to feed DNN. If the known inputs with known outputs previously classified with high confidence and low confidence suddenly begin to change, this can also be used as an indicator of the model being under attack.

Honeypot defenses for machine learning systems can be a practical addition to existing security architecture, but they are not suitable as a sole defense against adversarial attackers. Their primary function is to waste potential attackers' time while also providing some data on those attacker's behaviors; however, this requires the attacker to target and engage with the honeypot. While including obvious security flaws and "leaked" data can make the honeypot appealing to an attacker, such tactics do not guarantee that the attacker will interact with the decoy system for a long enough time to draw out the meaningful data.

8.4 POISONED DATA DEFENSE

Recent research into machine learning algorithms has shown that a potential attack vector is the training dataset used for the machine learning system, a so-called poisoned data attack. This compromise of the integrity of a machine learning system early into its operational lifetime can have deleterious effects, especially if the malicious alterations are subtle and difficult for human operators to detect. For example, a machine learning system trained to diagnose patient data could report more false positives than expected, reducing confidence in the system as well as worsening patient outcomes.

In their paper on the defensive techniques against poisoned data, Mozaffari-Kermani et al. presented general algorithm-independent approaches to using poisoning attacks in healthcare systems and countermeasures to those proposed generic attacks [5]. Since this chapter focuses on defensive measures against machine-learning attacks, the authors can only cover the generic attacks briefly.

The attack model for this poisoned data assumes that potential attackers have some knowledge of the training dataset, either through it being publicly available (population health data) or by eavesdropping on network traffic to gather specific data. These potential attackers' motivations may vary, but the general result of these attacks is the worsening model performance during particular classifications (denoted as the attacked class). For example, an attacker may wish to misdiagnose hypothyroidism as being normal for the patient: Once the initial misdiagnosis is made, either scenario is advantageous for a malicious attacker: (1) the diagnosis is corrected, but confidence in the ML system is reduced, or (2) the diagnosis is not corrected, and the patient suffers the health effects. For an adversarial attacker to accomplish this goal, it must add enough poisoned data to the training set that the machine learning system begins to identify perturbations incorrectly, mistaking the maliciously placed signals as actual identifiers of health conditions.

The countermeasure proposed by Mozaffari-Kermani et al. is the periodic construction and evaluation of a separate model using the training dataset, with threshold alarms being triggered if the accuracy statistics of the model changes significantly between constructions. By tracking the rate of correct classifications and Kappa statistic, perturbations in the model can be measured over time, and the timeframe of when malicious data entered the training set can be identified. By using both statistics, an alarm can be triggered even when malicious data does not directly decrease the classification ratio.

8.5 MIXUP INFERENCE AGAINST ADVERSARIAL ATTACKS

To reduce undesirable behaviors such as memorization and sensitivity to adverse examples, Zhang et al. introduced their approach of "mixup" during the training phase [5]. The main principle of *mixup* is to train the neural network on convex combinations of pairs of examples and their labels. In doing so, the neural network is regularized to favor simple linear behavior among training examples. The 2018 paper introducing the mixup concept showed that this principle improved the generalization of neural networks and increased the robustness of the networks against adversarial examples [5].

As an extension of this work, Pang et al. developed *mixup inference (MI)*, an inference principle that seeks to improve the defensive effects of mixup-trained models such that they can more actively fight against adversarial examples [6]. The core principle of a model using MI techniques is combining inputs with other random clean samples, shrinking and transferring any adversarial perturbations that may have existed in the dataset. This process of combining inputs seeks to prevent local perturbations from affecting the model's global predictions.

For their methodology, Pang et al. proposed an algorithm (Figure 8.3) for mixup-trained classifiers that takes an input x and performs a mixup upon it before feeding it to the prediction model. Theoretically (given enough computational time and sufficient clean samples), a well mixup-trained model F can be modeled as a linear function H on convex combinations of the clean examples. By adjusting the number of iterations performed during the prediction phase, the sensitivity of the model to adversarial examples and local perturbations can be changed post-training phase.

For the experimental results of their mixup-inference principle, training was done on CIFAR-10 and CIFAR-100 using ResNet-50 [6]. Pang et al. first evaluated the performance under oblivious-box attacks, in which the adversary is not aware of the defensive mechanism and generates adversarial examples based on general principles. In both targeted and untargeted modes, the MI method can improve the robustness of a trained model with global linearity. It is also compatible with training-phase defenses such as interpolated AT. Ablation studies of the MI method also demonstrated that it effectively exploits global linearities resulting from mixup-trained models, rather than inducing additional randomness to the process [7].

Inputs	Model input x; mixup-trained classifier F
Hyperparameters	Mixup ratio λ; number of executions N; sample distribution p
Algorithm	1. Sample a label y then select x s.t. $x \sim p(x|y)$.
	2. Mixup x with x^k from the sample distribution.
	3. Update prediction of $F(x)$ as running average of predictions.

FIGURE 8.3   Overview of algorithm proposed by Pang et al. [6], adapted.

Under white-box adaptive attacks in which the adversarial attacker is aware of the defense mechanisms, the MI method still performs strongly but does not scale as well as blind attacks. Additional iterations of the MI method offer more minuscule improvements for adaptive attacks than oblivious-box attacks.

8.6 CYBER-PHYSICAL TECHNIQUES

Cyber-physical systems control or monitor physical mechanisms via computer-based algorithms; the physical and software components are intertwined such that changes in one domain can prompt changes in the other domain [8]. Examples of cyber-physical systems include smart electrical grids, medical monitoring systems, and industrial control systems. A cyber-physical system's key distinguisher is a network design of physical inputs and outputs rather than distinct devices.

In their 2017 paper, Cheng et al. proposed a security methodology (Orpheus) for defending against data-oriented attacks [9]. Attacks of this type seek to alter a program's variables without interrupting control flow, causing a disconnect between what is physically occurring and what the CPS network believes to be occurring. The core approach behind Orpheus is an event-aware Finite State Machine (eFSA) to detect breaches via data-oriented attacks by evaluating state transitions and determining the accuracy of incoming data based on previous states. A proof-of-concept of this system was implemented on a Raspberry Pi platform to demonstrate the feasibility of integrating it into embedded control systems.

The first step of constructing an eFSA involves building a static FSA model that is then augmented via event identification and dependence analysis. Compared to a basic FSA model, the eFSA can check consistency between runtime behavior and program execution semantics, marking state transitions that deviate from the expected automaton or that fail to match the expected contextual data.

Once the eFSA is constructed, it can then be utilized during the system's runtime, cataloging and assessing potential anomalies. The first anomaly type is event-independent state transition – if a transition between states is not normally expected, the automaton is moved to a known good state. This can reduce the effectiveness of control-oriented attacks that seek to adjust the control flow of the model. The second anomaly type consists of event-dependent state transitions. Besides the previous static check, the eFSA is used to determine the consistency of the program behavior with given inputs; if valid input data is given that fails to match other physical contexts, an alarm can be set, and the program can return to a known good state. To overcome this check, an attacker would need to have full access to the system and simulate correct inputs to match each event's expected physical data.

8.7 INFORMATION FUSION DEFENSE

Information fusion (or data fusion) (Figure 8.4) integrates multiple distinct data sources to produce more helpful information than a single source can provide [10]. Depending on where this fusion takes place and the complexity of the data used, the fusion process can be placed on a continuum, with the low end corresponding to simple combinations of raw data and the high end corresponding to complex fusions of highly processed data streams. For machine learning defense, this concept of information fusion can be applied as a network-level mechanism for detecting and mitigating potential attacks against IoT devices, as demonstrated in the 2014 paper by Chen et al. [11].

In this paper, a fusion-based defense mechanism is proposed and evaluated where each node sends minimal feedback (one-bit) based on local decisions to the fusion center [11]. The fusion center then uses these feedback decisions to make attack inferences and deploy countermeasures. The core concept behind the fusion center is the offloading of more complex defense modules to a single powerful computation

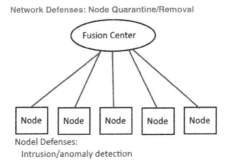

FIGURE 8.4   Network model of IoT fusion.

unit with each physical network node instead of making minimal defense decisions. Since the attacker and defender both tend to focus their efforts on a subset of nodes, abstracting the defensive mechanisms to a central fusion mechanism can reduce the per-node cost of defensive strategies and create a zero-sum game between the attacker and defender. From there, the zero-sum game equilibrium between the two parties can be used to model the payoff of various end states, allowing decisions to be made abstractly about potential adversarial scenarios.

At the network level, the fusion center monitors for high-level attacks (network breaches or routing changes) while also receiving and computing the feedback bits from all the network nodes. If a subset of the nodes comes under attack from an adversary, the defender seeks to mitigate these attacked nodes' damage potential on the rest of the network. Although the system model for this paper represents the network as a single layer, the general principles of the information fusion can be applied to more hierarchical networks with cascading feedback chains.

The worst-case scenario for the defender is an adversary with full knowledge of the nodal network and the ability to sabotage all nodes simultaneously, especially if the adversary is aware of the fusion center's thresholds and the importance of particular nodes. However, a uniform attack against every node is actually less effective than a targeted attack against a subset of nodes since the uniform attack will certainly trigger countermeasures by exceeding the fusion center thresholds. A more targeted intentional attack would instead seek to cripple the operation of the network without alerting the fusion center; this would mean fewer nodes are attacked, but the targets are more critical to network operation.

At the nodal level, each device employs simple anomaly detection and reports a minimal feedback bit to the fusion center. These feedback bits are used at the network level to decide which nodes should be quarantined, which nodes are producing false alarms, and when more aggressive countermeasures should be deployed (node and network resets). Since the operation of the network relies on some percentage of nodes operating correctly, the percolation-based connectivity of the network can be used as a measure of the effectiveness of attacks. If an attack successfully reduces the connectivity below the threshold (putting the network in a disconnected state), an attack is considered successful. In this model, false alarms have the same cost as successful node breaches since the result is the quarantine of a node from the network.

The two network types analyzed by Chen et al. were internet-oriented networks and CPS networks, since those topologies represent the largest share of networked node systems. For internet-oriented networks, the degree distribution can be modeled as a power law distribution, showing the vulnerability of such a network to intentional attacks that target critical nodes. If a major hub of connections is targeted and successfully removed from the network, the impact of the attack is much more successful than a random attempt. On the other hand, CPS systems tend to be more evenly connected and follow an exponential distribution for their connections. This growth in connections means that a sufficiently large network requires many simultaneous node failures before operational efficiency is impacted, improving its robustness against targeted adversarial attacks.

Based on their synthetic models and experimental results, a defense based on information fusion can enhance existing defensive mechanisms for both internet-oriented and CPS-oriented networks with minimal overhead to the primary nodes of the network. Since the feedback mechanism is constrained to send only minimal information to the fusion center, the communication overhead is low, and the computation overhead for each node is also kept low. While internet-oriented networks are still vulnerable to highly targeted attacks due to their power law distribution, a fusion center with well tuned thresholds can detect and mitigate attacks that might otherwise cripple such a network, especially if countermeasures such as rerouting can be deployed.

8.8 CONCLUSIONS AND FUTURE DIRECTIONS

As the previous sections indicated, attacks against ML systems have three primary goals: reducing the learning ability of the model, producing faults in model behavior, and acquiring sensitive information from the system. Attacks can be grouped by which phase (training or deployment) they target, and those groups can then be used to select defensive techniques such as those just presented. While no defense is absolute, an appropriately selected defensive schema can reduce attacker confidence and slow down their attempts at subverting the ML system's behavior, providing time for human operators to intervene.

As further research is conducted on potential attacks on machine learning systems, new defensive techniques and refinements of existing techniques will continue to be made. The most promising research seeks to find algorithmic approaches for detecting and correcting adversarial behavior in real time, enabling a deployed ML model to actively mitigate attacker damage. These techniques, like the mix up inference approach, rely on routine self-checks of model output but are intended to require less input from human operators to correct errors and attacker damage.

REFERENCES

[1] "The Three Types of Machine Learning Algorithms," *Pioneer Labs | Technology Strategists & Delivery Experts.* https://pioneerlabs.io/insights/the-three-types-of-machine-learning-algorithms (accessed Feb. 11, 2021).

[2] K. Eykholt *et al.*, "Robust Physical-World Attacks on Deep Learning Models," *ArXiv170708945 Cs*, Apr. 2018, Accessed: Feb. 11, 2021. [Online]. Available: http://arxiv.org/abs/1707.08945.

[3] "What is a Honeypot?," *usa.kaspersky.com*, Jan. 13, 2021. https://usa.kaspersky.com/resource-center/threats/what-is-a-honeypot (accessed Feb. 11, 2021).

[4] F. Younis and A. Miri, "Using Honeypots in a Decentralized Framework to Defend Against Adversarial Machine-Learning Attacks," in *Applied Cryptography and Network Security Workshops*, Cham, 2019, pp. 24–48, doi: 10.1007/978-3-030-29729-9_2

[5] H. Zhang, M. Cisse, Y. N. Dauphin, and D. Lopez-Paz, "Mixup: Beyond Empirical Risk Minimization," *ArXiv171009412 Cs Stat*, Apr. 2018, Accessed: Mar. 29, 2021. [Online]. Available: http://arxiv.org/abs/1710.09412.

[6] K. He, X. Zhang, S. Ren, and J. Sun, "Identity Mappings in Deep Residual Networks," *ArXiv160305027 Cs*, Jul. 2016, Accessed: Mar. 29, 2021. [Online]. Available: http://arxiv.org/abs/1603.05027.

[7] T. Pang, K. Xu, and J. Zhu, "Mixup Inference: Better Exploiting Mixup to Defend Adversarial Attacks," *ArXiv190911515 Cs Stat*, Feb. 2020, Accessed: Mar. 07, 2021. [Online]. Available: http://arxiv.org/abs/1909.11515.

[8] kristy.thompson@nist.gov, "Cyber-Physical Systems," *NIST*, Jun. 20, 2014. www.nist.gov/el/cyber-physical-systems (accessed Mar. 29, 2021).

[9]   L. Cheng, K. Tian, and D. (Daphne) Yao, "Orpheus: Enforcing Cyber-Physical Execution Semantics to Defend Against Data-Oriented Attacks," in *Proceedings of the 33rd Annual Computer Security Applications Conference*, New York, NY, USA, Dec. 2017, pp. 315–326, doi: 10.1145/3134600.3134640

[10] "Data Fusion – An Overview | ScienceDirect Topics." www-sciencedirect-com.libdata.lib.ua.edu/topics/computer-science/data-fusion (accessed Mar. 29, 2021).

[11] P. Chen, S. Cheng, and K. Chen, "Information Fusion to Defend Intentional Attack in Internet of Things," *IEEE Internet Things J.*, vol. 1, no. 4, pp. 337–348, Aug. 2014, doi: 10.1109/JIOT.2014.2337018

Defenses Against Deep Learning Attacks

9

Linsheng He[1] and Fei Hu[1]

[1]*Electrical and Computer Engineering, University of Alabama, USA*

Contents

9.1 INTRODUCTION

In recent years, technologies like machine learning and deep learning (DL) neural networks [1] have been widely used in real-world tasks such as image classification, speech recognition, autonomous driving, spam filtering, and intelligent antifraud on image-like data. As a result, DL systems are becoming commonplace and ubiquitous in our lives, which will open up new possibilities for adversaries to carry out attacks. Attackers attempt to achieve adversarial goals by altering the model's input features in various ways to bypass the detection of machine learning models in realistic tasks or by attacking the models directly to compromise their integrity.

For example, the initial adversarial attack is intended to construct an adversarial example by adding finely designed noise to the normal example that is not perceptible to humans, so as to drive the machine learning or DL model to misjudge the carefully constructed adversarial example without interfering with human cognition, as shown in Figure 9.1 [2]. Here the original image is recognized as "panda" by the image classification model with 57.7% confidence, while the adversarial image obtained after adding subtle perturbations is incorrectly recognized as "gibbon" with 99.3% confidence, yet the adversarial image is still recognized as a giant panda normally for a human.

Since such subtle perturbations are usually indistinguishable to the human eyes, they make the attack extremely stealthy and are sufficient to change the prediction results of the model. This is extremely harmful and thus poses a huge security threat to machine learning models actually deployed in real-world scenarios, especially risk-sensitive ones.

DOI: 10.1201/9781003187158-11

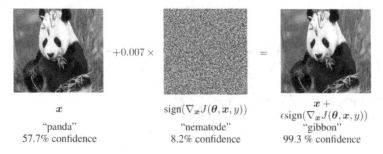

$$x \qquad \text{sign}(\nabla_x J(\theta, x, y)) \qquad \begin{array}{c} x + \\ \epsilon \text{sign}(\nabla_x J(\theta, x, y)) \end{array}$$

"panda"   "nematode"   "gibbon"

57.7% confidence   8.2% confidence   99.3 % confidence

FIGURE 9.1 Example of basic adversarial attack.

To explore defense schemes against DL models in depth, we briefly introduce the adversarial attack algorithms. The most important classification criterion for adversarial attacks is the degree of content acquisition of the DL model. Based on this, we can divide all adversarial attack models into two categories: white-box attacks and black-box attacks.

- *White-box attack*: White-box attack assumes that the attacker can fully obtain information such as the structure and parameters of the target model. Therefore, during the attack, the attacker can use the complete information of the model to solve the gradient information of the target model to guide the generation of adversarial examples.*Black-box attacks*: Unlike white-box attacks, black-box attacks assume that the attacker neither knows the training data and model structure nor can obtain the specific parameters of the model. It can only obtain the final decision result of the learning model. In this case, the attacker can only detect the sensitivity of the target model or estimate the gradient information of the model by manipulating the input of the model and using the final decision result to guide the adversarial sample. Therefore, compared with the white-box attack, the black-box attack uses less information, and the attack is more difficult to implement.

Adversarial attacks can also be classified according to different DL model tasks. Generally, DL models are mainly used to deal with real-world problems in images, data, speech, text, etc. For image and data tasks, DL has tasks such as segmentation, classification, and regression, and adversarial attacks usually use certain methods such as increasing perturbation and data poisoning. For speech and text tasks, DL mainly has tasks such as recognition, translation, prediction, etc. Adversarial attacks usually use methods such as adding high-frequency noise, keyword extraction, or synonym/antonym replacement when facing such tasks.

Once we have a general understanding of what an adversarial attack is, it is easy to see that defense scheme is also a much needed research topic in order to ensure that the widely used DL models are not disturbed by attackers and to enhance the robustness of DL models. In the following sections, we introduce methods of defenses with their particularities according to different application scenarios. At the end of this chapter, we propose several open issues that currently exist in adversarial attack and defense research.

9.2 CATEGORIES OF DEFENSES

When we classify the existing adversarial defenses, we can simply divide the level of defense into two categories: full defense and attack detection only. The full defense approach aims to enable the target network to achieve immunity from attacks on adversarial examples (e.g., attack targets). The classifier is able to predict the label of the adversarial attack accurately. On the other hand, the attack detection only approach may just add tags to potential adversarial examples in order to reject them in further processing.

We can also classify all defense methods into three types: modified training or modified input, modifying network architecture, and network add-on. The first category is not directly related to the learning model, while the other two categories are more concerned with the neural network itself. The difference between "modifying" the network and using "add-ons" is that the second changes the original deep neural network architecture/parameters during the training process. The third one keeps the original model intact and attaches an external model to the original model during testing.

9.2.1 Modified Training or Modified Input

Data Preprocessing

In image processing field, many adversarial perturbations look like high-frequency noise to an observer. Some researchers suggest the use of image preprocessing as a defense strategy against adversarial sample attacks, such as JPEG compression [3], total variance minimization (tvm) [4], image quilting [4], and bit-depth-reduction [5], etc.

For example, Dziugaite et al. [6] studied the effect of JPG compression on the perturbations computed by the fast gradient sign method (FGSM) [2]. It was reported that for FGSM perturbations, JPG compression can actually reverse the degradation of classification accuracy to a large extent. However, it is concluded that compression alone is far from being an effective defense, since JPG compression improves more for small perturbations and has less effect for large ones. FGSM is a typical white-box attack model. It is simple and effective and plays a major role in the field of image attacks, and many subsequent studies have been conducted based on this algorithm.

The feature squeezing algorithm also focuses on the data preprocessing to generate an effective defense. The input values of neural networks have many "redundant" features items, which make it more favorable for others to create adversarial examples. The authors in [5] provide a method of squeezing to reduce unnecessary features. Specifically, squeezing has the function of noise reduction. The authors proposed two squeezing methods:

1. *Decreasing the depth of each pixel*: The size of the channel of the image pixel is 2^8. It is found that using a compressing size of 2^i to represent the image does not affect the neural network's judgment of the image. It effectively reduces the possibility of being attacked; that is, the effect of noise added to the confrontation sample will be effectively reduced.
2. *Reduction in the spatial dimension*: In [5], the example of median smoothing was given, which is equivalent to pooling in convolutional neural networks. It takes the median of the $n \times n$ filter and replaces the value of the original region. Because the values of the pixels in the neighboring regions are similarly correlated, smoothing the whole map does not affect the final result but removes the noise added to it.

Although image preprocessing is effective in scenarios where the attacker is unaware of the defense method, it is almost ineffective in scenarios where the attacker is aware of the defense method. However, preprocessing is still an attractive class of defense methods because it can work with other defense methods to produce stronger defenses and can mitigate the harm of adversarial samples without knowing the target model.

Data Augmentation

Data augmentation is typically used to counter black-box attacks. It extends the original training set using generated adversarial examples and attempts to allow the model to see more data during training.

As an example, in [7], the text of adversarial examples by training on reading ability through genera-tive adversarial networks (GAN) algorithm is expanded [8]. They proposed two approaches to generate more data-diverse features: (1) knowledge-based words, i.e., replacing words with synonym/negative pairs from online database species; (2) neural network-based, i.e., using a seq2seq model to generate hypotheses containing examples and measuring the cross-entropy between the original and generated hypotheses by performing a loss function. In the training process, they used the GAN algorithm to train the discriminator and generator and merged the optimization steps of the adversarial exemplar discriminator in it.

Wang et al. [9] enhanced the complexity of the database, thus improving the resistance of the model. Moreover, its complexity conversion formula can update itself with the continuous detection of adversarial samples. Lee et al. [10] proposed the idea that a data augmentation module can be incorporated to simulate adversarial attacks when training DL models, which uses an uncertainty generator to train DL models. The generator can generate FGSM-like perturbations. After passing through the classifier of DL, it can yield either normal or interference results. The DL network learned autonomously by this data augmentation module has greater robustness against larger adversarial attacks based on FGSM perturbations.

9.2.2 Modifying Networks Architecture

Network Distillation

Distillation was introduced by Hinton et al. [11] as a training procedure to transfer knowledge of a more complex network to a smaller network. But this work only focuses on the task that ensembles models to allow the deployment to scale to a large number of users, especially if the individual models are large neural nets. The variant of the procedure introduced by Papernot et al. [12] essentially uses the knowledge of the network distillation to improve its own robustness. They argued that changing the network structure and retraining a more complex model when faced with an adversarial sample attack can certainly defend against such an attack. But it is too costly and requires more computational resources and time. Therefore, it is desirable to find a new defense method that can effectively deal with the attack without changing the network structure and affecting the accuracy of the model as much as possible. The probabilities of class vector of training data are extracted from initial DNN and fed back to train the original model in the dis-tilled network. Specifically, it uses the softmax output of the original DNN to train a second DNN that has the same structure as the original DNN. The sample model architecture is shown in Figure 9.2. The softmax of the original DNN is also modified by introducing a temperature parameter T:

$$q_i = \frac{\exp\left(z_i / T\right)}{\sum_j \exp\left(z_j / T\right)},$$

where z_i is the input of softmax layer. T controls the level of knowledge distillation. When $T = 1$, it turns back to the normal softmax function. If T is small, the function will output more extreme values. When it

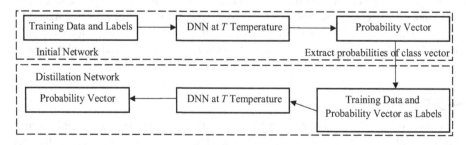

FIGURE 9.2 Sample model architecture of distillation DNN.

is large, q_i is close to a uniform distribution. Therefore, increasing T reduces the gradient of the model and makes it smoother, thus reducing the sensitivity of the model to perturbations.

Model Regularization

Model regularization enforces the generated adversarial examples as the regularizer and follows the form of:

$$\min\left(J\left(f\left(x\right),y\right)+\lambda J\left(f\left(x'\right),y\right)\right),$$

where λ is a hyperparameter. Following [2], the work [13] constructed the adversarial training with a linear approximation as follows:

$$-\log p\left(y|x+-\varepsilon g\, /\,\|g\|_2\, ,;\theta\right)$$

$$g = \partial_x \log p\left(y|x,;\theta\right),$$

where $\|g\|_2$ is the L_2-norm regularization, θ is the parameter of the neural model, and θ^* is a constant copy of θ. In [13], the authors compute the perturbation from FGSM proposed by Goodfellow to obtain the *adversarial loss* by the DL model. Such a loss is added to the classification loss (cross-entropy) of the DL model to obtain a new loss. By optimizing this loss, it is possible to achieve super performance on text classification tasks.

The word embedding after FGSM perturbation has a high probability of not mapping to the real word, in the same way that the perturbation of RGB in an image can produce antagonistic samples that are indistinguishable to the human eye. The authors argue that this adversarial training is more like a regularization method, allowing for better-quality word embedding, avoiding overfitting, and thus achieving excellent performance. It is experimentally proven that even though the computed *adversarial loss* is large, the loss is not involved in the gradient calculation, changes in the weight and bias of the model do not affect the adversarial loss, and the model can reduce it by changing the *word embedding weight*.

9.2.3 Network Add-on

Defense Against Universal Perturbations

Akhtar et al. [14] proposed a defense framework to counter adversarial attacks generated by using universal perturbations. This framework adds an additional "pre-input" layer before the target network and trains it to correct the perturbed images, without updating the parameters of the target network, ultimately allowing the classifier to make the same prediction as it would on an unperturbed version of the same image. The pre-input layer, called a scrambled rectification network (PRN), trains individual detectors by extracting features from the discrete cosine transform of the input–output differences of the PRN used to train the images. The test image is first passed through the PRN, and its features are then used to detect the disturbance. If adversarial perturbations are detected, the output of the PRN is used to classify the test image, and conversely the original image is used as the input to the model.

MegNet Model

Meng et al. [15] conceived that, for most AI tasks, the sample space is a a high-dimensional space but that the effective samples we need are actually in a manifold space with much lower dimensionality than the original sample space.

The authors implemented a two-pronged defense system, proposing the corresponding *detector* and *reformer* methods for two reasons: (1) Detector is used to determine whether a sample is far away from

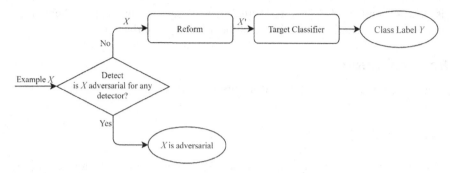

FIGURE 9.3 Workflow of MagNet.

the manifold boundary. For example, if the task is to classify handwritten digital images, all other images that do not contain numbers are adversarial samples that can lead to misclassification because the classifier of the target DL model has to classify that image. Input samples belonging to this class will be removed directly. (2) The adversarial sample is very close to the popular boundary for that task. If the generalization of that classification method is not good at that point, misclassification can also occur. Efforts are made to find a sample that is close to or just above the manifold for that task and then hand it over to the classifier. This is the second defense method conceived by the authors. Figure 9.3 demonstrates the workflow of MagNet.

Generally, the generation of functions to detect adversarial samples requires the knowledge of the specific adversarial sample generation process and therefore cannot generalize to other types of adversarial samples. Thus such a method can be easily bypassed by slightly modified attacks. In this paper, the authors proposed two new methods for detecting adversarial samples: reconstruction error-based detection and probabilistic divergence-based detection methods.

1. *Reconstruction error-based detection*: For the training set of all normal samples, the authors train an autoencoder such that the loss function of this training set is minimized, and its loss function is defined as

$$L\left(X_{\text{train}}\right) = \frac{1}{X_{\text{train}}} \sum_{x \in X_{\text{train}}} E(x),$$

   where X_{train} is the normal samples in training set. and $E(x)$ is the reconstruction error. The authors argue that, if the sample to be tested is a normal sample, its reconstruction error is small because the sample has the same generation process as the training data of autoencoder, and that, conversely, if the sample with detection is an adversarial sample, then the reconstruction error of that sample will be large.

2. *Probabilistic divergence-based detection*: The authors were able to find the adversarial input samples by comparing the results of two different sets of input data that went through the softmax layer output. They used the Jensen–Shannon scatter to measure the degree of disagreement between the results of the two sets of data.

The ideal reformer needs to satisfy the following two conditions: (1) It should not change the classification structure of the normal samples; (2) it should reconstruct the adversarial samples sufficiently so that the reconstructed samples are close to the normal samples. The authors trained the autoencoder to minimize the reconstruction errors on the training set and to ensure that it is well adapted to the validation set. Afterward, when given a normal input set, the autoencoder output a very similar example. However, when given an adversarial example, the autoencoder should output an example that approximates the adversarial one. In this way, MagNet can improve the classification accuracy while maintaining classification accuracy.

9.4 DISCUSSIONS AND OPEN ISSUES

The biggest challenges faced by current researchers can be summarized in two points.

1. *Effective defense against white-box attacks*: We still have not seen a defense that balances effectiveness and efficiency well. In terms of effectiveness, adversarial training shows the best performance but is computationally expensive. In terms of efficiency, many random and denoising-based defense algorithms can be configured in a few seconds, but simple defenses do not protect the DL model algorithm against attackers who already have the access to the DL model details.
2. *Scalability*: Since most defenses are unable to defend against adaptive white-box attacks, researchers have started to focus on provable defenses, which guarantee the performance of the defense to a certain extent regardless of the attacker's methods. But so far scalability is a common problem with most provable defenses today. For example, interval bound analysis is a recently popular proof-based defense method, but it does not scale to very deep neural networks and large datasets.

9.5 CONCLUSIONS

With the further development of deep learning research and the widespread application of DL technologies in real-world scenarios, deep learning security and privacy have become a new and promising research area, which attracts a large number of scholars from academia and industry to conduct in-depth research and achieve many remarkable research results. However, so far, DL security and privacy research is still in its infancy, and many critical scientific issues remain to be solved.

This chapter summarized adversarial defense strategies against the DL model. First, we introduced the targets of different adversarial attacks, for which we grouped all the defense strategies into three categories: modified training or modified input, modifying networks architecture, and network add-on. In the different categories, we also illustrated in detail the different defense algorithms, each of which has different advantages. However, the study of the adversarial defense of DL is still in its infancy, and we still face two major difficulties: how to combat white-box attacks and the pervasiveness of adversarial defense. We have no effective algorithms to defend against potentially harmful white-box attacks.

REFERENCES

[1] Ian Goodfellow, Yoshua Bengio, and Aaron Courville. 2016. *Deep Learning*. Vol. 1.
[2] Goodfellow, I. J., Shlens, J., & Szegedy, C. (2014). Explaining and harnessing adversarial examples. arXiv preprint arXiv:1412.6572.
[3] Das, N., Shanbhogue, M., Chen, S. T., Hohman, F., Li, S., Chen, L., ... & Chau, D. H. (2018, July). Shield: Fast, practical defense and vaccination for deep learning using jpeg compression. In *Proceedings of the 24th ACM SIGKDD International Conference on Knowledge Discovery & Data Mining* (pp. 196–204).
[4] Guo, C., Rana, M., Cisse, M., & Van Der Maaten, L. (2017). Countering adversarial images using input transformations. *arXiv preprint arXiv:1711.00117*.
[5] Xu, W., Evans, D., & Qi, Y. (2017). Feature squeezing: Detecting adversarial examples in deep neural networks. *arXiv preprint arXiv:1704.01155*.
[6] Dziugaite, G. K., Ghahramani, Z., & Roy, D. M. (2016). A study of the effect of jpg compression on adversarial images. *arXiv preprint arXiv:1608.00853*.

[7]   Kang, D., Khot, T., Sabharwal, A., & Hovy, E. (2018). Adventure: Adversarial training for textual entailment with knowledge-guided examples. *arXiv preprint arXiv:1805.04680.*

[8]   Goodfellow, I. J., Pouget-Abadie, J., Mirza, M., Xu, B., Warde-Farley, D., Ozair, S., ... & Bengio, Y. (2014). Generative adversarial networks. *arXiv preprint arXiv:1406.2661.*

[9]   Wang, Q., Guo, W., Zhang, K., Ororbia II, A. G., Xing, X., Liu, X., & Giles, C. L. (2016). Learning adversary-resistant deep neural networks. *arXiv preprint arXiv:1612.01401.*

[10] Lee, H., Han, S., & Lee, J. (2017). Generative adversarial trainer: Defense to adversarial perturbations with gan. *arXiv preprint arXiv:1705.03387.*

[11] Hinton, G., Vinyals, O., & Dean, J. (2015). Distilling the knowledge in a neural network. *arXiv preprint arXiv:1503.02531.*

[12] Papernot, N., McDaniel, P., Wu, X., Jha, S., & Swami, A. (2016, May). Distillation as a defense to adversarial perturbations against deep neural networks. In *2016 IEEE symposium on security and privacy (SP)* (pp. 582–597). IEEE.

[13] Miyato, T., Dai, A. M., & Goodfellow, I. (2016). Adversarial training methods for semi-supervised text classification. arXiv preprint arXiv:1605.07725.

[14] Akhtar, N., Liu, J., & Mian, A. (2018). Defense against universal adversarial perturbations. In *Proceedings of the IEEE conference on computer vision and pattern recognition* (pp. 3389–3398).

[15] Meng, D., & Chen, H. (2017, October). Magnet: a two-pronged defense against adversarial examples. In *Proceedings of the 2017 ACM SIGSAC conference on computer and communications security* (pp. 135–147).

Defensive Schemes for Cyber Security of Deep Reinforcement Learning

10

Jiamiao Zhao,[1] Fei Hu,[1] and Xiali Hei[2]

[1]Electrical and Computer Engineering, University of Alabama, USA

[2]School of Informatics, University of Louisiana at Lafayette, USA

Contents

10.1 INTRODUCTION

Machine learning (ML) can be divided into three categories: supervised, unsupervised, and reinforcement learning. In *supervised* learnings (e.g., SVM, decision trees, etc.), the training samples are labeled for finding the boundary among different clusters. On the other hand, if the training dataset does not have

DOI: 10.1201/9781003187158-12

ground-truth labels, its ML algorithm is called *unsupervised* learning, such as KNN, K-means, etc. They classify the training data with respect to the distance in hyperplane. *Reinforcement* learning (RL) uses the intelligent agent to learn from the history states/actions and can use the maximum cumulative reward to generate actions and interact with the environment in real time.

RL has been used for many decision-making problems in robotics, electric grids, sensor networks, and many other real-world applications. However, the RL algorithm may take a long time to converge to the optimal point (from a few hours to a few weeks depending on the state space size) [1]. Recently, a new algorithm combining deep learning (DL) with RL, called deep reinforcement learning (DRL), has been applied to handle high-dimensional inputs, such as camera images, big vector problems. DRL has shown great results in many complex decision-making processes, such as robotics [2], autonomous vehicles, smart city, and games such as Go [3], etc. DRL has shown its capabilities in dealing with critical real-world applications. An autonomous racing car, which is called AWS DeepRacer, has been designed by Amazon using DRL. It uses cameras to visualize the track and a DRL model to make decisions on throttle and direction.

Therefore, the security and privacy of DRL need to be fully investigated before deploying DRL in critical real-world systems. Recently, DRL is proved to be vulnerable to adversarial attacks [4]. Attackers insert perturbations into the input of DRL model and cause DRL decision errors. DRL utilizes a deep neural network (DNN) model to achieve high prediction, but it is not robust against input perturbations. Even a small change in the input may lead to dramatic oscillations in the output. This makes it necessary to have a comprehensive understanding of the types and features of adversarial attacks.

This chapter aims to introduce the security issues in DRL and current defensive methods to overcome adversary attacks. We discuss DRL's structure, its security problems, and the existing attack targets and objectives in DRL.

Organization of this chapter: Overviews of RL and DRL, as well as the privacy problems associated with them, are presented in section 10.2. Section 10.3 talks about a proposed defensive method to achieve a robust DRL against adversarial attacks. Section 10.4 is an extended version of section 10.3. It presents an upgraded DRL mechanism by finding a robust policy during the learning process. We conclude this chapter and discuss research challenges in designing a privacy-preserved and robust DRL in section 10.5.

10.2 BACKGROUND

In this section, we first briefly introduce RL. We especially discuss model-free RL. Then we show how we can combine reinforcement learning with deep learning to form the DRL process [1]. We summarize DRL techniques and their potential weaknesses in cyber security.

A general RL problem can be described as a Markov decision process (MDP). MDP contains the states, actions, reward, and policy of the intelligent agents. Here we first give the definitions of some terminologies and their functions in RL.

- *State* (S): An agent interacts and learns about states. At each time step, the agent interacts with the state space governed by the policy and obtains a reward that determines the quality of the action.
- *Policy* (π) defines how the agent will activate in an environment. It maps the present state to an action. Policies are divided into two types: deterministic policy and stochastic policy. When the action taken by the agent is *deterministic*, the policy is deterministic; if the action follows a probability distribution function for different states, the policy is *stochastic*.
- *On-policy algorithm* evaluates the exploration value.
- *Off-policy algorithm* does not evaluate the exploration value. Q-learning is an off-policy method.

- *Action* (A) is a behavior performed by the agent for interactions with environment.
- *Reward* (R): The agent receives the reward after making an action. The goal of the agent is to maximize the accumulative reward.
- *Value function*: The maximum expected accumulative reward an agent can get at a specific state. It can be calculated by

$$V_\pi(s) = E_\pi(G_t | S_t = s),$$

  where $G_t = R_{t+1} + \gamma R_{t+2} + \dots + \gamma^{T-1} R_T$, G_t is the sum of reward from states to the final point, and γ is a discount factor, $\gamma \in [0,1]$.
- *Exploration and exploitation*: *Exploration* is the process by which the agent tries to explore the environment by taking different actions at a given state. *Exploitation* occurs after exploration. The agent exploits the optimal actions to get the highest accumulated reward.

There are two major branches in RL: model-free and model-based RL algorithms. *Model-free* means no model is being used during the training. A model-free RL can learn directly from the interactions with the environment in real time. In *model-based* methods, there is a premature model (MDP model) before the training phase. The intelligent agent maintains this model and finds the optimized solution. The complexity of the optimizing model based on RL is proportional to the square of the number of states. This may require high computation overhead. In the real world, there is limited capability to enumerate all states, the transition conditions, and the transfer probability. Thus the majority of existing research works focus on model-free RL.

10.2.1 Model-free RL

Model-free RL focuses on calculating the value function directly from the interactions with the environment. It can be further divided into two types: policy optimizing and Q-learning.

1. *Policy optimizing*: Policy optimization is the process of directly searching for an optimal policy by using gradient-based or gradient-free schemes [1]. Gradient-based methods are the best fit for dealing with high dimensionality. Gradient-free methods are useful in low-dimensional problems. In policy optimization, the policy is denoted as $\pi_\theta(a|s)$, which means the probability distribution π over action a when state is s, and θ is neural network parameters for policy π. The policy gradient is defined as [5]

$$\nabla E_{\pi\theta} = E_{\pi\theta} r_t \nabla \log \pi_\theta(t).$$

2) *Q-learning*: This is an off-policy RL. The main aim of Q-learning is to maximize the Q-value, which is shown here. s_{t+1} is the next state based on the action taken in previous state s_t, and a_{t+1} is a possible action in s_{t+1}. α is the learning rate, i shows the iteration number, and γ is the discount factor:

$$Q_{i+1}(s_t, a_t) = Q_i(s_t, a_t) + \alpha(R_{i+1} + \gamma \max_{a_{t+1}}(Q_i(s_{t+1}, a_{t+1}) - Q_i(s_t, a_t)).$$

This equation clearly demonstrates that Q-learning decides the next move a_{t+1} and knows s_{t+1}. It uses ϵ-greedy policy and finds the action with the highest Q-value increment.

10.2.2 Deep Reinforcement Learning

In real-world applications, the input cannot be as simple as discrete states and can be represented by a value. The input data may have high dimension, like images in an autonomous vehicle. One solution is to extract the features from high-dimensional data. But this relies on the researcher's design. And it is intricate to extract the information of targets from the input images.

Another approach is to use a neural network to find the optimal policy value function or Q-function. This is the basic idea behind DRL. In 1992, Gerald Tesauro developed a computer backgammon program, called TD-Gammon, trained by neural network by using temporal-difference learning [6]. TD-Gammon is just slightly below the top human backgammon players of the time. It explores the strategies that no one has before and advances the theory of backgammon play.

When DL occurs, it behaves powerfully in image processing. DL can extract features from high-dimensional input images or videos, and those features are more generalized than artificial features.

There are three main approaches to solving RL by utilizing DL: methods based on Q-value, methods based on policy search, and a hybrid of both in actor-critic configurations.

Classic Q-networks have many problems when learning Q-values. First, the training data is not *iid* (independent and identically distributed) because successive samples are correlated [7]. Second, even slight changes to the Q-value can lead to rapid changes in the estimated policy. Finally, the gradients of Q-network can be sufficiently large so that Q-network is unstable [7].

These problems can be solved by using deep Q-network (DQN) [8] that employs experience replay, regulates the parameters of network, and normalizes the reward value to the range $[-1, +1]$ in 2015. Loss function is the difference between target value and current estimate of Q:

$$L_i(\theta_i) = E\left(\left(R + \gamma \max_{a'} Q(s', a', \theta) - Q(s, a, \theta)\right)^2.\right)$$

DQN is evaluated on Atari games. It gets the highest scores compare with other algorithms and beats human players in some games.

10.2.3 Security of DRL

DRL has different security challenges compared with those of ML. The major difference is that DRL's goal is to find the optimization policy from a sequence of problems, whereas most ML algorithms are trained to solve single-step prediction or cluster problems. And the training dataset for ML is a fixed distribution. On the other hand, intelligent agent in DRL is trained with a deterministic or stochastic policy.

RL problems have four parts (s, a, r, p), where s stands for state, a stands for action, r stands for reward, and p stands for the transient probability that an agent is in state s, takes action a, and switches to the next state s'. Hence an adversary has more attack targets in DRL than in ML. The attacks on DRL have four categories. Figure 10.1 shows adversarial attacks on DRL based on attack targets.

1. *Attacks targeting the reward*: In 2018, Hen et al. [9] evaluated the reaction of DRL in a software-defined network to adversary attacks. The adversary knows the structure and parameters of the model. By flipping the rewards and manipulating states, the attacker is able to mislead the intelligent node to choose suboptimal action. Pattamaik et al. [10] proposed three attack methods by adding perturbations to the observations. The evaluation results showed that the third attack, which uses stochastic gradient descent for creating cost function, can mislead the DRL agent to end up in a predefined adversarial state [1]. These adversaries are assumed to have the knowledge of the DRL, but Huang et al. [11] proved that even when the adversary has no information of the model, it can still fool the agent to perform a desired action.
2. *Attacks targeting the policy*: Trstschk et al. [12] added an adversarial transformer network (ATN) to impose adversarial reward on the policy network for DRL. The ATN makes the cumulative

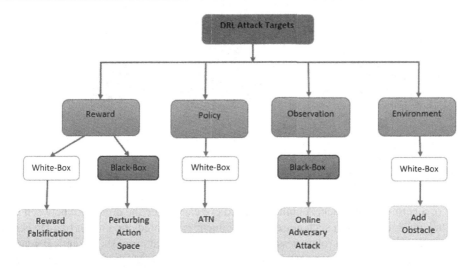

FIGURE 10.1 DRL attack targets and attack methods.

reward reach the maximum if the agent follows a sequence of adversarial inputs. Thus the DRL automatically follows the adversary's policy. But this method requires the attacker to have the complete information of the agent and the target environment (called a white-box attack).

3. *Attacks targeting the observation*: The adversary can also directly manipulate the sensory data. The research on real-time attacks to a robotic system in a dynamic environment has already been evaluated. Clark et al. [13] used a white-box adversarial attack on the DRL policy of an autonomous robot in a dynamic environment. The robot is intended to find a route in a maze while the adversary's goal is to mislead robot to the wrong routes. The adversary tampers with sensory data of the robot, so that the robot is deviated from the optimal route. Clark et al. [13] also observed that if the adversarial input is removed, the robot automatically corrects its actions and finds the correct route. Therefore, an adversary can tamper with the sensory data temporarily and leave behind very little evidence. Recently, Tu et al. [14] showed that the motion sensing and actuation systems in robotic VR/AR applications can be manipulated in real-time noninvasive attacks.

4. *Attacks targeting the environment*: The final way of performing an adversarial attack on DRL is compromising the environment. The core idea of environment attack is placing additional confusing obstacles to confuse the robot. Bai et al. [15] proposed a method of creating adversarial examples for deep Q-network (DQN), which is trained for pathfinding. The DQN model already has a complete solution to find an optimal path. The adversary extracts the model parameters, analyzes them, and finds the weakness of the Q-value. The adversary adds adversarial environmental examples to the robot. This method generates a successful attack and stops the robot from finding an optimal route in a maze.

10.3 CERTIFICATED VERIFICATION FOR ADVERSARIAL EXAMPLES

In DNN, a subtle perturbation to the input, known as adversarial example, can lead to incorrect prediction of the model. And the adversarial examples can cause danger in the real world, e.g., an autonomous vehicle may take a stop sign as an advertisement and ignore it, causing a collision. Figure 10.2 shows

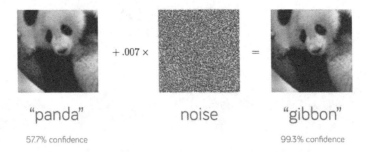

"panda"   noise   "gibbon"
57.7% confidence   99.3% confidence

FIGURE 10.2   Adversarial example.

one example of a perturbation attack. A human's naked eye cannot see the difference between the benign images and malicious images. But such noise highly influences the prediction result of a DNN model.

Lutjens et al. [16] proposed a robustness verification method that provides theoretical guarantees for improving robustness against adversarial attacks or noise. Lutjens et al. demonstrated that the existing methods to adversarial example, like adversarial training, defensive distillation, or model ensembles, do not have a theoretical algorithm that guarantees robustness improvement and are often ineffective for future adversarial attacks [17]. Verification methods can guarantee future robustness. It finds theoretically proven bounds on the maximum output deviation, given a bounded input perturbation [18]. But the current problem of finding exact verification bounds is NP-complete [18], thus it is infeasible to be run online on robots. Instead of finding the exact bound, Lutjens et al. [16] proposed a relaxed version of verification, which finds the lower bounds on the Q-value and chooses the optimal action under a worst-case deviation in the input space due to adversarial attacks or sensor noise. By utilizing the calculated lower bound, an intelligent agent can find all possible states lying inside the bounds and determine the optimal action.

10.3.1 Robustness Certification

In RL, the Q-value is expressed as $Q = E[\sum_{t=0}^{T} \gamma^t r_t]$, which is the expected reward at state s from $t = 0$ to $t = T$. The goal of Q-learning is to find the maximum cumulative Q-value under a worst-case perturbation of the observation by adversarial attacks or sensor noise. To calculate the certified lower bounds on DNN-predicted Q, Lutjens et al. [16] defined the certified lower bound of the state-action value, Q_L:

$$Q_L\left(s_{adv}, a_j\right) = \min_{s \in B_p\left(s_{sdv}, \varepsilon\right)} Q_l\left(s, a_j\right),$$

where s_{adv} is the perturbated state corrupted by the noise of adversary. Lutjens et. al. assumed that s_{adv} is bounded inside the ε-radius circle. a_j is the most robust action against the deviated input.

Lutjens et al. finally used a closed-form formula to calculate $Q_l\left(s, a_j\right)$ inside a DNN, in which ReLU is used as the activation function. H is the lower/upper bounding factor, W is the weight matrix, b is the bias in layer $\left(k\right)$ with r, j as indices:

$$Q_l\left(s, a_j\right) = A_{j,:}^{(0)}s + b_j^{(m)} + \sum_{k=1}^{m-1} A_{j,:}^{(k)}\left(b^{(k)} - H_{:,j}^{(k)}\right).$$

The matrix A contains the neural network weight and *ReLU* activation for each layer, $A^{(k-1)} = A^{(k)}W^{(k)}D^{(k-1)}$, and $A^{(m)} = 1$, an identity layer at the last layer.

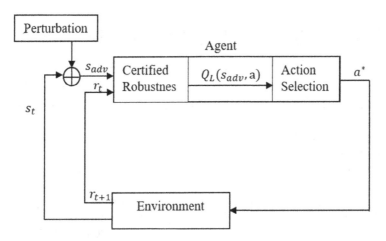

FIGURE 10.3   RL with certification add-on.

10.3.2 System Architecture

The certification system is similar to a standard model-free RL with added-on verification. Figure 10.3 depicts the system architecture. During training phase, agents use the DQL algorithm with uncorrupted state observations, s, and find the expected Q-value at each state.

At the online execution phase, the agents receive corrupted state observations and pass those to DNN. The certified DQN computes the lower bounds on Q with bounded perturbation of the input. Because Lutjens et al. assume that s_{adv}, the worst possible state observation, is bounded inside the ε-radius circle, centered at s_0, the true state, s_0 should be anywhere inside an ε-circle around s_{adv}. Therefore, the agent evaluates each Q-value of all the possible true states. The optimal action a^* is defined as the one with the highest Q-value under the perturbated observations:

$$a^* = \mathrm{argmax}_{a_j}\, Q_L\left(s_{adv}.a_j\right).$$

By follow this policy, the intelligent agents select the most robust action under the worst-case perturbation.

10.3.3 Experimental Results

The robustness metric is evaluated in simulations for collision avoidance among pedestrians and the cartpole domain. Pedestrian simulation is used to verify the collision avoidance for a robotic vehicle among pedestrians. This simulation uses a kinematic simulation environment for learning collision avoidance policy. In both cases, noise or adversarial attacks are added to the observations.

Nominal DQN policy is not robust against the perturbed inputs. Increasing the magnitude of adversarial or noise increases the average number of collisions and decreases the average reward.

Lutjens et al. ran the DQN with add-on defense, called certified adversarially robust reinforcement learning (CARRL). The results showed that, by adding certified defense to the DQL model, we can address existing misclassification under adversarial attacks and sensor noise. If the robustness parameter ε is increased, the number of collisions decrease when observations are corrupted. The reward increases with an increasing robustness parameter at $\varepsilon <\sim 0.1$. Also, the CARRL policy only marginally reduces the reward under no perturbation. It cannot significantly deteriorate the performance of DRL.

10.4 ROBUSTNESS ON ADVERSARIAL STATE OBSERVATIONS

Zhang et al. [19] also investigated the robust DRL algorithm under adversarial perturbation or noisy environment conditions (Figure 10.4). Zhang et al. [19] further demonstrated that the naïve adversarial training methods cannot improve DRL robustness significantly in noisy environment but instead make DRL training unstable and deteriorate agent performance. They concluded that naïvely applying techniques from supervised learning to RL is not appropriate since RL and supervised learning are two quite different problems.

Besides adversarial examples, a DRL agent may fail occasionally during regular training without adversarial attacks or noisy input. It is a challenging topic to debug the accident failure of DRL agents. Zhang et al. designed a few deep deterministic policy gradient (DDPG) agents that try to achieve high median reward. However, during 100 episodes of training, they observed some low reward runs [19]. These occasional failures can be dangerous in safety-critical applications like an autonomous vehicle.

To improve the robustness of DRL, Zhang et al. [19] proposed a robust policy regularizer that can decrease the sensitivity of the agent to observation. This policy regularizer can be used for multiple DRL algorithms: proximal policy optimization (PPO), deep deterministic policy gradient (DDPG), and deep Q-network (DQN). Zhang et al. [19] first formulated the perturbation on state observations as a modified Markov decision process (MDP) or state-adversarial MDP (SA-MDP). They proved that a stational and Markovian optimal policy may not exist for SA-MDP when agents are under adversarial attacks. Based on their SA-MDP theory, Zhang et al. proposed a theoretically principled robust policy regularizer that utilizes the total variation distance or KL-divergence on perturbed policies [19].

This section mainly presents the policy regularizer in DDPG and DQN, as well as the experimental results of the regulated DDPG and regulated DQN.

10.4.1 State-adversarial DRL for Deterministic Policies: DDPG

A MDP is defined as a 4-tuple, (S, A, R, p), as previously mentioned, and $\mathrm{p}: S \times A \to P(S)$ represents the transition probability of environment, where $P(S)$ is the set of all possible transition probability. Figure 10.4 displays RL with perturbed state observations. In SA-MDP, Zhang et al. [19] introduced an adversary $v(s): S \to S$. The adversary perturbs the observation of the agent, so that the action is taken as $\pi(a \mid v(s))$ instead of from true state $\pi(a \mid s)$. Since $v(s)$ is different from the true observations, the action $\pi(a \mid v(s))$ may be suboptimal. Zhang et al. [19] that assumed the adversarial attacks are dependent only on state and do not change over time. And then they defined a set $B(s)$, which consists of the "neighboring" states of s.

DDPG learns a deterministic policy $\pi(s) \in R^{|A|}$ to address the issue that the variation, D_{TV}, from the true state s may not overlap with the perturbed state $\hat{s}$ in DDPG. Zhang et al.[19] defined a smoothed version of the policy in DDPG, where independent Gaussian noise with variance σ^2 is added to each action: $\overline{\pi}(a \mid s) \sim N(\pi(s), \sigma^2 I_{|A|})$. Then the variation distance D_{TV} can be calculated by using the following theorem:

$$\textbf{Theorem}: D_{TV}\left(\overline{\pi}(\cdot \mid s), \overline{\pi}(\cdot \mid \hat{s})\right) = \frac{\sqrt{\frac{2}{\pi}}d}{\sigma} + O(d^3), \text{ where } d = \left\| \pi(s) - \pi(\hat{s}) \right\|_2.$$

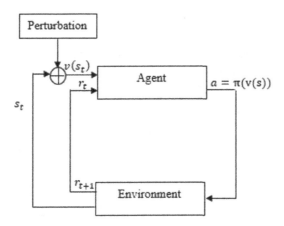

FIGURE 10.4   Reinforcement learning with perturbation from observation.

Thus the variation distance between two smoothed distributions is bounded. In DDPG, Zhang et al. parameterized the policy as a policy network π_θ. They deduced that the robust policy regularizer for DDPG is [19]

$$R_{DDPG}\left(\theta_\pi\right) = \sqrt{(2/\pi)}\left(\frac{1}{\sigma}\right)\sum_s \max \left\|\pi_{\theta_\pi}(s) - \pi_{\theta_\pi}(\hat{s})\right\|_2,$$

where $\hat{s} \in B(s)$, and $\|\cdot\|_2$ means the second norm. For each state s, the agent needs to solve a maximization problem, which can be solved using convex relaxations. This smoothing procedure can be done at test time. During training time, Zhang et al.'s goal is to keep $\max\left\|\pi_{\theta_\pi}(s) - \pi_{\theta_\pi}(\hat{s})\right\|_2$ small.

10.4.2 State-adversarial DRL for Q-Learning: DQN

Besides DDPG, Zhang et al. [19] also formulated the robust policy regularizer for DQN. Because the action space for DQN is finite, and Q-learning's policy is to maximize Q-value, then the total variation distance between true state s and adversarial state $\hat{s}$ in DQN is

$$D_{TV}\left(\pi(\cdot|s), \pi(\cdot|\hat{s})\right) = \begin{cases} 0 & \arg\max_a \pi(a|s) = \arg\max_a \pi(a|\hat{s}) \\ 1 & otherwise \end{cases}.$$

A hinge-like robust policy regularizer is used, where $a^*(s) = argmax_a Q_\theta(s,a)$ and c is a small positive constant:

$$R_{DQN}(\theta) = \sum_s \max\{\max_{\hat{s}\in B(s)}\max_{a\neq a^*} Q_.\left(\hat{s}, a\right) - Q_.\left(\hat{s}, a^*(s)\right), -c\}.$$

Zhang et al.[19] pointed out that, unlike DDPG, the regularized reward sum is similar to the robustness of classification tasks, if $a^*(s)$ is treated as the "correct" label. The maximization can be solved using projected gradient descent (PGD) or convex relaxation of neural networks [19].

10.4.3 Experimental Results

Similar to certificated verification, Zhang et al. [19] also assumed the set of adversarial states. $B(s)$ is a l_∞ -norm ball around s with radius ε. ε is also referred to as the perturbation budget.

Zhang et al. ran SA_DDPG in five different environments (Ant, Hopper, Inverted Pendulum, Reacher, and Walker2d) to test the robustness of SA-DDPG. Zhang et al. evaluated SA-DDPG with SGLD and convex relaxation. They measured the median, minimum, 25%, and 75% rewards in each training. Their results showed that SA-DDPG is able to consistently improve robustness: the statics of SA-DDPG is higher or equal to DDPG in five environments.

SA-DQN was implemented on four Atari games and one control problem: Pong, Freeway, BankHeist, RoadRunner, and Acrobot. Zhang et al. included vanilla DQN and adversarially trained DQN with 50% adversary as baselines. They observed that SA-DQN achieves much higher rewards under adversarial attacks in most environments and that naïve adversarial training is mostly ineffective under strong attacks.

10.5 CONCLUSION AND CHALLENGES

This chapter presented a survey of the security issues in deep reinforcement learning (DRL), which has been applied in many real-life applications. It is susceptible to adversarial attack, which corrupts the whole model, making DRL fail to achieve optimal rewards. We reviewed current adversarial attacks in DRL and a defense method that can be used in PPO, DDPG, and DQN. The overall defense method is derived from the relationship between the current state and action decided by the agent's policy. By deploying the adversarial defense method on DRL, the agent can choose an optimal action in the adversarial environment. Finally, the existing method to measure the robustness in DRL is still naïve because it assumes the true state is sitting inside a hypothetical circle. There is a need to properly define the deviation of DRL caused by adversarial attack.

ACKNOWLEDGMENT

This work is supported in part by the US NSF under grants OIA-1946231 and CNS-2117785. All the ideas presented here do not represent the opinions of NSF.

REFERENCES

[1]   I. Llahi, M. Usama, J. Qadir, M. U. Janjua, A. Al-Fuqaha, D. T. Hoang and D. Niyato, "Challenges and Countermeasures for Adversarial Attacks on Deep Resinforcement Learning," 2020. [Online]. Available: https://arxiv.org/abs/2001.09684.

[2]   S. Gu, E. Holly, T. Lillicrap and S. Levine, "Deep reinforcement learning for robotic manipulation with asynchronous off-policy updates," in *IEEE international conference on robotics and automation (ICRA)*, 2017.

[3]   D. Silver, A. Huang, C. J. Maddison, A. Guez, L. Sifre, G. v. d. Driessche, J. Schrittwieser and e. al, "Mastering the game of Go with deep neural networks and tree search," *Nature*, vol. 529, pp. 484–489, 2016.

[4]   V. Behzadan and A. Munir, "Vulnerability of deep reinforcement learning to policy induction attacks," in *International Coference on Machine Learning and Data Mining in Pattern Recognition*, 2017.

[5]   R. S. Sutton and A. G. Barto, "Reinforcement Learning: an Introduction," MIT Press, 2018.

[6]   G. Tesauro, "Temporal Difference Learning and TD-Gammon," *Communications of the ACM,* vol. 28, no. no. 3, pp. 58–68, 1995.

[7]   V. Behzadan and A. Munir, "The Fault in Our pi starts: Security Issues and Open challenges in Deep Reinforcement Learning," 2018. [Online].

[8]   V. Mnih, K. Kavukcuoglu, D. Silver, A. A. Rusu, J. Veness, M. G. Bellemare and e. al., "Human-level control through deep reinforcement learning," *Nature,* vol. 518, pp. 529–533, 2015.

[9]   Y. Han, B. I. Rubinstein, T. Abraham, T. Alpcan, O. D. Vel, S. Erfani, D. Hubczenko, C. Leckie and P. Montague, "Reinforcement Learning for Autonomous Defence in Software-Defined Networking," in *International Conference on Decision and Game Theory for Security*, 2018.

[10] A. Pattanaik, Z. Tang, L. Shujing, G. Bommannan and G. Chowdhary, "Robust Deep Reinforcement Learning with Adversarial Attacks," in *International Foundation for Autonomous Agents and Multiagent System*, 2018.

[11] Y. Huang and Q. Zhu, "Deceptive Reinforcement Learning under Adversarial Manipulations on Cost Signals," *arXiv,* 2019.

[12] E. Tretschk, S. J. Oh and M. Fritz, "Sequential Attacks on Agents for Long-Term Adversarial Goals," 5 Jul 2018. [Online].

[13] G. Clark, M. Doran and W. Glisson, "A Malicious Attack on the Machine Learning Policy of a Robotic System," in *17th IEEE International Conference On Trust, Security And Privacy In Computing And Communications/ 12th IEEE International Conference On Big Data Science And Engineering (TrustCom/ BigDataSE)*, New York, USA, 2018.

[14] Y. Tu, Z. Lin, I. Lee, X. Hei, "Injected and delivered: Fabricating implicit control over actuation systems by spoofing inertial sensors." In 27th USENIX Security Symposium (USENIX Security 18) 2018 (pp. 1545–1562).

[15] X. Bai, W. Niu, X. Gao and J. Liu, "Adversarial Examples Construction Towards White-Box Q Table Variation in DQN Pathfinding Training," in *2018 IEEE Third International Conference on Data Science on Cyberspace*, 2018.

[16] B. Lutjens, M. Everett and J. P. How, "Certified Adversarial Robustness for Deep Reinforcement Learning," in *3rd Conference on Robot Learning, Osaka*, Japan, 2019.

[17] N. Carlini and D. Wagner, "Adversarial Examples Are Not Easily Detected: Bypassing Ten Detection Methods," in *In Proceedings of the 10th ACM Workshop on Artificial Intelligence and Security*, 2017.

[18] K. Guy, C. Barrett, D. L. Dill, K. Julian and M. J. Kochenderfer, "Reluplex: An Efficient SMT Solver for Verifying Deep Neural Networks," in *Computer Aided Verification – 29th International Conference, CAV*, Heidelberg, Germany, 2017.

[19] H. Zhang, H. Chen, C. Xiao, B. Li, M. Liu, D. Boning and C.-J. Hsieh, "Robust Deep Reinforcement Learning against Adversarial Perturbations on State Observations," arXiv:2003.08938, 2020.

Adversarial Attacks on Machine Learning Models in Cyber-Physical Systems

11

Mahbub Rahman and Fei Hu

Electrical and Computer Engineering, University of Alabama, USA

Contents

11.1 INTRODUCTION

Modern technologies based on pattern recognition, machine learning, and data-driven artificial intelligence, especially after the advent of deep learning, have reported impressive performance in a variety of application domains, from classical pattern recognition tasks like speech and object recognition, used by self-driving cars and robots, to more modern cyber security tasks like spam and malware detection [1]. However, such technologies can be easily fooled by adversarial examples. For example, carefully perturbed input samples can mislead pattern detection at testing time. This has drawn considerable attentions since 2014, when Szegedy et al. [2] and subsequent work [3–5] showed that deep networks for object recognition could be fooled by the input images perturbed in an imperceptible manner. Recently, poisoning and

DOI: 10.1201/9781003187158-13

evasion attacks against clustering algorithms have also been formalized to show that malware clustering approaches can be vulnerable to well crafted attacks [6, 7].

Research in adversarial learning not only investigates the security properties of learning algorithms against well crafted attacks but also focuses on the development of more secure learning algorithms. For evasion attacks, this has been mainly achieved by explicitly embedding knowledge into the learning algorithm of the possible data manipulation that may be performed by the attacker, e.g., using theoretical game models for classification [8–10], probabilistic models of the data distribution drift under attacks [11, 12], and even multiple classifier systems [13–15]. Poisoning attacks and manipulation of the training data have been variously countered with data sanitization (i.e., a form of outlier detection) [16], multiple classifier systems [17], and robust statistics. Robust statistics has also been exploited to formally show that the influence function of SVM-like algorithms can be bounded under certain conditions [18], e.g., if the kernel is bounded. This ensures some degree of robustness against small perturbations of training data, and it is desirable to improve the security of learning algorithms against poisoning [19].

In this chapter, we investigate the vulnerability of several popular machine learning algorithms (like KNN, random forest, and SVMs) to a variety of attacks, namely evasion attack, membership attack, data poisoning attack. These attacks can occur during the training and testing phases by manipulating the training data and labels, injecting foreign test sets, etc. These attacks can increase the misclassification rate and devastate the CPS system.

11.2 SUPPORT VECTOR MACHINE (SVM) UNDER EVASION ATTACKS

Machine learning (ML) is commonly used in multiple disciplines and real-world applications, such as information retrieval, financial systems, health, biometrics, and online social networks. However, their security profiles against deliberate attacks have not often been considered. Sophisticated adversaries can exploit specific vulnerabilities exposed by classical ML algorithms to deceive intelligent systems. It is critical to perform a thorough security evaluation in terms of the potential attacks against the ML techniques before developing novel methods to guarantee that machine learning can be securely applied in adversarial settings. This section presents a survey of different attacks on linear or nonlinear SVM.

In [20], an optimization method to generate adversarial examples for SVM and neural networks was proposed. The method has a regularization method that allows the model to generate adversarial examples separately in the black-box and white-box settings. They apply the method for malware detection in PDF files [21].

Given a classification algorithm $f\colon X{\mapsto}Y$ that is assigned to a feature $x \in X$ with a label $y \in Y = \{-1, +1\}$, where $(+1)$ represents the legitimate class and (-1) is the adversarial or malicious class. The label $y^c = f(x)$ given by a classifier is obtained by thresholding a discriminant function $g\colon X{\mapsto}R$, so we assume $f(x) = -1$ if $g(x) < 0$, and +1 otherwise.

11.2.1 Adversary Model

- *Adversary's goal*: Find a sample x such that $g(x){<}{-}\epsilon$ for every $\epsilon{>}0$. This means that the sample just crosses the decision boundary. It can also be asked to find a sample in which the classifier is highly confident on a wrong decision.
- *Adversary's knowledge*: Depending on information about the classifier, data, or training model, the adversary may find different strategies.
- *Adversary's capability*: The scenarios are restricted to those where the adversary can only modify the testing data.

11.2.2 Attack Scenarios

- *Perfect knowledge (PK)*: The adversary knows everything about the classifier: architecture, data, training.
- *Limited knowledge (LK)*: The authors assume the adversary can generate a new dataset $D' = \{(\hat{x}, \hat{y})\}$ on which the adversary can train a new classifier $\hat{f}$ and discriminant $\hat{g}$ function that approximates the original discriminant function g of the classifier f.

11.2.3 Attack Strategy

For an adversarial target sample x^0, an optimal attack strategy aims to find sample x^* to minimize g or its estimate $\hat{g}$, subject to a bound on its distance from x^0:

$$x^* = \arg\min_x \hat{g}(x),$$

$$\text{s.t} \, d(x, x^0) \leq d_{\max}.$$

But this is hard, so they regularize it:

$$\arg\min_x F(x) = \hat{g}(x) - \frac{\lambda}{n} \sum_{i|yc=-1} K\left(\frac{x - xi}{h}\right),$$

$$\text{s.t.} \, d(x, x^0) \leq d_{\max}.$$

They approach this via gradient descent.

11.3 SVM UNDER CAUSALITY AVAILABILITY ATTACK

If the target of an attack is to damage a classifier's generalization and usability by poisoning the training data, we call this kind of attack a causative availability attack [22]. In such an attack, the attacker injects some carefully selected foreign data $Df = (Xf, Yf)$ into the training data Dtr. However, the learner is unaware of the changes and learns a model ΘD by using the tainting data $Dt = Dtr \cup Df$, rather than using Dtr. This attack may cause the classifier to lose its generalization capability, e.g., to cause many misclassification errors including false positive (FP) and false negative (FN). In the following example, we select a support vector point (x_0, y_0) as a poisoning attacking point. The attacking goal is to add a foreign data item to Dtr to maximize the convex QP problem.

In [22], the SVM attack strategy was explained. It first identifies the support vectors by training the SVM model on the original training dataset Dry and randomly selects one of the support vectors without flipping its label as the starting attack point. It is worth noting that one can still flip the initial support vector's label as an attack strategy. However, since the labeling process is usually conducted by an oracle, their experiment study has kept the original label during the attack process. Therefore, the experimental setting is more reasonable from a practical point of view. Furthermore, they calculated the gradient using parameters of the SVM model trained on Dtr $\cup$ (X_0, Y_0). Afterward, we update the initial attacking point according to the gradient ascent method. Then they retrain the SVM model and optimize the attacking point iteratively until it reaches the termination condition.

In this study, they use the gradient ascending technique. The solution is in the negative direction of the gradient. L_1 is a convex objective function, so it will never reach a convergence point. In this respect, they use L_2 to constrain the iterative steps. The algorithm terminates either when there are no further changes regarding the hinge loss, $L_2(Xo)$, or the margin cannot be maximized anymore. When the value returned by $L_2(Xo)$ does not change, we have $L_2^{t-1} = L_2^t$, meaning that the attacking point will not cause further hinge loss because of the fault-tolerant nature of the SVM. Therefore, there is no need to further update the attacking point.

11.4 ADVERSARIAL LABEL CONTAMINATION ON SVM

Attacks against learning methods can be carried out either during the training or testing stage.

1. Attacks at the *testing* stage exploit the characteristics of the underlying classes that can be modified without affecting the true classification. But it deleteriously impacts on the discriminative model learned from the training data [23–25]. Given knowledge of invariances in the task, the effect of potential testing attacks can be somewhat mitigated by the learning algorithm. However, this comes at a cost of increased complexity of the training problem.
2. Attacks at the *training* stage attempt to exert a long-lasting impact on learning by modifying the training data. For example, important points in the training data may be changed, which makes the learning problem more complex. Such attacks introduce feature noise to the training points. Another possibility is to flip the labels of certain points (label noise), which has a similar effect on the learning problem. While some previous work has developed some methods that are robust against feature noise [26, 27], little is known on how learning algorithms are affected by adversarial (rather than random) label noise.

In [28], the authors proposed a model for the analysis of label noise in support vector learning and developed a modification of the SVM formulation that indirectly compensates for the noise. The model is based on a simple assumption that any label may be flipped with a fixed probability. They demonstrated that the noise can be compensated for by correcting the kernel matrix of SVM with a specially structured matrix, which depends on the noise parameters. They adopt two different strategies for contaminating the training sets through label flipping: random or adversarial label flips.

11.4.1 Random Label Flips

In the first case, they randomly select a number of samples from the training data (chosen according to a fixed percentage of data that can be manipulated by the adversary) and flip their labels. This can be regarded as a nonadversarial kind of noise, since it is not dependent on the given classifier.

11.4.2 Adversarial Label Flips

In the second case, given a number of allowed label flips, the adversary instead aims to find the combination of label flips to maximize the classification error on the untainted testing data. However, the problem of finding the optimal (worst-case) combination of label flips to attack the SVM learning algorithm is not trivial. Thus they resort to a heuristic approach, which has shown to be quite effective in a set of experiments. The idea behind the adversarial label flip attack is to first flip the labels of samples with nonuniform probabilities, depending on how well they are classified by the SVM learned on the untainted training set. Then it repeats this process a number of times, eventually retaining the label flips to maximally decrease performance. In particular, it increases the probability of flipping labels of samples that are classified with very high confidence (i.e., nonsupport vectors) and decreases the probability of flipping labels of support vectors and error vectors (inversely proportional to the α value). The reason is that the former (mainly, the nonsupport vectors) are more likely to become support vectors or error vectors when the SVM is learned on the tainted training set, and consequently the decision hyperplane is closer to them. This reflects a considerable change in the SVM solution and potentially in its classification accuracy. Furthermore, the labels of samples in different classes can be flipped in a correlated way in order to force the hyperplane to rotate as much as possible.

11.5 CONCLUSIONS

Adversarial examples show that many modern machine learning algorithms can be broken in surprising ways. These failures of machine learning demonstrate that even simple algorithms can behave very differently from what their designers intend. Thus it is imperative for machine learning researchers to design methods for preventing adversarial examples in order to close this gap between what designers intend and how algorithms behave.

REFERENCES

[1] J. Gu, Z. Wang, J. Kuen, L. Ma, A. Shahroudy, B. Shuai, T. Liu, X. Wang, G. Wang, J. Cai, T. Chen, *Recent advances in convolutional neural networks, Pattern Recognition 77* (2018) 354–377.

[2] C. Szegedy, W. Zaremba, I. Sutskever, J. Bruna, D. Erhan, I. Goodfellow, R. Fergus, *Intriguing properties of neural networks*, in: ICLR, 2014.

[3] I. J. Goodfellow, J. Shlens, C. Szegedy, *Explaining and harnessing adversarial examples*, in: ICLR, 2015.

[4] A. M. Nguyen, J. Yosinski, J. Clune, Deep neural networks are easily fooled: High confidence predictions for unrecognizable images., in: *IEEE CVPR*, 2015, pp. 427–436.

[5] S.-M. Moosavi-Dezfooli, A. Fawzi, P. Frossard, Deepfool: a simple and accurate method to fool deep neural networks, in: *IEEE CVPR*, 2016, pp. 2574–2582.

[6] B. Biggio, I. Pillai, S.R. Bulò, D. Ariu, M. Pelillo, F. Roli, Is data clustering in adversarial settings secure?, in: *Proceedings of the 2013 ACM Workshop on Artificial Intelligence and Security, ACM*, NY, USA, 2013, pp. 87–98.

[7] B. Biggio, S.R. Bulò, I. Pillai, M. Mura, E.Z. Mequanint, M. Pelillo, F. Roli, Poisoning complete-linkage hierarchical clustering, in: P. Franti, G. Brown, M. Loog, F. Escolano, M. Pelillo (Eds.), *Joint IAPR International Workshop on Structural, Syntactic, and Statistical Pattern Recognition, Lecture Notes in Computer Science*, vol. 8621, Springer Berlin Heidelberg, Joensuu, Finland, 2014, pp. 42–52.

[8] A. Globerson, S.T. Roweis, Nightmare at test time: robust learning by feature deletion, in: W.W. Cohen, A. Moore (Eds.), *Proceedings of the 23rd International Conference on Machine Learning*, vol. 148, ACM, Pittsburgh, Pennsylvania, USA, 2006, pp. 353–360.

[9] C.H. Teo, A. Globerson, S. Roweis, A. Smola, Convex learning with invariances, in: J. Platt, D. Koller, Y. Singer, S. Roweis (Eds.), *NIPS 20*, MIT Press, Cambridge, MA, 2008, pp. 1489–1496.

[10] M. Brückner, C. Kanzow, T. Scheffer, Static prediction games for adversarial learning problems, *J. Mach. Learn. Res. 13* (2012) 2617–2654.

[11] B. Biggio, G. Fumera, F. Roli, Design of robust classifiers for adversarial environments, in: *IEEE International Conference on Systems, Man, and Cybernetics*, 2011, pp. 977–982.

[12] R.N. Rodrigues, L.L. Ling, V. Govindaraju, Robustness of multimodal biometric fusion methods against spoof attacks, *J. Vis. Lang. Comput. 20* (2009) 169–179.

[13]   A. Kolcz, C.H. Teo, Feature weighting for improved classifier robustness, in: *6th Conference on Email and Anti-Spam, Mountain View*, CA, USA, 2009.

[14]   B. Biggio, G. Fumera, F. Roli, Multiple classifier systems under attack, in: N.E. Gayar, J. Kittler, F. Roli (Eds.), *9th International Workshop on Multiple Classifier Systems, Lecture Notes in Computer Science*, vol. 5997, Springer, Cairo, Egypt, 2010, pp. 74–83.

[15]   B. Biggio, G. Fumera, F. Roli, Multiple classifier systems for robust classifier design in adversarial environments, *Int. J. Mach. Learn. Cybern.* 1 (2010) 27–41.

[16]   G.F. Cretu, A. Stavrou, M.E. Locasto, S.J. Stolfo, A.D. Keromytis, Casting out demons: sanitizing training data for anomaly sensors, in: *IEEE Symposium on Security and Privacy, IEEE Computer Society, Los Alamitos*, CA, USA, 2008, pp. 81–95.

[17]   B. Biggio, I. Corona, G. Fumera, G. Giacinto, F. Roli, Bagging classifiers for fighting poisoning attacks in adversarial environments, in: C. Sansone, J. Kittler, F. Roli (Eds.), *10th International Workshop on Multiple Classifier Systems, Lecture Notes in Computer Science*, vol. 6713, Springer-Verlag, Naples, Italy, 2011, pp. 350–359.

[18]   A. Christmann, I. Steinwart, On robust properties of convex risk minimization methods for pattern recognition, *J. Mach. Learn. Res.* 5 (2004) 1007–1034.

[19]   www.princeton.edu/~abhagoji/files/iitm_seminar_04_2018.pdf

[20]   B. Biggio, I. Corona, G. Fumera, G. Giacinto, F. Roli, "Evasions attacks against machine learning at test time"

[21]   https://vitalab.github.io/article/2021/02/26/Evasion-attacks.html

[22]   Shigang Liu et.al., 'A Data-driven Attack against Support Vectors of SVM', ASIACCS'18, June 4–8, 2018, Incheon, Republic of Korea.

[23]   A. Globerson and S. T. Roweis. Nightmare at test time: robust learning by feature deletion. In William W. Cohen and Andrew Moore, eds, *ICML*, vol. 148 of ACM Int'l Conf. Proc. Series, pp. 353–360, 2006.

[24]   O. Dekel, O. Shamir, and L. Xiao. Learning to classify with missing and corrupted features. *Machine Learning*, 81:149–178, 2010.

[25]   C. H. Teo, A. Globerson, S. Roweis, and A. Smola. Convex learning with invariances. In J.C. Platt, D. Koller, Y. Singer, and S. Roweis, eds, *Advances in Neural Information Processing Systems 20*, pp. 1489–1496, 2008.

[26]   A. Christmann and I. Steinwart. On robust properties of convex risk minimization methods for pattern recognition. *J. of Machine Learning Research*, 5:1007–1034, 2004.

[27]   H. Xu, C. Caramanis, and S. Mannor. Robustness and regularization of support vector machines. *J. of Machine Learning Research*, 10:1485–1510, 2009.

[28]   Battista Biggio, et.al. 'Support Vector Machines Under Adversarial Label Noise.' *JMLR: Workshop and Conference Proceedings* 20 (2011) 97–112. Asian Conference on Machine Learning.

Federated Learning and Blockchain: An Opportunity for Artificial Intelligence with Data Regulation

12

Darine Ameyed,[1] Fehmi Jaafar,[2] Riadh ben Chaabene,[1] and Mohamed Cheriet[1]

[1]*École de technologie supérieure, Montreal, Canada*
[2]*Quebec University at Chicoutimi, Quebec, Canada*

Contents

DOI: 10.1201/9781003187158-14

12.1 INTRODUCTION

In recent years, we have seen a tremendous increase in the focus and concerns of policy makers and consumers regarding data collection, data privacy, and data usage. Two of the most popular examples that had affected the data market are the General Data Protection Regulation (GDPR) [3] that took place on 2018 in Europe. It provided multiple regulations related to data usage and data manipulation with a focus on data security. This affected most of the European Union enterprises, making them more mindful with their collection, usage, storing, and transferring of client data. The second example is the California Consumer Privacy Act (CCPA) [4] in the United States. It provided the possibility for citizens to demand the disclosure and the deletion of their own data possessed by enterprises.

Those regulations and policies have affected the artificial intelligence field in the field of either machine learning or deep learning. The process of training and testing artificial intelligence models requires a sufficient amount of data to work properly and to provide better results and accuracy. But sometimes these models require personal data to be trained, and, with such legislation, using personal data has become either difficult or impossible.

Since data security and privacy became a critical concern, finding new artifice intelligence methodologies has become a necessity. Federated learning, which is a machine learning technique, has been developed for this purpose. Privacy is at the core of federated learning since it provides a decentralized approach to training machine learning models while respecting the privacy of the data. No data is collected during the process of training; the model is sent to the data, rather than the traditional way, in which the data is sent to the model. But with that arrangement come other challenges and other security attributes that we need to address, and blockchain technology could be a solution to provide more security options to federated learning.

In this chapter we discuss the major data security attributes and the regulations and policies regarding personal data, as well as how federated learning contributes to AI models regarding those regulations. Then we provide a full description of federated learning, its types, and its techniques. That is followed by the current challenges and opportunities facing federated learning. In this section, we discuss how blockchain technology could be a factor to address those challenges. Finally, we provide our use case and the conclusion of the chapter.

12.2 DATA SECURITY AND FEDERATED LEARNING

Data security [5] is defined as the process of securing and protecting private and public sectors data and preventing them from data loss, data tampering, and unauthorized access. This represents the CIA triad, which consists of data confidentiality, integrity, and availability.

1. *Confidentiality:* Confidentiality refers to the effort to keep data private or secret from unauthorized parties. To ensure the confidentiality of the data, Methods to control the data confidentiality include role-based access control (RBAC), volume/file encryption, file permissions, and encryption of data.
2. *Integrity:* This is a fundamental part of the CIA triad and is intended to protect data from deletion or alteration from any unapproved access, and it guarantees that, when an approved individual

provides an incorrect update, the harm can be remedied. So it is very important to ensure that the data is secured from both ends, in terms of modification and deletion so that the data can be trusted.

3. *Availability:* This principle refers to the availability of the data, systems, and applications to authorized parties. Availability should be always consistently and readily accessible, whether impacted by either hardware, software, network failure, human error, or cyber attack. Redundancy for the servers and network, system upgrading, backup and recovery systems, and fault tolerance are methods that could be used to ensure data and applications. Intrusion detection systems, hashing, encryption authentication systems, digital signature, and access control are all methods that can be used to maintain data integrity.

Seven principles of the GDPR specify the rules that each company or individual needs to follow in order to deal with customers' personal data. Following [6], the principals are as follows:

1. *Lawfulness, fairness, and transparency*: The term *lawfulness* refers to the reasons for processing personal data, meaning that all information related to the usage of the data should be well-defined and specific. Fairness and transparency go hand and hand with lawfulness, meaning that organizations shouldn't withhold information about what data they are collecting or why. They should be clear to the data subjects about their identity and why and how they are processing their data.

2. *Purpose limitation*: Data should be collected only for a specific reason. The purpose of processing the data needs to be well-defined to the data subjects.

3. *Data minimization*: Data users should only collect the smallest amount needed of data to complete the processing. There should be a reason to ask for specific data. For example, there is no need to ask for a phone number or address if it's not related to the data processing.

4. *Accuracy*: It's up to the organization to deal with data accuracy. Any data stored incorrectly or incompletely should be corrected, updated, or erased.

5. *Storage limitation*: There is a need to justify the time span during which the data is being stored. Data retention shouldn't occur if there is no longer a need for the data.

6. *Integrity and confidentiality*: Like the CIA triad, the GDPR require all data users to maintain the integrity and confidentiality of the collected data and to ensure its security from unauthorized access.

7. *Accountability*: This last principle means that the GDPR require a level of accountability for all organizations. It signifies that documentation and proofs should be associated with the data processing principles. At any time, supervisory authorities can ask for this documentation.

The CCPA principles [4] are as follows:

- Consumers have the right to access all the data that an organization collects about them.
- Consumers can choose to not have their information sold to third parties.
- Consumers can demand that an organization delete their personal data.
- Consumers have the right to know to whom their data has been sold.
- Consumers have the right to know the reason for the data collection.
- Consumers can take legal action without proof of damages if they are subjected to a beach of privacy.

In all those regulations, federated learning [7] was introduced. Privacy is one of the essential properties of federated learning. This requires security models and analysis to provide meaningful privacy guarantees. The interest of federated learning is to mutually train a global model without directly sacrificing data privacy. In particular, it has provided significant privacy advantages compared to a data center training in a dataset.

12.3 FEDERATED LEARNING CONTEXT

In the classic centralized learning, we have data centralized in a server that is sent to a model for training. We have spent decades using that technique until we were presented with federated learning. Federated learning (FL) [8] is a machine learning method where distributed clients, such as edge devices or organizations, mutually train a model under the coordination of a central server (e.g., service provider), while keeping the training data decentralized. It embodies the principles of focused collection and data minimization and can mitigate many of the systemic privacy risks and costs resulting from traditional, centralized machine learning. Federated learning conducts machine learning in a decentralized approach [9]. It aims to provide the model to the data rather than the traditional way, in which we create a model and then provide the data. The data is distributed among different users in their edge mobiles. Each one provides its own data to train the model, leading to multiple unique training datasets rather than just one. Only the metadata is returned to the centralized model or the global model using encrypted communication. At the same time, this provides knowledge performance for the model and enhances data privacy, since all the calculation happen within the owner's device without any exchange of its data to a centralized server.

It is important to not confuse federated learning with distributed learning. They both follow a same schema for training models by using a distributed node for the data and a centralized server for the model. However, the main differences arise from the data assumptions: Distributed learning uses parallelism to enhance computational power for additional training productivity, whereas federated learning, using encryption, focuses more on conserving data privacy.

12.3.1 Type of Federation

Federated learning is divided into mainly two types based on the most used approach: data driven and model driven (Figures 12.1 and 12.2).

12.3.1.1 Model-centric Federated Learning

This the more common of the two types [10]. In model-centric, the model is hosted in a cloud service, and its API is preconfigured (layers, weight, etc.). Each individual downloads the model, enhances it, and uploads a newer version. But this could happen over a long period of time, frequently weeks and months. A great example is Google's GBoard mobile app. It learns users' typing preferences over time without sending any private data. The learning is distributed and happens remotely, updating a central model through a suitable federation technique.

12.3.1.2 Data-centric Federated Learning

Less common than model-centric but more suitable for experimentation science, data-centric is hosted inside a cloud server not as a model but as a dataset whose API is preconfigured (Schema, Attributes, etc.). Users can use those datasets to locally train their models in a specially appointed experimental way. To be more specific, in data-centric federated learning, the idea is to have a super or global oracle that is centralized in a server and that contains multiple datasets added by contributors. Any model owner wanting to use that data can send its model and a request for the usage of a specific dataset for its training. If the request is accepted, the oracle searches for the requested dataset and starts the training of the model. As soon as the training is finished, the model is provided back to the model owner. The most likely scenario for data-centric is where a person or organization wants to protect data (instead of hosting the model,

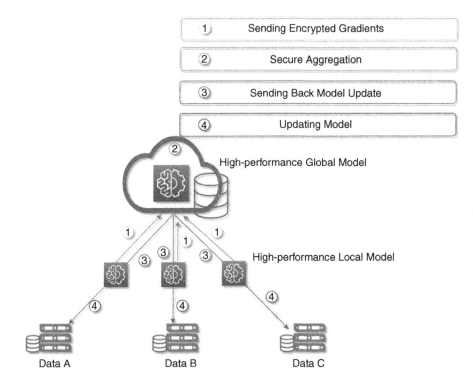

FIGURE 12.1   Model-centric federated learning.

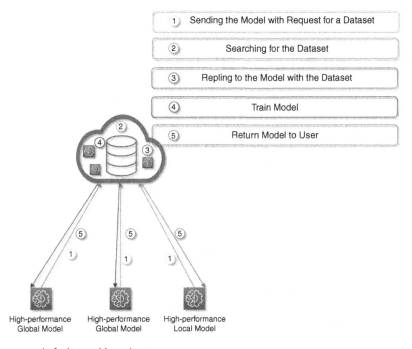

FIGURE 12.2   Data-centric federated learning.

they host data). This allows a data scientist, who is not the data owner, to make requests for training or reference against that data.

12.3.2 Techniques

In federated learning, understanding your data, how it is split, and its technical and practical challenges is really important in defining the terms of implementation.

12.3.2.1 Horizontal Federated Learning

Horizontal federated learning, or sample-based federated learning, is used in cases where multiple data sets share the same features but differ on instances. We also call it a *homogeneous federated learning approach*, since we are dealing with similarity for the usage of same features.

12.3.2.2 Vertical Federated Learning

Vertical federated learning [11] is the opposite of the horizontal approach. It refers to the use of multiple data sets having many samples that intersect and few intersected features. In this case, we have a *heterogeneous federated learning*, on account of differing feature sets. It uses different datasets of different feature spaces to jointly train a global model.

12.4 CHALLENGES

In this section we discuss the current issues and challenges that federated learning is facing and that it's required to focus on in order to obtain higher efficiency [12].

12.4.1 Trade-off Between Efficiency and Privacy

When using secure multicomputation (SMPC) [13] and differential privacy (DP) [14], the privacy measures of federated learning become very efficient. However, obtaining a high level of privacy comes with a trade-off. With the use of SMPC, the encryption of the model parameters needs to occur before sending it back to the global model for aggregation. This requires the use of additional computation resources and power, which can affect the efficiency of the training. On the other hand, the usage of DP requires adding noise to the data and the model and, in turn, a trade-off in accuracy and privacy since the use of the additional noise affects the accuracy of the model. Finding a suitable trade-off of SMPC and DP in federated learning is very challenging.

12.4.2 Communication Bottlenecks [1]

The communication cost for bringing the models to the device should be moderately low since it may impact the FL environment where one device can be crippled due to communication bottlenecks, which in turn stall the federated training process. There are several works on addressing communication bottlenecks, such as dropping stragglers (devices that fail to compute training within the specified time window), and model compression/quantization to reduce the bandwidth cost.

12.4.3 Poisoning [2]

Poisoning has two forms. In *data poisoning*, malicious users send fake or malicious data during the process training, which affects the overall performance of the model. Using federated learning, it is very difficult to detect such users. *Model poisoning* is the opposite of the first type. Malicious users tamper with the received model by modifying its gradient/parameters before the training. So during aggregation, the global model is then highly affected with false gradients during the process.

12.5 OPPORTUNITIES

Some concerns still need to be dealt with in federated learning in order to guarantee greater security without the cost of lost performance. An idea is to merge blockchain technology with data-centric federated learning. Blockchain is a "peer-to-peer" (P2P) decentralized ledger technology [15]I that provides a method to record and distribute information about transactions publicly on a peer-to-peer system of computers through crypto protocol. The database is scattered according to the rule that each duplicate of new information isn't just put away on a solitary server but is additionally shipped off to all clients in the chain or framework. To change any bit of the data set, programmers need to change 51% of the duplicates for passage on the framework, and every one of these duplicates needs to contain the entire past exchanges of this information. This convention has effectively wiped out outsiders, while guaranteeing security and adaptability with high intuitiveness [16].

The concept includes two types of blockchain, public and private. However, other variations consist of consortium and hybrid blockchains. All of them consist of clusters of nodes functioning on a peer-to-peer network system in which each node possesses a replica of the shared ledger with timely updates. Also, they all verify, initiate, and recover transactions while also creating new blockchain. Nevertheless, all these types differ from one other in various ways.

12.5.1 Leveraging Blockchain

Leveraging blockchain technology [17] permits two types of interaction with data-centric federated learning. First, it offers new security layers for data-centric federated learning such as immutability, traceability, transparency, and a level of trust for all participants. Second, it offers a genuine execution of an incentive-based system that encourages contributors to share data and to train the ML model to raise performance. The data structure in a blockchain is a back-linked list of records. In this way, the information can't be modified or erased. Hence, we are able to ensure the immutability of our data and model. Each transaction and each block in the blockchain are time-stamped. Thus, assuming data is lost, there is no chance that it cannot be recovered. The transactions that happen are straightforward, and participants having permission can see any transaction, leading to a full transparency. The beginning of any record can be traced along the chain to its starting place. To guaranteed full interaction transparency, we add public models in a smart contract. Now data providers are able to be fully aware about what training their data will be involved in. Smart contracts offer agreed-on specification insurance, which helps the model to be updated in the blockchain environment as well as in a centralized server. It is impossible to tamper with the smart contract code, and, being evaluated by many machines, it guarantees the full functionality of the model specifications. Smart contracts [18] are unmodifiable and evaluated by many machines, helping to ensure that the model does what it is specified to do. The changeless nature and perpetual record of smart contracts additionally permit us to dependably register and convey prizes for good data contributors.

12.6 USE CASE: LEVERAGING PRIVACY, INTEGRITY, AND AVAILABILITY FOR DATA-CENTRIC FEDERATED LEARNING USING A BLOCKCHAIN-BASED APPROACH

In this section, we present a proof of how the blockchain opportunity can enhance model and data security and integrity in federated learning.

Federated learning architecture was the reference approach in creating our own architecture. Whereas in federated learning we distribute the model to the data, in our case we do the opposite by distributing the data to the model. The usage of blockchain for data-centric federated learning is shown in Figure 12.3.

In our case, the model and the data are stored in a decentralized environment that is the public blockchain, not in a centralized server that is accessible by all network participants. Those same participants provide the data to the model in a collaborative approach, making the data available for model owners. The personal data is fully controlled by its owners, i.e., they can control the amount of data they want to provide. With every input of data, we have a data aggregation, but before any update to the model, the data needs to be verified. To ensure the integrity of the data, we implement an incentive mechanism to encourage contributors and to oblige each user to insert a deposit in order to add data to train the model, making it costly for malicious users to disturb model efficiency. The process starts by the model owner providing a deposit, reward, and time-out function, thus establishing the foundation of our incentive mechanism. The *deposit* corresponds to the monetary amount that a user needs to make in order to have access to the model smart contract, so that it can add data to train the model inside the blockchain. The *reward* represents the amount that a user receives when checking the integrity of the data and restores it. The *time-out function* defines the time between a return of deposit and validation of the owner. When the approved data is sent to the model for training, the model is updated; for each beneficial deposit, a refund is made, and for each beneficial verification of the inputs, a reward is offered. Access to the contributor's data is necessary since the data will only be used to train the model within the smart contract. Those features will provide an enormous improvement in terms of data security and machine learning.

12.6.1 Results

We explored in this approach different security aspects with respect to data regulation policies while at the same time enhancing the efficiency of AI models. Using federated learning, we are able to create a model

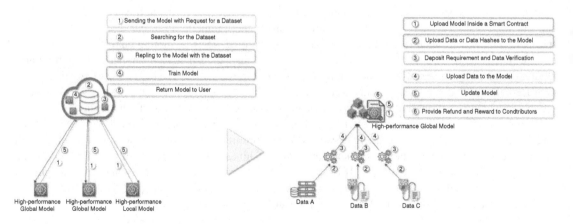

FIGURE 12.3 Blockchain for data-centric federated learning.

for training while attending to privacy concerns for personal data, and with blockchain, we provided contributors with full anonymity, control, and access in providing the data that they want the model to assess, while having full transparency of the training process. Those features lead to the growth of the training data set by an average of 500 additional data points per day. Also, with the usage of the incentive mechanism, we are able to avoid malicious contribution since the data is verified by the contributors before the training phase. Since we are using a deposit/refund mechanism, each input requires monetary deposit. After a period of time, we saw that the account balance of malicious users went all the way to zero because all data was verified and deleted when it led to no refund to the contributor. Simultaneously the account balance of good contributors went up, since they were rewarded for their verification. This mechanism help avoids the poisoning problem described previously. As for availability, the decentralized infrastructure of the blockchain helps in reaching countless users, who can provide their data in a secure and private environment.

Despite all those security features, the accuracy and the efficiency of the model increased by 5%. This is due to the fact that we don't consider adding noise to the data to enhance privacy, such as we do in differential privacy. Thus the model is being trained with quality data received from contributors

12.7 CONCLUSION

In this chapter, we have shown that, despite new data regulation and policies, we are still able to provide quality and secure model training for the AI model, while complying with regulatory rules. And this is possible due to federated learning and blockchain technology. As we saw in our use case, where we merged federated learning and blockchain, multiple data security attributea were addressed while enhancing a AI model. This combination offers a specific approach to contribute to the CIA triad; i.e., by using federated learning, we manage to provide privacy and confidentiality to the data during training. At the same time, the blockchain offers traceability, integrity, and transparency. Those attributes also answer to the GDPR principals and CCPA regulations.

REFERENCES

1. Filipe Betzel, Karen Khatamifard, Harini Suresh, David J. Lilja, John Sartori, and Ulya Karpuzcu. Approximate Communication: Techniques for Reducing Communication Bottlenecks in Large-Scale Parallel Systems. *ACM Computing Surveys*, 51(1):1–32, April 2018.
2. Minghong Fang, Xiaoyu Cao, Jinyuan Jia, and Neil Zhenqiang Gong. Local Model Poisoning Attacks to Byzantine-Robust Federated Learning. page 19.
3. Michelle Goddard. The EU General Data Protection Regulation (GDPR): European Regulation that has a Global Impact. *International Journal of Market Research*, 59(6):703–705, November 2017.
4. Lydia de la Torre. A Guide to the California Consumer Privacy Act of 2018. *SSRN Electronic Journal*, 2018.
5. Dorothy Elizabeth Robling Denning. *Cryptography and Data Security*. AddisonWesley, Reading, Mass, 1982.
6. Oliver Radley-Gardner, Hugh Beale, and Reinhard Zimmermann, editors. *Fundamental Texts On European Private Law*. Hart Publishing, 2016.
7. Xin Yao, Tianchi Huang, Chenglei Wu, Ruixiao Zhang, and Lifeng Sun. Towards Faster and Better Federated Learning: A Feature Fusion Approach. In *2019 IEEE International Conference on Image Processing (ICIP)*, pages 175–179, Taipei, Taiwan, September 2019. IEEE.
8. Qiang Yang, Yang Liu, Tianjian Chen, and Yongxin Tong. Federated Machine Learning: Concept and Applications. *arXiv:1902.04885 [cs]*, February 2019. arXiv: 1902.04885.
9. Kelvin. Introduction to Federated Learning and Challenges, November 2020.

10. Huawei Huang and Yang Yang. WorkerFirst: Worker-Centric Model Selection for Federated Learning in Mobile Edge Computing. In *2020 IEEE/CIC International Conference on Communications in China (ICCC)*, pages 1039–1044, August 2020. ISSN: 2377–8644.

11. Jiankai Sun, Xin Yang, Yuanshun Yao, Aonan Zhang, Weihao Gao, Junyuan Xie, and Chong Wang. Vertical Federated Learning without Revealing Intersection Membership. *arXiv:2106.05508 [cs]*, June 2021. arXiv: 2106.05508.

12. Peter Kairouz, H. Brendan McMahan, Brendan Avent, Aur´elien Bellet, Mehdi Bennis, Arjun Nitin Bhagoji, Kallista Bonawitz, Zachary Charles, Graham Cormode, Rachel Cummings, Rafael G. L. D'Oliveira, Hubert Eichner, Salim El Rouayheb, David Evans, Josh Gardner, Zachary Garrett, Adria' Gasco´n, Badih Ghazi, Phillip B. Gibbons, Marco Gruteser, Zaid Harchaoui, Chaoyang He, Lie He, Zhouyuan Huo, Ben Hutchinson, Justin Hsu, Martin Jaggi, Tara Javidi, Gauri Joshi, Mikhail Khodak, Jakub Koneˇcny´, Aleksandra Korolova, Farinaz Koushanfar, Sanmi Koyejo, Tancr'ede Lepoint, Yang Liu, Prateek Mittal, Mehryar Mohri, Richard Nock, Ayfer Ozgu¨r, Rasmus Pagh, Mariana Raykova, Hang Qi, Daniel Ra-¨mage, Ramesh Raskar, Dawn Song, Weikang Song, Sebastian U. Stich, Ziteng Sun, Ananda Theertha Suresh, Florian Tram'er, Praneeth Vepakomma, Jianyu Wang, Li Xiong, Zheng Xu, Qiang Yang, Felix X. Yu, Han Yu, and Sen Zhao. Advances and Open Problems in Federated Learning. *arXiv:1912.04977 [cs, stat]*, March 2021. arXiv: 1912.04977.

13. Lifei Wei, Haojin Zhu, Zhenfu Cao, Xiaolei Dong, Weiwei Jia, Yunlu Chen, and Athanasios V. Vasilakos. Security and Privacy for Storage and Computation in Cloud Computing. *Information Sciences*, 258:371–386, February 2014.

14. Cynthia Dwork. Differential Privacy: A Survey of Results. In Manindra Agrawal, Dingzhu Du, Zhenhua Duan, and Angsheng Li, editors, *Theory and Applications of Models of Computation*, pages 1–19, Berlin, Heidelberg, 2008. Springer Berlin Heidelberg.

15. Hyesung Kim, Jihong Park, Mehdi Bennis, and Seong-Lyun Kim. Blockchained On-Device Federated Learning. *IEEE Communications Letters*, 24(6):1279–1283, June 2020.

16. Quoc Khanh Nguyen and Quang Vang Dang. Blockchain Technology for the Advancement of the Future. In *2018 4th International Conference on Green Technology and Sustainable Development (GTSD)*, pages 483–486, Ho Chi Minh City, November 2018. IEEE.

17. Konstantinos Christidis and Michael Devetsikiotis. Blockchains and Smart Contracts for the Internet of Things. *Ieee Access*, 4:2292–2303, 2016.

18. Bin Hu. A Comprehensive Survey on Smart Contract Construction and Execution: Paradigms, Tools, and Systems. *OPEN ACCESS*, page 51.

PART III

Using AI/ML Algorithms for Cyber Security

Using Machine Learning for Cyber Security: Overview

13

D. Roshni Thanka, G. Jaspher W. Kathrine, and E. Bijolin Edwin

Karunya Institute of Technology and Sciences, Coimbatore, Tamil Nadu, India

Contents

DOI: 10.1201/9781003187158-16

13.1 INTRODUCTION

The facilities of the internet and its ease of use have created a platform for the movement of data from anywhere to everywhere. Huge data is accumulated by the various data sources, and they grow day by day and even by second to second. Protecting all this data from malicious users in the real world, where an attack is pertinent at each corner of the network, is the main focal point. Cyber security has a significant role in this context of maintaining security. Brute-force attack, credential stuffing, phishing, malware attacks, botnets denial-of-service, instant messaging abuse, worms, Trojans, intellectual property theft, rootkit, password sniffing, etc. are just some of the attacks commonly faced by any organization or by users of the internet. To detect and prevent these kinds of attacks, to assist analysts in dealing with them, and to prevent the abuse of data in the modern era, artificial intelligence categories like machine learning and deep learning play a vital role. With the use of machine learning techniques, patterns are developed and applied to algorithms to preemptively prevent various unforeseen attacks and bolster the security system. To generate these patterns, relevant, potentially rich-quality data is essential. Models developed using machine learning techniques automate computers and assist analysts in preserving valuable data without the explicit presence of the analysts. In any kind of attack, the very first thing is to know about the goal of the attack [2], which may be under any of the categories shown in Figure 13.1. A second important dimension is to deflate an adversary's ability to execute malicious activities.

13.2 IS ARTIFICIAL INTELLIGENCE ENOUGH TO STOP CYBER CRIME?

In today's emerging technology, data all over the world is evolving more and more, squeezing and bringing the essence is the major challenge. These squeezing reveal threats. In this aspect use of machine learning improves cyber security by predicting the future with the support of human expertise.

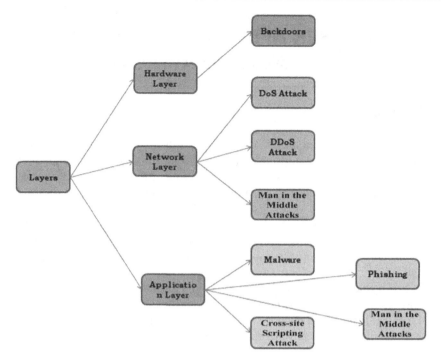

FIGURE 13.1 Types of attack.

13.3 CORPORATIONS' USE OF MACHINE LEARNING TO STRENGTHEN THEIR CYBER SECURITY SYSTEMS

The corporations that use machine learning techniques for their operation and enhanced security implementation [3] are Microsoft, Chronicle, Splunk, Blackberry, and Demisto, to name a few.

For cyber security, Microsoft uses the Windows Defender Advanced Threat Protection (ATP), which automatically spots threats. Microsoft also has launched a machine learning cyber attack threat matrix to engage security threats against ML.

In Chronicle, Backstory, which was released by Alphabet (a parent company of Google), provides instant security analysis and context for actual susceptibilities and uses machine learning to summarize information insights.

Splunk software includes a number of applications such as IT operations, analytics, and cyber security. This software is designed to categorize customer's weak points, computerize fissure inquiries, and respond to malware attacks. Products like Splunk Enterprise Security and Splunk User Behavior Analytics use machine learning to identify attacks and to assist in rapidly excluding them.

BlackBerry provides cyber security resolutions that practice AI and machine learning in order to thwart cyber security attacks and industrialize customers' threat response competences.

Demisto utilizes machine learning to monitor and highlight security alerts.

13.4 CYBER ATTACK/CYBER SECURITY THREATS AND ATTACKS

13.4.1 Malware

Intruders have developed different kinds of software to destroy computer systems, devices, and networks in order to steal sensitive information, greatly disrupting users. The modalities of malware come in the form of email attachments, infected files, file-sharing software, etc. Some of the variants of malware include botnets, crypto jacking, virus, worms, malvertising, ransomware, Trojans, and spyware, to name a few.

Botnets are networks of infected computers that mainly cause denial of service attacks. Crypto jacking is a kind of malicious software that mines for cryptocurrencies through malicious links in an email. These are financially motivated, may finally overload the affected system, and can lead to physical damage.

Viruses are executable files that attach themselves in a clean code or in an automated process and spread across the network very quickly, inflicting great damage on the functionalities of the system by corrupting and locking them. Worm malware is a type of self-replicative attacks and may or may not be attached with existing files or programs that start from one system but spread through the entire network very quickly.

A seemingly legitimate advertisement that turns into an attack is malvertising. Here, an advertisement in a legitimate network and website is used to spread the malware. The ransomware/scareware malware blocks or places a hold on some or all of the computer system until a certain amount of money is paid to the attackers. This creates an expensive loss to the organization since they cannot access their system.

The Trojan malware acts as legitimate software and creates backdoors for other malware to attack easily. Spyware is a malware that hides in the background of a system, by which attackers obtain sensitive information like credit card numbers, passwords etc. from the infected systems.

13.4.2 Data Breach

A data breach is gaining access to unauthorized information such as credit card information, passwords, eavesdropping, recording keystrokes etc. Phishing is a kind of data breach done by sending apparently legitimate emails or messages and gaining access to the valuable data. Whale-phishing attacks and spear-phishing attacks are some of its variants.

13.4.3 Structured Query Language Injection (SQL-i)

This is mainly to gain access to a SQL server database in the form of a SQL query in order to get valuable information.

13.4.4 Cross-site Scripting (XSS)

Similar to SQL-I, JavaScript code is injected as a browser side script instead of as a query to achievement website vulnerabilities.

13.4.5 Denial-of-service (DOS) Attack

This attack renders the resource inaccessible by flooding internet traffic that cannot be handled by server, thus preventing needful users from getting their online services.

13.4.6 Insider Threats

People inside the organization itself sometimes endanger the organization to gain restricted privileges to make significant changes or even to change security policies and obtain valuable data.

13.4.7 Birthday Attack

Here the attacker creates a similar hash identical to the one sent by the original sender; this can mislead the receiver, which identifies the fake as the real one. To prevent this type of attack, a longer hash can be used, which significantly reduces the effectiveness of such attacks.

13.4.8 Network Intrusions

Safeguarding the network from numerous security gaps is more crucial. Network intrusion leads to the stealing of valuable network resources.

13.4.9 Impersonation Attacks

Impersonation attacks take the form of fraud in which attackers may pose as a known contact and entice a user into sharing sensitive information or sharing login credentials. They also can lead to transferring money to a fraudulent account.

13.4.10 DDoS Attacks Detection on Online Systems

DDoS attacks create significant problems in many organizations. Identifying them in their early stages mitigates their serious effects on the organization. Using such machine learning algorithms as MLP, RF, KNN, and SVM in preprocessing helps in the identification various floods, such as UDP flood, ICMP flood, HTTP flood, etc. and provides good accuracy [37].

13.5 DIFFERENT MACHINE LEARNING TECHNIQUES IN CYBER SECURITY

Artificial intelligence and machine learning have already brought incredible transformations in the most significant parts of many industries. In order to provide an efficient cyber security in the current technology domain, machine learning is the only possible way. Because of the changing behavior of attack patterns, this can be achieved only through the proper comprehension of available quality data from their respective environments. Cyber security analysis, malware analysis, and other kinds of threat analysis can

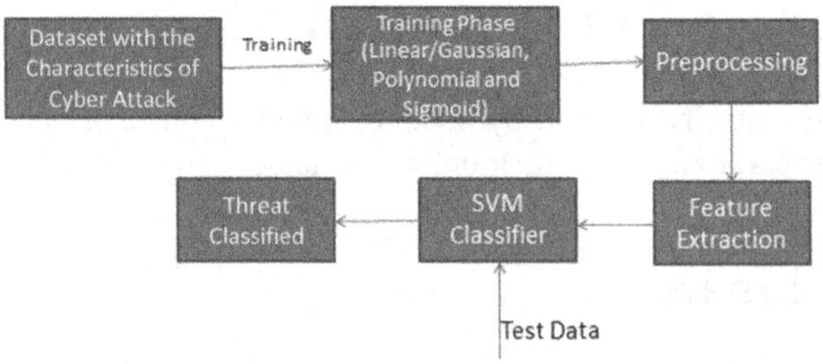

FIGURE 13.2 SVM architecture.

be put to widespread use in conjunction with deep learning techniques. This will increase the productivity time of an organization. Researchers say that, by 2025, more than 30% of investment will be based on AI/ML for cyber security, for assisting in identifying, preventing, and bringing solutions to the problems.

13.5.1 Support Vector Machine (SVM)

SVM, a classification of machine learning, is a supervised algorithm that plays a major role in perceiving cyber attacks based on the pattern of normal and abnormal attacks [9]. Here, misclassification can be reduced by increasing the marginal space in spatial resolution [4]. The performance can be further improved by modifying the Gaussian kernel. The SVM algorithm runs in the background when the user starts the browser, monitors each and every activity happening in the pages visited by the user, matches it with vulnerability types, and classifies the cyber attack depicted in Figure 13.2.

13.5.2 K-nearest Neighbor (KNN)

A strong security system is essential to make a system protected. Using the KNN classifier algorithm, the abnormal nodes whose feature is different are segregated from the normal node by their behavior in the intrusion system [8] depicted in Figure 13.3. When many of the nearest neighbor nodes are associated with a particular classification, then the data points belong to same classification. The closest data points are found through Euclidean distance or Manhattan distance. The Euclidean distance is calculated by:

$$d(x,y) = \sqrt{\sum_{i=1}^{n}(x_i - y_i)^2}.$$

Here n denotes the total quantity of data points, and x_i and y_i specify the ith and jth feature elements of instances x and y. The objective of any intrusion detection system is the quick detection of intrusion, mitigating the false alarm and providing fast activation of the protective system [5, 6, 7]. But the problem with KNN is that is is a lazy learner and requires high computational time. To overcome this, fast KNN is implemented to provide less computational time with better accuracy.

13.5.3 Naïve Bayes

Network anomalies such as anomaly detection and misuse detection can be detected by several data mining tasks. When a system has integrity, confidentiality, or availability, then the system is

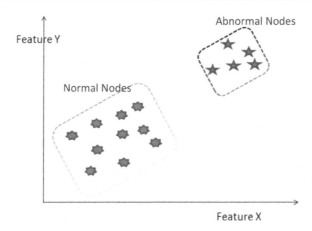

FIGURE 13.3   KNN intrusion detection.

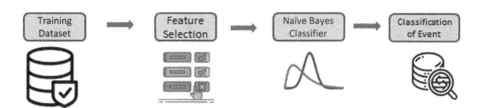

FIGURE 13.4   Naïve Bayes classifier.

considered to be secure. Naïve Bayes classifier is a simplified Bayesian probability model used in the field of uncertainty depicted in Figure 13.4. Here the possibility of one attribute does not affect the possibility of another model.

The naïve Bayes algorithm calculates and saves the probabilities of theft by providing certain attributes during the training phase. This is done for each attribute, and the time it takes to calculate the necessary probability for each attribute is recorded. The time it takes to calculate the probability of a given class for each sample in the worst scenario is proportional to n, the number of attributes, during the testing phase. In the worst-case scenario, the testing phase takes the same amount of time as the training phase.

13.5.4 Decision Tree

As communication technologies increase day by day, it's crucial to improve the security of computer network systems and protect them from intrusions. Intrusions in the system can be detected by changes in the activities from the normal profile.

The decision tree is a classification algorithm that mainly assists in finding the association between the predictor variable and target variable. Every data point in the training dataset is divided into subdomains by categorizing all the similar target valued under one subdomain. The subdomain further categorizes the target value of the newly arrived data point depicted in Figures 13.5 and 13.6. The path from source to destination of the decision tree specifies the pattern of the predictor variable to analyze the target variable. The predictor variable specifies the event, and the target variable specifies the event type. Every event occurring in the system is analyzed to categorize it as a normal or abnormal event and thereby detect intrusion.

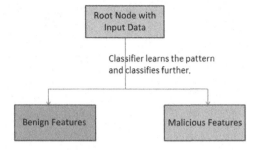

FIGURE 13.5   Example of decision tree.

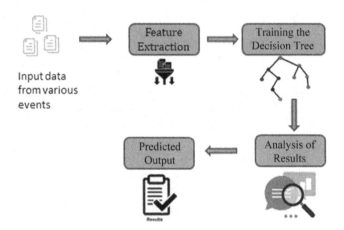

FIGURE 13.6   Processing steps of decision tree.

13.5.5 Random Forest (RF)

Random forest is an ensemble method that combines numerous models for better accuracy, depicted in Figure 13.7. The use of technology against communication intrusion activities is also increased. Intrusion can be in the data level in the computer or in the device used at the network level during communication. Random forest includes a number of decision tree that provide far less classification error. Randomization is used to choose the right node to divide when making a distinct number of trees [12]. RF arbitrarily selects the best node to split and generates noisy trees, which may reduce the accuracy of the model. Feature selection is applied to find the features that play a substantial role in predicting the output class.

13.5.6 Multilayer Perceptron (MLP)

MLP is grounded on a neural network where all the neurons on all the layers are fully connected, as in Figure 13.8. This algorithm is more efficient in classifying massive amount of records and in identifying the patterns in the complex dataset [30]. The error in the output is calculated by the difference between the actual output and the expected output.

To rectify this error, back-propagation is done in the backward direction. This starts from the output layer to the hidden layer, and it is transmitted toward the input layer. During this transmission, the weight is updated to obtain accurate results in the network. This also achieves better accuracy in the intrusion detection system [29, 31].

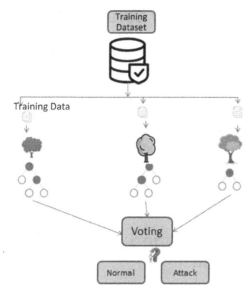

FIGURE 13.7   Random forest classifier.

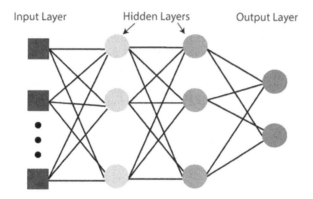

FIGURE 13.8   MLP.

13.6 APPLICATION OF MACHINE LEARNING

13.6.1 ML in Aviation Industry

Antidrone systems can detect drones using video, audio, thermal imaging, radar, or radio frequency technology. Protector schemes have the capacity to interfere with a drone's functionality. In detection, machine learning techniques are widely used. The three most common attacks on UAVs are jamming, meaconing, and spoofing. In a simulated setting, they create a GPS spoofing threat recognition method depending on a one class support vector machine algorithm. They put a malicious path into the replicated UAV in an endeavor to divert it from the preprogrammed course [26].

13.6.2 Cyber ML Under Cyber Security Monitoring

In cyber security monitoring, data collected from a UCI machine repository generates the training, validation, and test sets to get the samples on an anomalous network and without an anomalous network. The user states the time period of the data assortment and which device is to be monitored by mentioning its IP address. The record values are regularized and are incorporated into training, validation, and testing sets.

The user also specifies whether the data has to be worked with a multiclass algorithm or a one-class classification algorithm. A multiclass algorithm makes use of anomalous and nonanomalous data for all sets. Anomalous data is used only in the testing set. For designing the model and for performance evaluation, the machine learning algorithm is performed using the collected data. This model, as shown in Figure 13.9, is used for identifying the network status [27].

13.6.3 Battery Energy Storage System (BESS) Cyber Attack Mitigation

For obtaining sufficient detection accuracy, the quality and the dataset of appropriate size are essential. A battery management system (BMS) is used for managing BESS. For preventing the attack against a battery's state of charge, predicting approaches can be applied. Depending on the state estimation forecast with the sensing data, a state-of-the-art FDIA against the electric grid detection method is done. If there is a variance between the estimation and measurements, it means the residual surpasses the given threshold. For residual-based method implementation, the measurement prediction is necessary due to the constant data exchange between the BESS and the electric grid; cyber attacks influence the integrity of commands that the BESS receives, as depicted in Figure 13.10. Distributed methods have been used in decentralized systems for detecting cyber attacks. The main objective of the technique is to form a residual signal. It decides on a residual assessment function that is associated with the predefined threshold.

FIGURE 13.9   ML in cyber security monitoring.

FIGURE 13.10   BESS cyber threat.

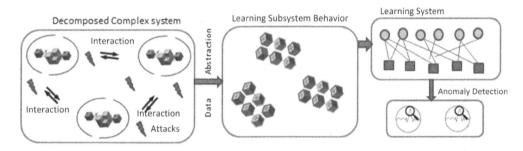

FIGURE 13.11   Cyber attack detection using unsupervised learning.

Once the attack is detected, its impact on the BESS operation has to be removed. An unobserved FDIA might threaten the chronological data applied for the training purposes and corrupt the prediction. For that reason, we use pseudo-measurements to feel the gap generated by a cyber-attack. Pseudo-measurements generation and SE forecast are the two major mitigation schemes [28].

13.6.4 Energy-based Cyber Attack Detection in Large-Scale Smart Grids

Dynamic Bayesian networks (DBN) are a probabilistic graphical model used to simplify the structure. Directed acyclic graph (DAG), mutual information (MI) to obtain the features, and restricted Boltzmann machine (RBM), which trains the data and learns the pattern in the context behavior with the help of unsupervised DBN model, are used in the process of cyber attack detection depicted in Figure 13.11.

These methods are computationally effective in identifying anomalies by extracting the features and time-based activaties from the behaviors of the subsystem. DBN and RBM identify the unnoticed attacks depending on the free energy [42].

13.6.5 IDS for Internet of Vehicles (IoV)

To provide a safe driving security for IoV, VANET is more essential for secure vehicle-to-vehicle communication. Different types of sensors and embedded controls are available in the vehicle to gather information such as traffic conditions, parking facilities, road accidents, etc. This information is communicated among nearby vehicles in order to have a comfortable journey. These communications are sometimes compromised by attackers that affect the whole network. MLP based on neural network plays a major role in identifying such intrusions and attacks in the IoV network [35]. This security can also be achieved using the deep belief network of deep learning [36, 38, 43].

13.7 DEEP LEARNING TECHNIQUES IN CYBER SECURITY

These days, there is a huge development in various technologies such as IoT, cloud computing, big data, requiring cyber security to protect applications and networks [44]. Technology giants like Microsoft, Google, Salesforce, Facebook, etc. have deep learning in their deliverables [45]. Deep learning that is grounded on artificial neural networks with more hidden layers is the next level of machine learning. Deep learning when compared with machine learning is more proficient due to its deep layers structure and its capability to learn by itself and produce better results.

13.7.1 Deep Auto-encoder

Deep auto-encoder monitors and learns the usage pattern of normal users' behavior, as depicted in Figure 13.12. It is an unsupervised learning that recreates the output based on the input [15]. When the usage pattern of the user changes, that is measured as an anomaly. Here the resemblance among present behavior $V_i(t)$ and normal behavior $V_i(0)$ is measured [13]. A threshold value k is fixed by the admin, when the difference goes beyond the predefined threshold value that is considered abnormal activity.

$$d\left(V(t) - V(0)\right) = \frac{1}{n} \sum_{i=1}^{n} \left(ci - pi\right)^2.$$

$(if\ V(t) - V(0) > k, \text{it means 'abnormal'.})$

13.7.2 Convolutional Neural Networks (CNN)

As the usage of the internet increases, malware is also increasing. CNN is one of the deep learning methods that depend on neural network. This processes the input in the form of an array, generally used for processing two-dimensional and three-dimensional array. The main three components of CNN are (1) the convolutional layer named after the convolution operation in mathematics; (2) the pooling layer used in the dimensionality reduction; and (3) the classification layer used for the finally classification. One-dimensional convolutional layers with single-dimensional filters of 3×1 size slide at one stride along with rectified linear unit activation are used, followed by two fully connected dense layers [18, 21], as shown in Figure 13.13. Here, there is no need to reduce input size because max pooling layers are not used. Dropout layers are used to avoid overfitting.

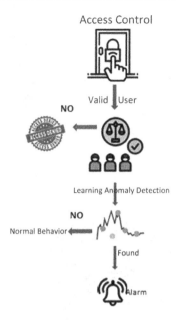

FIGURE 13.12   Flow of intrusion detection system.

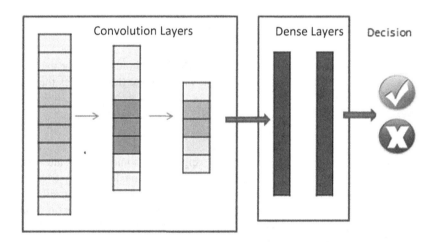

FIGURE 13.13   Sample 1 dimension CNN architecture.

This one-dimensional CNN is an intrusion detection mechanism that provides accurate results [16, 17]. CNN-based intrusion detection outperforms other intrusion detection mechanisms due to its deep layers and reduced training time.

13.7.3 Recurrent Neural Networks (RNNs)

In the world of advanced technology with its enormous data, intrusion and malware become major threats for small to big companies through any means, such as email, downloadable file, software, etc. Such intrusions require efficient systems to prevent these activaties and to protect valuable data and the network itself. Recently, deep learning algorithms have played a major part in cyber security. RNN is a

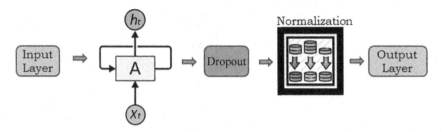

FIGURE 13.14   Recurrent neural network.

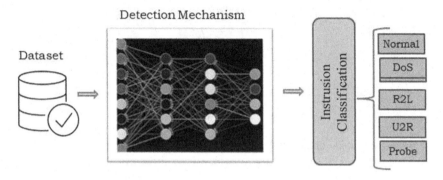

FIGURE 13.15   Deep neural network (DNN).

feed-forward neural network base on a deep learning method that holds not only the current state but also the previous state, so that is is called recursive. The hidden layer that includes the previous state is represented as follows:

$$S_t = \phi \, (wx_t + Hx_{t-1}).$$

where S_t represents the hidden state at time t, w is the weight, H is some hidden state, and ϕ represents some activation function, which can be a sigmoid or tanh function [19]. This RNN can solve any supervised or unsupervised task and outperforms traditional machine learning algorithm in intrusion or fraud detection. The number of layers and the epochs varies depending on the task on which it acts [20]. This RNN includes a recurrent layer with neurons that maintain the previous information along with the current time information, depicted in Figure 13.14. Followed by a recurrent layer, batch normalization and regularization are done to preclude overfitting and to speed up the training rate. Finally, classification is done with a particular activation function based on the output class.

13.7.4 Deep Neural Networks (DNNs)

An intrusion detection system is the need of the hour in today's world due to the rapid development of information technology. DNN takes on the major role in finding security breaches without accessing the information inside the packet, thus maintaining data privacy [23]. The model created by DNN includes multiple hidden layers, and the neurons in the layers are interconnected. Forward-propagation with the appropriate activation function for the perceptron is depicted in Figure 13.15.

To produce the most accurate results, back-propagation with a suitable optimizer and loss function is used to adjust the weights and bias [22]. Here the classification can be binary classification, denoting

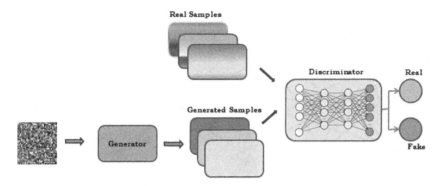

FIGURE 13.16   Generative adversarial networks architecture.

normal and attack, or multiclassification, such as user to root attack, denial of service attack, remote to user attack, probe, and normal.

13.7.5 Generative Adversarial Networks (GANs)

The digital data available needs to be secured to protect it from unknown attacks. Many companies, healthcare organizations, and public institutions have lot of secure data to be protected. GAN can generate new data that resembles the original data by which the original data can be safeguarded; a sample is shown in Figure 13.16. The new data generated by GAN has a high correlation with the original data, thus improving the security of the original data.

Another application of GAN is steganography where the sensitive information is hidden inside the images. Also GAN is used to hide a grayscale image inside the color image with high security. Sensitive data can also be stored inside an image container more securely using deep convolutional generative adversarial networks and auxiliary classifier generative adversarial networks. Neural network (NN), combined with steganography called neural cryptography, plays a significant role in securing the information. Here synchronizing the output weights present in NN generates a secret key. PassGAN is used for quality password guesses [24].

13.7.6 Restricted Boltzmann Machine (RBM)

The exponential use of technologies in our daily lives has increased the volume of data exchange among people and organizations, leading to new and different attacks in the network. Many organizations have initiated security measures to protect their networks. ThRBM algorithm is beneficial in different applications such as regression, dimensionality reduction, feature learning, classification, and many more. RBM comes with two biases: The forward pass is supported by a hidden bias, and the backward pass supported by visible layer's bias for reconstruction. Here no two neurons on same layer are connected. It is said to be restricted because no intralayer connection or communication is present. Figure 13.17 shows the graphical depiction of RBM with an m visible and an n hidden node [25]. Here, there is no connection between visible and visible nodes or between hidden nodes; the only permissible connection is visible to hidden or hidden to visible.

Each node represents a feature from the dataset that is multiplied with the weight and added with the bias passed through the activation function, giving the output of the particular node. Also, multiple inputs can be combined and given to one particular neuron by multiplying the multiple input with its weight, summing it up and adding the bias, and then sending it through the activation function to generate the

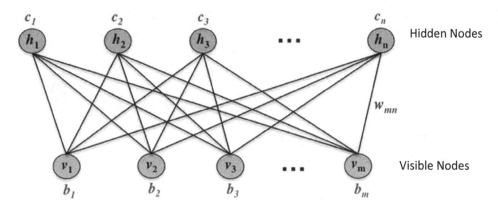

FIGURE 13.17   Graphical representation of RBM.

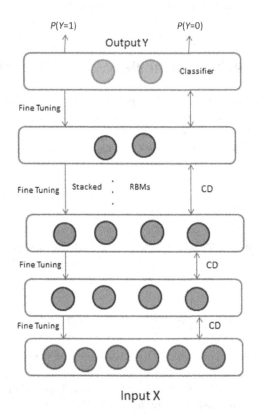

FIGURE 13.18   DBN structure.

output. Training makes it possible to produce a compact depiction of input. RBM helps in differentiating normal and anomalous traffic.

13.7.7 Deep Belief Network (DBN)

DBN is an unsupervised deep learning algorithm with a generative graphical model. This is a composite of stacked restricted Boltzmann machines, trailed by a softmax classification layer or auto encoder with both a directed and an undirected graphical network. The parameters of RBM are trained by a contrastive divergence (CD) algorithm. Figure 13.18 shows the structure of DBN [32].

Each hidden layer in a particular subnetwork acts as a visible layer for the next, thus improvising the training method due to its pretraining and fine-training. DBN does require labeled data, as in a deep neural network, but it works with unsupervised data, and the weights are updated and fine-tuned by back-propagation. This helps many in the industry that lack large amounts of data. Also, DBN gains good insight from the data and, when combined with CNN, produces higher accuracy. This method also performs better in malicious URLs identification [33], as well as in detection of security breaches in the IoT [34].

13.8 APPLICATIONS OF DEEP LEARNING IN CYBER SECURITY

13.8.1 Keystroke Analysis

To protect the user's credentials from the attacker's username and pin code are not enough. It is more essential to move one step forward to silently and constantly monitor the typing style of a regular user with that of another user. Many of the deep learning algorithms having enough hidden layers with prominent feature extraction capability play a major role in capturing the keystroke features of different users versus those of regular users, thereby protecting user credential information [39].

13.8.2 Secure Communication in IoT

In the emerging new technological world, many things are connected through the internet, and they exchange data. To enable security systems in this task and to avoid security breaches and losses, deep learning algorithms are used to detect threats and malware in the IoT environment [40].

13.8.3 Botnet Detection

Given the enormous usage of the internet, botnets pose a major security threat. Botnets compromise the basic requirements of network security, such as integrity, confidentiality, and availability. To detect and resist this type of attack, deep algorithms play a significant role and provide better performance and accuracy [41].

13.8.4 Intrusion Detection and Prevention Systems (IDS/IPS)

These systems protect users against security breaches by identifying malicious activities and alerting users to attackers. The traditional machine learning algorithms used in this method sometimes gives false alerts, which fatigue users [47]. This flaw can be overcome with deep learning algorithms [46, 49]. Network traffic is also monitored for malicious activity. Also, normal and attackers' network IDs can be identified.

13.8.5 Malware Detection in Android

It is rare to see people without android devices. Here malware detection is one of the big challenges. Deep learning with NLP helps in analyzing the static and dynamic features categorized by DBN and provides improved performance [47].

13.8.6 Cyber Security Datasets

For intrusion detection, dataset plays a major role, as shown in Table 13.1.

13.8.7 Evaluation Metrics

After implementing different modeld to provide cyber security, measuring their accuracy and performance is more significant. The metrics are accuracy, F1-score, precision, recall, false positive rate, and area under the curve, which is based on a confusion matrix that gives information in actual and predicted classes. (See Table 13.2.)

TABLE 13.1 Roles of Datasets in Intrusion Detection

DATASET	DESCRIPTION
NSL-KDD	Includes data extracted from military network environment, which includes the categories of DOS Attack, R2L Attack, U2R Attack, and Probing Attack [10, 11]. It is an improved form of KDDcup99.
KDD-CUP'99	Two million data for training and testing available with 41 features for intrusion detection system. It has four main output classes: DOS, probing, U2R, R2L [48].
ISCX 2012	An intrusion detection evaluation dataset, which includes the format of full packet payload [50].
DARPA	Big number of intrusion detection systems is evaluated using this popular dataset. This includes the communication between source and destination IPs [51].
CIDDS-001	Coburg Intrusion Detection Data Set produced by [14] assists in anomaly-based IDS. This dataset includes five classes: normal, suspicious, unknown, attacker, and victim.
UNSWNB15	This intrusion detection dataset has nine classes, backdoors, worms, DoS [52].
Microsoft BIG Malware Dataset	This dataset categorizes the malware depending on the content and features of the file [53].
ISCX	Information Security Center of Excellence (ISCX) dataset [54].
CTU- 13	This contains botnet, background traffic, and normal [55].
CSIC 2010 HTTP Dataset	Testing the web attack protection system [56].

TABLE 13.2 Confusion Matrix

		PREDICTED CLASS	
		ATTACK	NORMAL
Actual class	Attack	True positive (TP)	False negative (FN)
	Noraml	False positive (FP)	True negative (TN)

$$\text{Pr}ecision = \frac{TP}{TP + FP},$$

where *TP* represents the data points that are rightly predicted as attack. *TN* represents the data points classified as normal. *FN* are those wrongly classified as normal. *FP* represents data points wrongly said to be attack. The different evaluation metrics are as follows:

$$\text{Recall} = \frac{TP}{TP + FN}$$

$$F\text{-measure} = 2\left(\frac{\text{Precision Recall}}{\text{Precision} + \text{Recall}}\right)$$

Precision refers to the rightly identified threats to all the samples, i.e., said to be attacks.

Recall specifies the data points rightly classified as attacks.

This measures the accuracy by considering precision and recall. Accuracy is the ratio of correctly classified data points to the total number of data points.

13.9 CONCLUSION

Advancement in internet and information technology has brought an enormous intensification in network size and its related data. Every day the challenge to identify attacks on networks and on sensitive data is heightened. Even though the intrusion detection system plays a vital role in identifying and preventing intrusion in order to provide the fundamental requirements for networks and data, such as confidentiality, integrity, and availability, such systems still face many challenges in providing accurate results while minimizing false alarms. Here enter the artificial intelligent techniques, such as machine learning and deep learning algorithms implemented in the network, to identify intrusions successfully and show progress in this field. In this chapter, the ML and DL algorithms are discussed with their different applications with emphasis on the role of artificial intelligence in the intrusion detection system. In the comparison of ML algorithms with DL algorithms, DL-based algorithms provide better solutions in terms of feature extraction and prompt fitting to the model. The major challenge faced by this system is its complexity in terms of processing power and storage facilities. Addressing these requirements can meet the current need in intrusion detection systems.

REFERENCES

[1] Hadis Karimipour, Ali Dehghantanha, Reza M. Parizi, Kim-Kwang Raymond, and Henry Leung, "A Deep and Scalable Unsupervised Machine Learning System for Cyber-Attack Detection in Large-Scale Smart Grids", IEEE Access, Special Section On Digital Forensics Through Multimedia Source Inference, Volume 7, 2019, pg. No. 80778–80788.

[2] R. Geetha, T. Thilagam, "A Review on the Effectiveness of Machine Learning and Deep Learning Algorithms for Cyber Security", Springer, Archives of Computational Methods in Engineering, CIMNE, Barcelona, Spain 2020

[3] "Machine Learning Cybersecurity: How It Works and Companies to Know", https://builtin.com/artificial-intelligence/machine-learning-cybersecurity

[4]   Kirankumar Yadav, Tirupati Pai, Ravi Rane, "Classification of Cyber Attacks Using Support Vector Machine", Imperial Journal of Interdisciplinary Research, Vol 3, Issue 5, 2017, ISSN: 2454-1362.

[5]   B. Basaveswara Rao, and K. Swathi, "Fast kNN Classifiers for Network Intrusion Detection System", Indian Journal of Science and Technology, Vol 10(14), DOI: 10.17485/ijst/2017/v10i14/93690, April 2017.

[6]   Jamshidi Y, Nezamabadi-pour H. "A Lattice based nearest neighbor classifier for anomaly intrusion detection". Journal of Advances in Computer Research. 2013 Nov; 4(4):51–60.

[7]   Wenchao Li, Ping Yi, Yue Wu, Li Pan, and Jianhua Li, "A New Intrusion Detection System Based on KNN Classification Algorithm in Wireless Sensor Network", Hindawi Publishing Corporation Journal of Electrical and Computer Engineering Volume 2014, Article ID 240217, 8 pages.

[8]   Diwakar Reddy M, Bhoomika T Sajjan, Anusha M, "Detection of Cyber Attack in Network using Machine Learning Techniques", Vol. 1 No. 2 (2021): International Journal of Advanced Scientific Innovation-May 2021.

[9]   Kinan Ghanem, Francisco J. Aparicio-Navarro, Konstantinos Kyriakopoulos, Jonathon Chambers, "Support Vector Machine for Network Intrusion and Cyber-Attack Detection", December 2017 DOI: 10.1109/SSPD.2017.8233268, Conference: 2017 Sensor Signal Processing for Defence Conference (SSPD) http://dx.doi.org/10.1155/2014/240217

[10]   Nsl-kdd Data Set for Network-Based Intrusion Detection Systems, Available on: http://nsl.cs.unb.ca/KDD/NSLKDD.html, Last Visited, 2009.

[11]   Bhavsar Y. and Waghmare K., "Intrusion Detection System Using Data Mining Technique: Support Vector Machine", International Journal of Emerging Technology and Advanced Engineering, vol. 3, no. 3, pp. 581–586, 2013.

[12]   Nabila Farnaaz, M.A. Jabba, "Random Forest Modeling for Network Intrusion Detection System", Elsevier, Multi-Conference on Information Processing-2016, pp. 213–217.

[13]   Samir Fenanir, Fouzi Semchedine, Saad Harous, Abderrahmane Baadache, "A Semi-supervised Deep Auto-encoder Based Intrusion Detection for IoT", International Information and Engineering Technology Association, Vol. 25, No. 5, October, 2020, pp. 569–577.

[14]   Ring, M., Wunderlich, S., Grüdl, D., Landes, D., Hotho, A. (2017). Flow-based benchmark data sets for intrusion detection.

[15]   Kishwar Sadaf and Jabeen Sultana, "Intrusion Detection Based on Autoencoder and Isolation Forest in Fog Computing", IEEE Access, Volume 8, 2020, pp 167059–167068.

[16]   Zeliang Kan, Haoyu Wang, Guoai Xu, Yao Guo, and Xiangqun Chen, "Towards light-weight deep learning based malware detection", In 2018 IEEE 42nd Annual Computer Software and Applications Conference (COMPSAC). IEEE, Jul 2018.

[17]   Bojan Kolosnjaji, Apostolis Zarras, George Webster, and Claudia Eckert. Deep learning for classification of malware system call sequences. In AI 2016: Advances in Artificial Intelligence, pages 137–149. Springer International Publishing, 2016.

[18]   Arindam Sharma, Pasquale Malacaria, MHR Khouzani, "Malware Detection Using 1-Dimensional Convolutional Neural Networks", IEEE European Symposium on Security and Privacy Workshops, 2019.

[19]   Mohammed Harun Babu R, Vinayakumar R, Soman KP, "RNNSecureNet: Recurrent neural networks for Cyber security use-cases", arXivLabs, arXiv:1901.04281, Jan 2019.

[20]   T.T. Teoh; Graeme Chiew; Yeaz Jaddoo; H. Michael; A. Karunakaran; Y.J. Goh, "Applying RNN and J48 Deep Learning in Android Cyber Security Space for Threat Analysis", IEEE Xplore, International Conference on Smart Computing and Electronic Enterprise, 2018.

[21]   Chia-Mei Chen; Shi-Hao Wang; Dan-Wei Wen; Gu-Hsin Lai; Ming-Kung Sun, "Applying Convolutional Neural Network for Malware Detection", IEEE Xplore, International Conference on Awareness Science and Technology, 2019.

[22]   Mohammed Maithem, Dr.Ghadaa A. Al-sultany, "Network intrusion detection system using deep neural networks", Journal of Physics: Conference Series, 2020.

[23]   Rahul K. Vigneswaran; R. Vinayakumar; K.P. Soman; Prabaharan Poornachandran, "Evaluating Shallow and Deep Neural Networks for Network Intrusion Detection Systems in Cyber Security", IEEE Xplore, International Conference on Computing and Networking Technology, 2018.

[24]   Indira Kalyan Dutta, Bhaskar Ghosh, Albert H. Carlson, Magdy Bayoumi, "Generative Adversarial Networks in Security: A Survey", Intelligent Systems Modelling and Simulations Lab, 2020.

[25]   Tamer Aldwairi, Dilina Pereraa, Mark A. Novotny, "An evaluation of the performance of Restricted Boltzmann Machines as a model for anomaly network intrusion detection", Elsevier, Computer Networks, Volume 144, 24 October 2018, pp. 111–119.

[26] Ali Rahim Taleqani, Kendall E. Nygard, Raj Bridgelall, Jill Hough, "Machine Learning Approach to Cyber Security in Aviation", IEEE Xplore, International Conference on Electro/Information Technology, 2018.

[27] Yogita Goyal, Anand Sharma, "A Semantic Machine Learning Approach for Cyber Security Monitoring", IEEE Xplore, International Conference on Computing Methodologies and Communication, 2019.

[28] Nina Kharlamova, Seyedmostafa Hashemi, Chresten Traeholt, "Data-driven approaches for cyber defense of battery energy storage systems", Elsevier, Energy and AI, 2021.

[29] Sumit Soni; Bharat Bhushan, "Use of Machine Learning algorithms for designing efficient cyber security solutions", IEEE Xplore, International Conference on Intelligent Computing, Instrumentation and Control Technologies, 2019.

[30] Francisco Palenzuela, Melissa Shaffer, Matthew Ennis, Tarek M. Taha, "Multilayer perceptron algorithms for cyberattack detection", IEEE Xplore, IEEE National Aerospace and Electronics Conference (NAECON) and Ohio Innovation Summit, 2016.

[31] Flora Amato, Giovanni Cozzolino, Antonino Mazzeo, Emilio Vivenzio, "Using Multilayer Perceptron in Computer Security to Improve Intrusion Detection", International Conference on Intelligent Interactive Multimedia Systems and Services, May 2018.

[32] Wangyan Feng, Shuning Wu, Xiaodan Li, Kevin Kunkle, "A Deep Belief Network Based Machine Learning System for Risky Host Detection", arXiv.org, Computer Science, Cryptography and Security, 2017.

[33] ShymalaGowri Selvaganapathy, Mathappan Nivaashini & HemaPriya Natarajan, "Deep belief network based detection and categorization of malicious URLs", Information Security Journal: A Global Perspective, April 2018, DOI: 10.1080/19393555.2018.1456577

[34] Nagaraj Balakrishnan, Arunkumar Rajendran, Danilo Pelusi, Vijayakumar Ponnusamy, "Deep Belief Network enhanced intrusion detection system to prevent security breach in the Internet of Things", Elsevier, Internet of Things, Volume 14, June 2021.

[35] Ayesha Anzer; Mourad Elhadef, "A Multilayer Perceptron-Based Distributed Intrusion Detection System for Internet of Vehicles", IEEE Xplore, International Conference on Collaborative Computing: Networking, Applications and Worksharing, 2018.

[36] Rasika S. Vitalkar, Samrat S. Thorat, "A Review on Intrusion Detection System in Vehicular Ad- hoc Network Using Deep Learning Method", International Journal for Research in Applied Science & Engineering Technology, Volume 8 Issue V May 2020.

[37] Dilek Başkaya, Refi Samet, "DDoS Attacks Detection by Using Machine Learning Methods on Online Systems", International Conference on Computer Science and Engineering, 2020.

[38] Boumiza, Safa, Braham, Rafik, "An Anomaly Detector for CAN Bus Networks in Autonomous Cars based on Neural Networks", International Conference on Wireless and Mobile Computing, Networking and Communications, 2019.

[39] Bernardi, Mario Luca, Cimitile, Marta, Martinelli, Fabio, Mercaldo, Francesco, "Keystroke Analysis for User Identification Using Deep Neural Networks", International Joint Conference on Neural Networks, 2019.

[40] Danish Javeed, Tianhan Gao, Muhammad Taimoor Khan, Ijaz Ahmad, "A Hybrid Deep Learning-Driven SDN Enabled Mechanism for Secure Communication in Internet of Things (IoT)", MDPI, Sensors, https://doi.org/10.3390/s21144884, 2021.

[41] Nugraha, B., Nambiar, A., Bauschert, T., "Performance Evaluation of Botnet Detection using Deep Learning Techniques", International Conference on Network of the Future, 2020.

[42] Sravan Kumar Gunturia, Dipu Sarkar, "Ensemble machine learning models for the detection of energy theft", Elsevier, Electric Power Systems Research, 2021.

[43] Qiyi He, Xiaolin Meng, Rong Qu, Ruijie Xi, "Machine Learning-Based Detection for Cyber Security Attacks on Connected and Autonomous Vehicles", MDPI, Mathematics, 2020.

[44] Priyanka Dixit, Sanjay Silakari, "Deep Learning Algorithms for Cybersecurity Applications: A Technological and Status Review", Elsevier, Computer Science Review, 2021.

[45] Using the Power of Deep Learning for Cyber Security, Available: www.analyticsvidhya.com/blog/2018/07/using-power-deep-learning-cyber-security (July 5, 2018)

[46] N. Shone, T. N. Ngoc, V. D. Phai, Q. Shi, A deep learning approach to network intrusion detection, IEEE Transactions on Emerging Topics in Computational Intelligence 2 (1) (2018) 41–50.

[47] Mohammed Harun Babu R, Vinayakumar R, Soman KP, "A short review on Applications of Deep learning for Cyber security", arXiv, 2018.

[48] KDD Cup 1999 Data, Available: http://kdd.ics.uci.edu/databases/kddcup99/kddcup99.html

[49]   R. Vinayakumar, K. Soman, P. Poornachandran, Applying convolutional neural network for network intrusion detection, in: Advances in Computing, Communications and Informatics (ICACCI), 2017 International Conference on, IEEE, 2017, pp. 1222–1228.

[50]   Intrusion detection evaluation dataset (ISCXIDS2012), Available: www.unb.ca/cic/datasets/ids.html

[51]   1999 Darpa Intrusion Detection Evaluation Dataset, Available: www.ll.mit.edu/r-d/datasets/1999-darpa-intrusion-detection-evaluation-dataset

[52]   The UNSW-NB15 Dataset, Available: https://research.unsw.edu.au/projects/unsw-nb15-dataset

[53]   Microsoft Malware Classification Challenge (BIG 2015), Available: www.kaggle.com/c/malware-classification

[54]   Information Centre of Excellence for Tech Innovation, Available: www.iscx.ca/datasets/

[55]   The CTU-13 Dataset. A Labeled Dataset with Botnet, Normal and Background traffic. Available: www.stratosphereips.org/datasets-ctu13

[56]   HTTP DATASET CSIC 2010, Available: www.tic.itefi.csic.es/dataset

[57]   https://onlinelibrary.wiley.com/doi/full/10.1002/ett.4150

Performance of Machine Learning and Big Data Analytics Paradigms in Cyber Security

14

Gabriel Kabanda

Zimbabwe Academy of Sciences, University of Zimbabwe, Harare, Zimbabwe

Contents

DOI: 10.1201/9781003187158-17

14.1 INTRODUCTION

14.1.1 Background on Cyber Security and Machine Learning

Cyber security consolidates the confidentiality, integrity, and availability of computing resources, networks, software programs, and data into a coherent collection of policies, technologies, processes, and techniques in order to prevent the occurrence of attacks [1]. Cyber security refers to a combination of technologies, processes, and operations that are framed to protect information systems, computers, devices, programs, data, and networks from internal or external threats, harm, damage, attacks, or unauthorized access, such as ransomware or denial of service attacks [2]. The rapid advances in mobile computing, communications, and mass storage architectures have precipitated the new phenomena of big data and the Internet of Things (IoT). Outsider and insider threats can have serious ramifications on an institution, for example, failure to provide services, higher costs of operations, loss in revenue, and reputational damage [2, 3]. Therefore, an effective cyber security model must be able to mitigate cyber security events, such as unauthorized access, zero day attack, denial of service, data breach, malware attacks, social engineering (or phishing), fraud, and spam attacks, through intrusion detection and malware detection [4, 5].

The researcher identifies the primary variable that this study seeks to investigate as cyber security. Prior literature notes that the definition of cyber security differs among institutions and across nations [6]. However, in basic terms cyber security may be defined as a combination of technologies and processes that are set up to protect computer hosts, programs, networks, and data from external and external threats, attacks, harm, damage, and unauthorized access [2]. The major cyber security applications are intrusion detection and malware detection. The transformation and expansion of the cyber space has resulted in an exponential growth in the amount, quality, and diversity of data generated, stored, and processed by networks and hosts [7]. These changes have necessitated a radical shift in the technology and operations of cyber security to detect and eliminate cyber threats so that cyber security remains relevant and effective in mitigating costs arising from computers, networks, and data breaches [2].

A cyber crime is a criminal activity that is computer related or that uses the internet. Dealing with cyber crimes is a big challenge all over the world. Network intrusion detection systems (NIDS) are a category of computer software that monitors system behaviour with a view to ascertain anomalous violations of security policies and that distinguishes between malicious users and the legitimate network users [8]. The two taxonomies of NIDS are anomaly detectors and misuse network detectors. According to [9], the components in intrusion detection and prevention systems (IDPSs) can be sensors or agents, servers, and consoles for network management.

Artificial intelligence (AI) emerged as a research discipline at the Summer Research Project of Dartmouth College in July 1956. Genetic algorithms are an example of an AI technique, which imitates

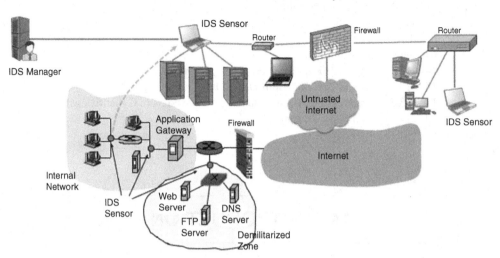

FIGURE 14.1 Typical Intrusion detection and prevention system.

the process of natural selection and is founded on the theory of evolutionary computation. Data over networks may be secured through the use of antivirus software, firewall, encryption, secure protocols, etc. However, hackers can always devise innovative ways of breaking into network systems. An intrusion detection and prevention system (IDPS), shown on Figure 14.1, is placed inside the network to detect possible network intrusions and, where possible, prevent cyber attacks. The key functions of the IDPSs are to monitor, detect, analyze, and respond to cyber threats.

The strength of the overall security in cyber security is determined by the weakest link [10]. Access controls and security mechanisms should be encapsulated in the company objectives. It is common practice to give employees access to only what they are entitled to, with the appropriate compartimentalized security and access privileges to mitigate against any possible security breaches [10]. Firewall protection has proved to be inadequate because of gross limitations against external threats [11]. However, operating systems hardening can supplement the firewall protection, and this involves installing and patching the operation system and then hardening and/or configuring the operating system to adequately provide acceptable security measures.

Machine learning is a field within artificial intelligence (AI) that is the science and engineering of making intelligent machines. Machine learning algorithms require empirical data as input and then learn from this input. The amount of data provided is often more important than the algorithm itself. Computers are instructed to learn through the process called machine learning (ML). Machine learning (ML) algorithms, inspired by the central nervous system, are referred to as artificial neural networks (ANNs). Deep learning (DL), as a special category of ML, brings us closer to AI. ML algorithms as part of artificial intelligence (AI) can be clustered into supervised, unsupervised, semisupervised, and reinforcement learning algorithms. The main characteristic of ML is the automatic data analysis of large data sets and the production of models for the general relationships found among data. The three classes of ML are as illustrated on Figure 14.2 [12]:

1. *Supervised learning*: Where the methods are given inputs labeled with corresponding outputs as training examples
2. *Unsupervised learning*: Where the methods are given unlabeled inputs
3. *Reinforcement learning*: where data is in the form of sequences of observations, actions, and rewards

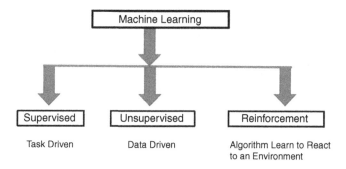

FIGURE 14.2   Three levels of machine learning.

Source: [13].

Machine learning uses intelligent software to enable machines to perform their jobs proficiently, which is the backbone of machine intelligence [14]. Most industries are embracing artificial intelligence and machine learning in their customer relations strategies to ensure that their client's experience is the best it can be. To succeed, one needs an investment strategy that adapts to deal with the new paradigm in markets.

Supervised learning models are grounded on generating functions that map big datasets (features) [15] into desired outputs [16]. Artificial intelligence is the simulation of human intelligence in machines through programming computers to think and act like human beings [17]. Big data analytics [18] has emerged as a discipline of ways to analyze, systematically extract/mine information from [76], or otherwise deal or work with enormous or complex datasets that are too large for traditional data-processing methodologies to deal with them [17]. Unsupervised learning is seen as a mathematical process of minimizing redundancy or categorizing huge datasets based on likeness [17]. Big data analytics is not only about the size of data but also hinges on volume, variety, and velocity of data [19]. It is important to note that machine learning is a technique of big data analytics that includes programming analytical model construction [20]. The output of a machine learning model often includes perceptions and/or decisions.

The greater risk of cyber attacks has stemmed mainly from the monotonic increase in the use of internet-connected systems [1]. The cyber attacks are becoming increasingly sophisticated with zero-day exploits and malware that evade security measures. Furthermore, commercial interest in cyber attacks has increased. The transformation and expansion of the cyber space have led to the generation, use, storage, and processing of big data, that is, large, diverse, complex, multidimensional, and usually multivariate datasets [21]. The characteristics of big data are volume, velocity, variety, veracity, vocabulary, and value [21]. *Volume* denotes big data as massive; *velocity* denotes the high speed of big data; *variety* denotes the diversity of big data; *veracity* denotes the degrees of trustworthiness in big data; *vocabulary* denotes conformity of big data to different schema, models, and ontologies; and *value* denotes the cost and worth of big data [21]. Big data has necessitated the development of big data mining tools and techniques widely referred to as big data analytics [22, 23]. Big data analytics [24] refer to a combination of well-known tools and techniques, for example machine learning, and data mining, that are capable of leveraging useful data usually hidden in big data [25] and creating an interface in the form of linear and visual analytics [21].

An artificial intelligence system uses computer technologies to perform tasks based on natural human problem-solving techniques. People's daily lives and work have been impacted by the rapid development of computing technology and the internet. As some of its advantages, AI can discover new and sophisticated changes in attack flexibility. AI can handle the volume of data, and this means AI can enhance network security by developing autonomous security systems to detect attacks and respond to breaches. AI can offer assistance by expanding the monitoring and detection of suspicious activities. Some of the relevant common applications of AI include the following:

- Self-driving cars
- Google Translate
- SIRI/intelligent personal assistants
- Facial recognition
- Data analysis
- Automated weaponry
- Game playing (Chess, Go, etc.)
- News generation
- Fraud Detection
- Oncology, treatment, and diagnosis

14.1.2 Background Perspectives to Big Data Analytics and Cyber Security

Big data analytics makes use of analytic techniques such as data mining, machine learning, artificial learning, statistics, and natural language processing. In an age of transformation and expansion in the Internet of Things (IoT), cloud computing services, and big data, cyber attacks have become enhanced and complicated [6], and therefore cyber security events have become difficult or impossible to detect using traditional detection systems [4, 5]. For example, malicious programs may be capable of continuously modifying their behavior so as to hide their malicious actions from traditional detection systems [4]. These challenges necessitate more robust cyber security processes with advanced monitoring capabilities and improved threat assessment such as big data analytics that are able to correlate, analyze, and mine large datasets for meaningful patterns [4].

According to [26], big data refers to the flood of digital data from many digital sources. The data types include images, geometries, texts, videos, sounds, and combinations of each. In [27], big data was explained as the increase in the volume of data that offers difficulty in storage, processing, and analysis through traditional database technologies. Big data is shortly described in three phrases according to [28] astoo big, too fast, and too hard. *Too big* means the quantum of data is very large. *Too fast* means the data coming through from different angles has to be processed quickly and results have to be obtained fast. *Too hard* means the data obtained might not be easily processed by existing resources. Organizations benefit a lot from big data as it enables them to gather, to store, to manage, and to manipulate these vast amounts of data at the right speed and right time in order to gain the right insights (value). Big data has also been defined according to the 5 Vs as stipulated by [28] where

- *volume* refers to the amount of data gathered and processed by the organization;
- *velocity* refers to the time required to process the data;
- *variety* refers to the type of data contained in big data;
- *value* refers to the key important features of the data, defined by the added value that the collected data can bring to the intended processes; and
- *veracity* means the degree to which the leaders trust the information to make a decision.

Big data came into existence when the traditional relational database systems were not able to handle the unstructured data generated by organizationd, social media [79], or any other data generating sourced [29]. It is easy to "predict" the inexorably increasing availability of big data due to ongoing technology evolution. Specifically, environmental sensor data and social media data will become increasingly available for disaster management due to the advances of many kinds of capable sensors [30]. Characterising big data reveals that it is one place where organizations can derive very useful data that can aid them in informed decision making. The challenge with developing countries is that, due to various reasons, they lag far behind when it comes to technological innovations that support the use of current inventions. There is a lack of infrastructure to support such innovations, a lack of skilled data scientists, and a lack of policies or legislation that promote such innovations.

Some researchers have added the fourth V, that is, veracity. to stress the importance of maintaining quality data within an organization. In such a situation, BDA is largely supported by Apaches Hadoop framework, which is an open-source, completely fault-tolerant, and highly scalable distributed computing paradigm. Compared to traditional approaches, security analytics provides a "richer" cyber security context by separating what is "normal" from what is "abnormal," i.e., separating the patterns generated by legitimate users from those generated by suspicious or malicious users. Passive data sources can include computer-based data, for example geographical IP location; computer security health certificates; and keyboard typing and clickstream patterns. Mobile-based data may include, for example, GPS location, network location, WAP data. Active (relating to real-time) data sources can include credential data, e.g., username and password, one-time passwords for, say, online access and digital certificates.

Big data analytics (BDA) can offer a variety of security dimensions in network traffic management, access patterns in web transactions [31], configuration of network servers, network data sources, and user credentials. These activities have brought a huge revolution in the domains of security management, identity and access management, fraud prevention and governance, risk and compliance. However, there is also a lack of in-depth technical knowledge regarding basic BDA concepts, Hadoop, predictive analytics, and cluster analysis, etc. With these limitations in mind, appropriate steps can be taken to build on the skills and competences of security analytics.

The information that is evaluated in big data analytics includes a mixer of unstructured and semistructured data, for instance, social media content [32, 33, 34], mobile phone records, web server logs, and internet click stream data. Also analysed includes text from survey responses, customer emails, and machine data captured by sensors connected to the Internet of Things (IoT).

14.1.3 Supervised Learning Algorithms

The supervised machine learning algorithm, which can be used for both classification or regression challenges, is called the support vector machine (SVM). The original training data can be transformed into a higher dimension where it becomes separable by using the SVM algorithm, which searches for the optimal linear separating hyperplane. Estimations of the relationships among variables depend mainly on the statistical process of regression analysis. The independent variables determine the estimation target. The regression function can be linear as in linear regression or a common sigmoid curve for the logistic function.

The easiest and simplest supervised machine learning algorithm which can solve both classification and regression problems is the K-nearest neighbors (KNN) algorithm. Both the KNN and SVM can be applied to finding the optimal handover solutions in heterogeneous networks constituted by diverse cells. Given a set of contextual input cues, machine learning algorithms have the capability to exploit the user context learned. The hidden Markov model (HMM) is a tool designed for representing probability distributions of sequences of observations. It can be considered a generalization of a mixture-based model rather than being independent of each other. The list of supervised learning algorithms is shown on Table 14.1.

Common examples of generative models that may be learned with the aid of Bayesian techniques include but are not limited to the Gaussians mixture model (GM), expectation maximization (EM), and hidden Markov models (HMM) [41, p. 445]. In Table 14.1, we summarize the basic characteristics and applications of supervised machine learning algorithms.

14.1.4 Statement of the Problem

The fact is that the most network-centric cyber attacks are carried out by intelligent agents such as computer worms and viruses; hence, combating them with intelligent semiautonomous agents that can

TABLE 14.1   Supervised Machine Learning Algorithms

CATEGORY	LEARNING TECHNIQUES	KEY CHARACTERISTICS	APPLICATION IN 5G
Supervised learning	Regression models	• Estimate the variables' relationships • Linear and logistics regression	Energy learning [8]
	K-nearest neighbor	• Majority vote of neighbors	Energy learning [8]
	Support vector machines	• Nonlinear mapping to high dimension • Separate hyperplane classification	MIMO channel learning [35]
	Bayesian learning	• A posteriori distribution calculation • GM, EM and HMM	• Massive MIMO learning [36] • Cognitive spectrum learning [37–39]

Source: [40].

detect, evaluate, and respond to cyber attacks has become a requirement [8]. Agility security involves the designing and devising of cyber defense strategies that seek to preserve the computer resources, networks, programs, and data from unauthorized access, change, and even destruction. With the rapid development of computing and digital technologies, the need to revamp cyber defense strategies has become a necessity for most organizations [42]. As a result, there is an imperative for security network administrators to be more flexible and adaptable and to provide robust cyber defense systems in the real-time detection of cyber threats. The key problem is to evaluate machine learning (ML) and big data analytics (BDA) paradigms for use in cyber security. According to the National Institute of Standards and Technology [43], the NIST Cybersecurity Framework [44] is based on standards such as the International Organization for Standardization (ISO), Control Objectives for Information and Related Technologies (COBIT), and International Electrotechnical Commission (IEC).

14.1.5 Purpose of Study

The research is purposed to evaluate machine learning and big data analytics paradigms for use in cyber security.

14.1.6 Research Objectives

The research objectives are to:

- evaluate machine learning and big data analytics paradigms for use in cyber security;
- develop a cyber security system that uses machine learning and big data analytics paradigms.

14.1.7 Research Questions

The main research question was:

Which machine learning and big data analytics paradigms are most effective in developing a cyber security system?

The subquestions are:

- how are the machine learning and big data analytics paradigms used in cyber security?
- how is a cyber security system developed using machine learning and big data analytics paradigms?

14.2 LITERATURE REVIEW

14.2.1 Overview

Computers, phones, internet, and all other information systems developed for the benefit of humanity are susceptible to criminal activity [10]. Cyber crimes consist of offenses such as computer intrusions, misuse of intellectual property rights, economic espionage, online extortion, international money laundering, non-delivery of goods or services, etc. [11]. Today, "cyber crimes" has become known as a very common phrase, yet its definition has not been easy to compose as there are many variants to its classification. In their varying and diverging definitions, the majority of the authors agree on two of the most important aspects: that it's usually an unauthorized performance of an act on a computer network by a third party and that, usually, that party seeks to bring damage or loss to the organization.

On a daily basis, the amount of data processed and stored on computers as well as other digital technologies is increasing and human dependence on these computer networks is also increasing. Intrusion detection and prevention systems (IDPS) include all protective actions or identification of possible incidents and analysing log information of such incidents [9]. In [11], the use of various security control measures in an organization was recommended. Various attack descriptions from the outcome of the research by [37] are shown on Table 14.2.

However, there is a clear distinction between cyber security and information security, as illustrated in Figure 14.3.

The monotonic increase in an assortment of cyber threats and malwares amply demonstrates the inadequacy of the current countermeasures to defend computer networks and resources. These issues demand the development of efficient solutions to cyber security. To alleviate the problems of the classic techniques of cyber security, research in artificial intelligence and more specifically in machine learning is sought after [1]. Machine learning teaches the distinction between good and malicious files. Deep learning is a submodule of machine learning and is often referred to as deep neural networks (DNNs) [41]. Deep learning doesn't need any domain expert knowledge assistance to understand the importance of new input.

TABLE 14.2 Various Attack Descriptions

ATTACK TYPE	DESCRIPTION
DoS	Denial of service: an attempt to make a network resource unavailable to its intended users; temporarily interrupt services of a host connected to the Internet
Scan	A process that sends client requests to a range of server port addresses on a host to find an active port
Local access	The attacker has an account on the system in question and can use that account to attempt unauthorized tasks
User to root	Attackers access a user account on the system and are able to exploit some vulnerability to gain root access to the system
Data	Attackers involve someone performing an action that they may be able to do on a given computer system but that they are not allowed to do according to policy

Source: [37].

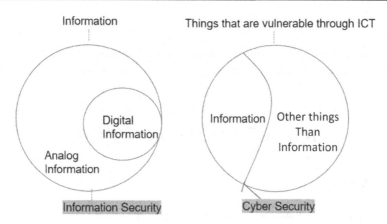

FIGURE 14.3   Differences between information security and cyber security.
Source: [45].

To enhance the malware and cyber attack detection rate, one can apply deep learning architectures to cyber security [35]. Experiments with classical machine learning algorithms are run on CPU-enabled machines, and deep learning architectures are run on GPU-enabled machines.

14.2.2 Classical Machine Learning (CML)

Machine learning is a field in artificial intelligence where computers learn like people. We present and briefly discuss the most commonly used classical machine learning algorithms [46].

14.2.2.1 Logistic Regression (LR)

As an idea obtained from statistics and created by [47], logistic regression is like linear regression, yet it averts the misclassification that may occur in linear regression. Unlike linear regression, logistic regression results are basically either 0 or 1. The efficacy of logistic regression is mostly dependent on the size of the training data.

14.2.2.2 Naïve Bayes (NB)

Naïve Bayes (NB) classifier is premised on the Bayes theorem, which assumes the independence of features. The independence assumptions in naïve Bayes classifier overcome the curse of dimensionality.

14.2.2.3 Decision Tree (DT)

A decision tree has a structure like flow charts, where the root node is the top node and a feature of the information is denoted by each internal node. The algorithm might be biased and may end up unstable since a little change in the information changes the structure of the tree.

14.2.2.4 K-nearest Neighbor (KNN)

K-nearest neighbor (KNN) is a nonparametric approach that uses similarity measure in terms of distance function classifiers other than news cases. KNN stores the entire training data, requires larger memory, and so is computationally expensive.

14.2.2.5 AdaBoost (AB)

The AdaBoost (AB) learning algorithm is a technique used to boost the performance of simple learning algorithms used for classification. AdaBoost constructs a strong classifier using a combination of several weak classifiers. It is a fast classifier and at the same time can also be used as a feature learner. This may be useful in tasks that use imbalanced data analysis.

14.2.2.6 Random Forest (RF)

Random forest (RF), as an ensemble tool, is a decision tree derived from a subset of observations and variables. The random forest gives better predictions than an individual decision tree. It uses the concept of bagging to create several minimal correlated decision trees.

14.2.2.7 Support Vector Machine (SVM)

Support vector machine (SVM) belongs to the family of supervised machine learning techniques that can be used to solve classification and regression problems. SVM is a linear classifier, and the classifier is a hyper plane. It separates the training set with maximal margin. The points near the separating hype plane are called support vectors, and they determine the position of the hyper plane.

14.2.3 Modern Machine Learning

Deep learning is modern machine learning with the capability to take raw inputs and learn the optimal feature representation implicitly. This has performed well in various long-standing artificial intelligence tasks [41]. The most commonly used deep learning architectures are discussed next in detail.

14.2.3.1 Deep Neural Network (DNN)

An artificial neural network (ANN) is a computational model influenced by the characteristics of biological neural networks. The family of ANN includes the feed-forward neural network (FFN), convolutional neural network, and recurrent neural network (RNN). FFN forms a directed graph in which a graph is composed of neurons group as a mathematical unit. Each neuron in the ith layer has a connection to all the neurons in $i + 1th$ layer.

Each neuron of the hidden layer denotes a parameter h that is computed by

$$h_i(x) = f(w_iT\,x + b_i), \tag{14.2.1}$$

$$h_{ii}: R_{d_i-1} \rightarrow R_{d_i}, \tag{14.2.2}$$

$$f: R \rightarrow R, \tag{14.2.3}$$

where $w_i \in R_{d\times d\ i-1}$, $b_i \in R_{d_i}$, d_i denotes the size of the input, f is a nonlinear activation function $ReLU$.

The traditional examples of machine learning algorithms include linear regression, logistic regression, linear discriminant analysis, classification and regression trees, naïve Bayes, K-nearest neighbor (KNN, K means clustering), learning vector quantization (LVQ), support vector machines (SVM), random forest, Monte Carlo, neural networks and Q-learning.

- *Supervised Adaptation*: Adaptation is carried out in the execution of system at every iteration.
- *Unsupervised Adaptation:* This follows the trial-and-error method. Based on the obtained fitness value, the computational model is generalized to achieve better results in an iterative approach.

14.2.3.2 Future of AI in the Fight against Cyber Crimes

This section briefly presents related work and some ANN as computational mechanisms that simulate the structural and functional aspects of neural networks existing in biological nervous systems. Experiments have shown that NeuroNet is effective against low-rate TCP-targeted distributed DoS attacks. In [48], presented the intrusion detection system using neural network based modeling (IDS-NNM) was presented, which proved to be capable of detecting all intrusion attempts in the network communication without giving any false alerts [14].

The characteristics of NIC algorithms are partitioned into two segments: swarm intelligence and evolutionary algorithm. The swarm intelligence-based algorithms (SIA) are developed based on the idea of the collective behaviors of insects in colonies, e.g., ants, bees, wasps, and termites. Intrusion detection and prevention systems (IDPS) include all protective actions or identification of possible incidents and analyzing log information of such incidents [9].

14.2.4 Big Data Analytics and Cyber Security

14.2.4.1 Big Data Analytics Issues

Big data analytics requires new data architectures, analytical methods, and tools. Big data environments ought to be magnetic, which accommodates all heterogeneous sources of data. Instead of using mechanical disk drives, it is possible to store the primary database in silicon-based main memory, which improves performance. The HDFS storage function in Hadoop provides a redundant and reliable distributed file system, which is optimized for large files, where a single file is split into blocks and distributed across cluster nodes. According to [49], there are four critical requirements for big data processing, the first two of which are fast data loading and fast query processing. The Map function in Hadoop accordingly partitions large computational tasks into smaller tasks and assigns them to the appropriate key/value pairs.

Nowadays, people don't want to just collect data; they want to understand the meaning and importance of the data and use it to aid them in making decisions. Data analytics are used to extract previously unknown, useful, valid, and hidden patterns and information from large data sets, as well as detect important relationships among the stored variables. Forecasts in big data analytics can increasingly benefit the manufacturing, retail, and transport and logistics industries.

14.2.4.2 Independent Variable: Big Data Analytics

The researcher identifies big data analytics (BDA) as the main independent variable influencing the effectiveness of cyber security in an age of big data, growth in cloud computing services [50], and the Internet of Things (IoT) [51]. Big data analytics comes in handy in providing the data mining tools and techniques that are used to extract useful data, usually hidden in large datasets, in a timely and efficient way, for the purposes of informing the formulation and implementation of effective cyber security strategies [4, 52]. The application of big data analytics to cyber security encompasses behavioral analytics, forensics analytics, forecast analytics, and threat intelligence analytics. Behavioral analytics provides information about the behavioral patterns of cyber security events or malicious data [2]. Forensics analytics locates, recovers, and preserves reliable forensic artifacts from specifically identified cyber security events or attacks [53]. Forecast analytics attempts to predict cyber security events using forecast analytics models and methodologies [52]. Threat intelligence helps to gather threats from big data, analyze and filter information about these threats, and create an awareness of cyber security threats [2].

14.2.4.3 Intermediating Variables

The researcher identified cyber framework as the other variable worthy of investigation in this study. The effectiveness of big data analytics in mitigating cyber security problems is moderated by the architecture, that is, the network architecture, computer architecture, and big data architecture used by the organization.

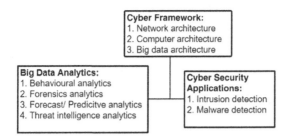

FIGURE 14.4 Conceptual framework for cyber security and big data analytics.

14.2.4.4 Conceptual Framework

In developing a conceptual framework, the researcher brings the dependent and independent variables to interaction, as shown in Figure 14.4.

14.2.4.5 Theoretical Framework

14.2.4.5.1 Game Theory

Prior research has attempted to use game theory in which two major players, the classifier (data miner) and an adversary (initiator), engage in a cost-sensitive game [4]. In game theory, the adversary seeks to maximize his return through a set of sequential actions, while the classifier or data miner seeks to minimize the cost from misclassifying data. However, where the adversary is able to continuously transform its performance, the initial successes of the classifier (data miner) may not necessarily guarantee subsequent successes against the adversary [4]. Game theory is important in explaining why weak big data analytics models will not succeed in detecting and eliminating malicious cyber threats that continuously modify their behaviors in such a way as to hide their actions from detection by the cyber security framework [4]. Game theory in the context of insider threats brings further complications in that the adversary has intimate knowledge about the organization or institution for example employees, partners, and suppliers who are associated with the organization or institution [3].

14.2.4.5.2 Situation Awareness Theory

The situation awareness theory postulated by [54] postulates that the success of a cyber security domain depends on its ability to obtain real-time, accurate, and complete information about cyber security events or incidents [14]. The situation awareness model consists of situation awareness, decisions, and action performance, as shown in Figure 14.5.

Situational awareness involves the perception, comprehension, and projection of environmental elements prior to decision making [14]. Decision making involves the making of informed decisions about the capabilities of the cyber security system and cyber security mechanisms, among other things, which lead to action performance [54]. The situational awareness theoretical models allow cyber security experts to ask and answer questions about what is happening and why, what will happen next, and what the cyber security expert can do about it [14].

14.2.4.6 Big Data Analytics Application to Cyber Security

There is consensus in prior literature that cyber security has evolved to become a problem for big data analytics. This is due to the understanding that the transformation and expansion of the cyber space [4] has rendered traditional intrusion detection and malware detection systems obsolete. Further, even the data mining models that have been used in the past are no longer sufficient for the challenges in cyber security [4]. For this reason, there has been a proliferation of new big data analytics models for cyber security [4].

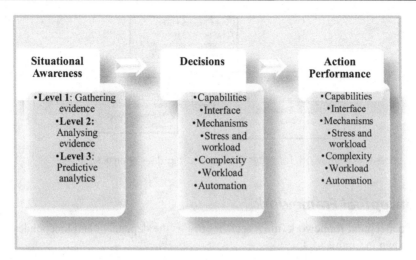

FIGURE 14.5   Simplified theoretical model based on [54] situation awareness.
Source: Adapted from [14].

Prior research provides evidence of failures in prior data analytics models in cyber security especially in the face of malicious adversaries that are constantly modifying their behavior in order to evade detection [4]. In other words, a big data analytics model for cyber security can be evaluated on the basis of its agility and robustness [4].

Data analytics models for cyber security differ among institutions; for example, the data analytics model for cyber security in private health care institutions will not necessarily be the same as the model used in financial institutions or those used in the public sector, although they may share similar frameworks.

14.2.4.7 Big Data Analytics and Cyber Security Limitations

The authors of [55] indicated that Big Data has to do with large-volume, complex, growing data sets coming from multiple and autonomous sources. As [56] underlined, the main characteristics of big data are volume, velocity and variety [57] cited in [58]. In [59], big data was defined as data that has volume, variety, velocity, value. According to [60], big data is defined not only by the amount of the information that it delivers but also by its complexity and by the speed at which it is analyzed and delivered. With reference to [61], big data can be defined as multifaceted data that combines the following characteristics: veracity, variety, volume, velocity, and value.

Cyber security ensures security issues, especially data security, and privacy protection issues remain the primary concern for its adoption. Cloud computing is the process of sharing resources through the internet [62]. Cloud computing finds applications in many industries. The primary benefits of cloud computing making it necessary for modern businesses are security, cost reduction, accessibility, and greater efficiency.

There have been several attacks on critical infrastructures in different countries, and it is therefore necessary for critical systems to have periodic cyber security vulnerability assessments so that attacks that could be used against them can be mitigated beforehand. Cyber attacks can range from the theft of personal information to extortion attempts for individual targets, according to McAfee security. The broad scope of what cyber security covers for both individuals and corporate organizations are

- critical infrastructure security,
- network security,
- cloud security,
- application/system security,

- user security, and
- Internet of Things security.

14.2.4.8 Limitations

According to [63], there are three main three major limitations of big data.

14.2.4.8.1 Outlier Effect

The first such limitation on big data is that outliers are common. Technology is said to be not at a point yet where the methods for gathering data are completely precise. As an example, Google Flu Trends famously highlighted the limitations of big data a few years ago. Thus the program was created to provide real-time monitoring of global flu cases by taking data from search queries related to the flu.

14.2.4.8.2 Security Limitations

The results of a data leak can include, for starters, lawsuits, fines, and a total loss of credibility. Security concerns can greatly inhibit what companies are able to do with their own data. For instance, having data analyzed by a third-party group or agency can be difficult because the data may be hidden behind a firewall or private cloud server. This can create some big hiccups if a company is trying to share and transfer data to be analyzed and acted on consistently.

14.2.4.8.3 Misinterpretation of Data

Data can reveal the actions of users. However, it cannot tell you why users thought or behaved in the ways they did. A company may not have the resources, capabilities, or staff to sift through data reports to analyze actionable insights for both short-term and long-term business strategies.

14.2.4.8.4 Other Cloud Computing Limitations

Everything at the cloud depends on internet availability. If the cloud server faces some issues, so will the application. Plus, if a company use an internet service that fluctuates a lot, cloud computing is not for it. Sometimes, a vendor locks customers in by using proprietary rights so that they cannot switch to another vendor. For example, there is a chance that a vendor does not provide compatibility with Google Docs or Google Sheets. As customers or employees of the customer advance in their expertise, the business may have a crucial need for these features. So the contract with the provider must meet the organization's terms, not the vendor's.

So if one owns a business where a single miss or corruption of data is not at all acceptable, that owner should think 100 times before going for the cloud, especially for a large-scale business. But for small business owners, the question here is whether you will be able to provide more security levels to the applications than a cloud service provider does. Because it may happen that, to be compatible with cloud, the business applications need to be rewritten. Moreover, if the business demands huge data transfers, every month company will face huge bills as well. If the business faces heavy traffic every day and heavier on weekends, then a quick fix is always a top priority.

14.2.5 Advances in Cloud Computing

14.2.5.1 Explaining Cloud Computing and How It Has Evolved to Date

14.2.5.2 Definition

Cloud computing is about using the internet to access someone else's software running on someone else's hardware in someone else's data center [64]. Cloud computing is essentially virtualized distributed processing, storage, and software resources and a service, where the focus is on delivering computing as an

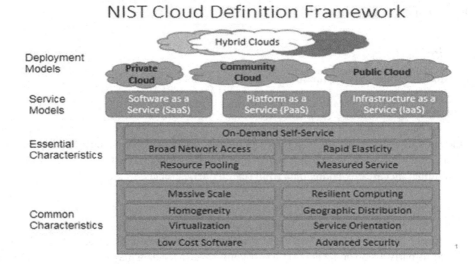

FIGURE 14.6   NIST visual model of cloud computing definition.
Source: [66].

on-demand, pay-as-you-go service. In [1], similar definitions were provided that put more emphasis on the service being highly scalable and subscription based, the pay-per-use aspect and delivery channel being the Internet. In [26],cloud computing's role in driving IT adoption in developing economies was highlighted. According to [65], a *cloud* is a type of parallel and distributed system consisting of a collection of interconnected and virtualized computers that are dynamically provisioned.

The characteristics of a cloud include the following:

- Pay-per-use
- Elastic capacity
- Illusion of infinite resources
- Self-service interface
- Resources that are abstracted or virtualized
- Provision of development tools and API to build scalable applications on their services

The architecture of the NIST Visual Model of Cloud Computing is shown in Figure 14.6. The NIST Cloud Computing framework states that cloud computing is made up of five essential characteristics, three service models, and four deployment models [66].

14.2.6 Cloud Characteristics

The five essential cloud characteristics are:

1. On-demand self-service
2. Broad network access
3. Resource pooling, i.e., location independence
4. Rapid elasticity
5. Measured service

The essential characteristics of cloud computing are briefly explained as follows:

1. *On-demand self-service*: A consumer can unilaterally and automatically provision computing capabilities such as server time and network storage as needed, without requiring human interaction with a service provider. This means one can get computing capabilities as needed automatically.
2. *Broad network access*: Heterogeneous client platforms available over the network come with numerous capabilities that enable the provision of network access.
3. *Resource pooling*: Computing resources are pooled together in a multitenant model depending on the consumer demand. Resource pooling means
   * location independence,
   * provider resources pooled to server multiple clients.
4. *Rapid elasticity*:
   * This is when capabilities are rapidly and elastically provisioned to quickly scale out and rapidly released to quickly scale in.
   * Unlimited capabilities can be purchased by the consumer in any quantity at any time as these are available for provisioning.
   * Rapid elasticity means the ability to quickly scale service in/out.
5. *Measured service*:
   * A metering capability can be be automatically controlled and optimized in cloud systems at some level of abstraction appropriate to the type of service.
   * Transparency for both the provider and consumer of the service can be demonstrated by resource usage that is monitored, controlled, and reported.
   * Measured service is about control and optimizing services based on metering.

14.2.7   Cloud Computing Service Models

Service delivery in cloud computing comprises three cloud service models: software as a service (SaaS), platform as a service (PaaS), and infrastructure as a service (IaaS). These three models are shown on Figure 14.7 and are discussed here.

14.2.7.1 *Software as a Service (SaaS)*

The provider's applications running on a cloud infrastructure provide a capability for the consumer for use. It utilizes the internet to deliver applications to the consumers (e.g., Google Apps, Salesforce, Dropbox, Sage X3, and Office 365) [67]. Applications range widely from social to enterprise applications such as email hosting, enterprise resource planning, and supply chain management. The underlying cloud infrastructure from the cloud provider is configured to manage network, servers, operating systems, storage, or even individual application capabilities. The consumer only handles minimal user specific application configuration settings. This is the most widely used service model [66, 68].

SaaS provides off-the-shelf applications offered over the internet. Examples are

* Google Docs,
* Aviary,
* Pixlr,
* Microsoft Office Web App.

14.2.7.2 *Platform as a Service (PaaS)*

PaaS pertains to the consumer infrastructure for third-party applications. Just as in SaaS, the consumer does not manage or control the underlying cloud infrastructure, including the network, servers, operating systems, or storage, but does have control over the deployed applications and possibly configuration

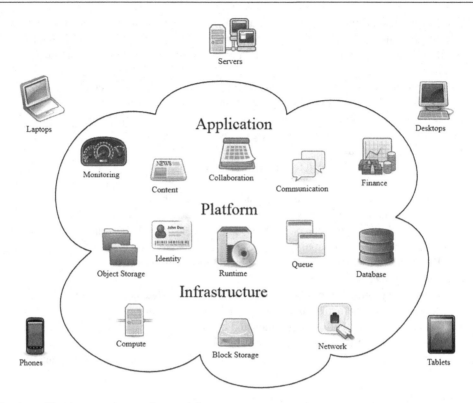

FIGURE 14.7   Cloud computing service models.

settings for the application-hosting environment [66, 68]. Examples include Windows Azure, Apache Stratos, Google App Engine, CloudFoundry, Heroku, AWS (Beanstalk), and OpenShift [67, 62]. PaaS provides faster and more frequent delivery of functionality for the sake of direct support for business agility.

PaaS provides an enabling environment for a consumer to run applications. PaaS is a cloud computing service that offers a computing platform and solution stack for users, and this may include

- language,
- operating system (OS),
- database,
- middleware,
- other applications.

Features to look for in Paas include the following:

1. Programming languages (Python, Java, .Net Languages, Ruby)
2. Programming frameworks (Ruby on Rails, Spring, Java EE, .Net)
3. Persistence options

A PaaS Cloud should be able to support various programming models for different types of programming. For example,

- programming large datasets in clusters of computers (MapReduce) [69, 70],
- development of request-based web services and applications,

- orchestration of a business process in the form of workflows (WorkFlow Model),
- high-performance distributed execution of tasks.

The commonly used PaaS examples include Google Apps Engine, Windows Azure Platform, and Force.com.

The pros for PaaS are

- rapid deployment,
- low cost,
- private or public deployment.

The cons for PaaS are

- not much freedom,
- limited choices of tools,
- vendor lock-in.

14.2.7.3 Infrastructure as a Service (IaaS)

IaaS provisions processing, networks, storage, and other essential computing resources on which the consumer is then able to install and run arbitrary software, which can include operating systems (virtual machines [VM], appliances, etc.) and applications [66, 68]. The customer in essence can run and control its own virtual infrastructure without the overheads of cost and maintenance from running its own hardware. However, this comes at a cost. Common global examples are Amazon Web Services (AWS), Cisco Metapod, Microsoft Azure, and Rackspace, and the local ones include TelOne cloud services and Dandemutande [67].

IaaS is a cloud service that allows existing applications to run on its hardware.

- It rents out resources dynamically wherever they are needed.
- Services include:
  - compute servers,
  - data storage,
  - firewall,
  - load balancer.

The distinguished features of IaaS are as follows:

1. Geographical presence
   a. Responsiveness
   b. Availability
2. User interfaces and access to servers
   a. Providing means of accessing their cloud
      i. GUI
      ii. CLI
      iii. Web services
3. Advance reservation of capacity
   a. Time-frame reservations
4. Automatic scaling and load balancing
   a. Elasticity of the service
   b. One of the most desirable features of an IaaS cloud
   c. Traffic distribution

5. Service-level agreement
    a. As with all services, parties must sign an agreement.
    b. Metrics
        • Uptime, performance measures
    c. Penalties
        • Amazon
    d. Hypervisor and operating system choice
        • Xen
        • VMWare, vCloud, Citric Cloud Center

14.2.8 Cloud Deployment Models

The three commonly used cloud deployment models are private, public, and hybrid. An additional model is the community cloud. However, this is less commonly used. In a cloud context, the term *deployment* basically refers to where the software is made available – in other words, where it is running.

14.2.8.1 Private Cloud

The private cloud is normally either owned or exclusively used by a single organization. The services and infrastructure are permanently kept on a private network, and the hardware and software are dedicated solely to the particular organization. The service provider or the particular organization may manage the physical infrastructure. The major advantage of this model is improved security as resources are not shared with others, thereby allowing for higher levels of control and security [71].

14.2.8.2 Public Cloud

The cloud infrastructure is provisioned for use by the general public, i.e., any organization or individual can subscribe. In [26], it was further clarified that cloud services are provided on a subscription basis to the public, typically ,on a pay-per-use model. The advantages include lower costs, near-unlimited scalability, and high reliability [71]. Examples include Amazon (EC2), IBM's Blue Cloud, Sun Cloud, Google App Engine, and Windows Azure [72]. The public cloud is sold to the public as a megascale infrastructure and is available to the general public.

14.2.8.3 Hybrid Cloud

A hybrid cloud model is a mix of two or more cloud deployment models such as private, public, or hybrid [73, 74]. This model requires determining the best split between the public and private cloud components. This model is commonly used by organizations starting the cloud adoption journey. The advantages include control over sensitive data (private cloud), flexibility (i.e., ability to scale to the public cloud whenever needed), and ease of transitioning to the cloud through gradual migration [71]. The use of standardized or proprietary technology allows for data and application portability [75].

14.2.8.4 Community Cloud

This model is provisioned for exclusive use by a particular community of consumers bound by shared interests (e.g., policy and compliance considerations, mission and security requirements). Technically speaking, a community cloud is a multitenant platform that is provisioned only for a particular subset of customers. A community cloud shares computing resources among several organizations and can be managed by either organizational IT resources or third-party providers [66]. A typical example is the US-based exclusive IBM SoftLayer cloud, which is dedicated for use by federal agencies only. This approach builds confidence in the platform, which cloud consumers use to process their sensitive workloads [72].

The community cloud infrastructure is

- shared for a specific community,
- intended for several organizations that have shared concerns,
- managed by an organization or a third party.

14.2.8.5 Advantages and Disadvantages of Cloud Computing

14.2.8.5.1 Cloud Computing in General

According to [76], the cloud symbol is typically used to symbolize the internet. Cloud computing services are delivered when they are needed in the quantity needed at a certain time. Cloud computing is analogous to the rent-a-car model. Cloud computing has many benefits for the organizations, which include cost savings (low capital outlay), scalability, anytime-anywhere access, use of latest software versions, energy savings, and quick rollout of business solutions [77]. However, security, privacy, regulatory, and technological readiness concerns have had an impact on the pace of adoption.

The cost-effectiveness and efficiency of the cloud platforms is tempting most organizations to migrate to the cloud and enjoy a wide range of general benefits [78], which according to [79], include

- free capital expenditure,
- accessibility from anywhere at anytime,
- no maintenance headaches,
- improved control over documents as files are centrally managed,
- dynamic scalability,
- device independence,
- instantaneity,
- cost-efficiency,
    - task-centrism,
    - private server cost.

Cloud computing has the potential to significantly bolster economic growth through the provision of cost-effective advanced IT solutions and improved efficiencies. Also, economies of scale can substantially reduce the IT costs of the organization. The use of shared computing facilities improves efficiencies in infrastructural usage as organizations effectively only pay for what they use and effectively lowers their operational costs [80]. Cloud computing enables rapid provisioning of IT infrastructure and access to applications, thereby reducing the deployment period. The shared hosting environment reduces the initial capital outlay, which is often an impediment to technology adoption in most organizations [81, 82]. The major pros of cloud computing include the following:

- Lower-cost computers for end users
- Improved performance on users' PCs
- Lower IT infrastructure and software costs
- Fewer maintenance issues
- Instance software updates
- Unlimited storage capacity
- Increased data safety
- Easier group collaboration
- Universal access to data/documents

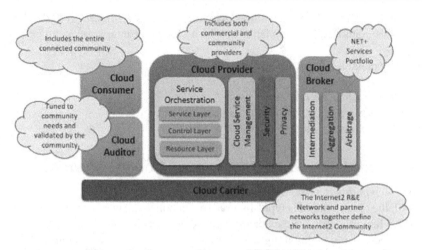

Cloud Computing: NIST Framework

FIGURE 14.8   NIST cloud computing definition framework.

The NIST Cloud Computing Definition Framework is shown in Figure 14.8.

The business and technical benefits of cloud computing are summarised here. The business benefits of cloud computing include

- almost zero upfront infrastructure investment,
- just-in-time infrastructure,
- more efficient resource utilization,
- usage-based costing,
- reduced time to market.

The technical benefits of cloud computing are

- automation – "scriptable infrastructure,"
- auto-scaling,
- proactive scaling,
- more efficient development life cycle,
- improved testability,
- disaster recovery and business continuity.

However, the major issues of concern and cons for cloud computing are that

- it requires a constant internet connection;
- it doesn't work well with low-speed connections;
- it can be slower than using desktop software;
- its features might be more limited;
- stored data might not be secure;
- ff the cloud loses your data, that's a big problem.

Other key concerns have to do with

- privacy,
- security,

- availability,
- legal issues,
- compliance,
- performance.

Who benefits from cloud computing (CC)? CC should not be used by the internet-impaired, offline workers, the security conscious, and anyone married to existing applications, e.g., Microsoft Office. The most commonly used CC services (SaaS) include

- calendars, schedules, and task management,
- event and contact management,
- email,
- project management,
- word processing, spreadsheets, and presentations,
- databases,
- storing and sharing files,
- sharing digital photographs,
- sharing songs and videos.

For most businesses today when they plan to modernize their computing and networking architectures, cloud-native architectures have become the principal target environments [83]. Cloud computing is now a vital part of corporate life, bringing a significant opportunity to accelerate business growth through efficient utilization of IT resources and providing new opportunities for collaboration.

Several benefits are associated with cloud adoption, and these are highlighted in this section. In [84] as cited in [85], the attributes of cloud computing include:

- *Scalability and on-demand services*: Cloud computing offers resources and services for users on demand. The scalability allows for services to be expanded or shrunk with ease. Cloud computing efficiently scales up and down, and this enables it to meet sudden spikes in user demand and create temporary digital environments as required. This has the potential to reduce operating costs [86].
- *Guaranteed quality of service (QoS)*: Cloud computing guarantees high QoS (hardware/CPU performance, bandwidth, memory capacity, and uptime) for users.
- *Computing power*: Massive computing power comes at a marginal cost [87].
- *Autonomous system*: There is seamless management of the cloud platform such that most platform changes happening at the back end are hardly noticeable to the end user.
- *Pricing*: In most instances, cloud computing requires little or zero capital expenditure as it uses the pay-as-you-go pricing model [86].
- *Easy access*: Most cloud-based services are easily accessible from any modern end user device such as desktop, laptop, tablet, and mobile phone [86]. In [73], this was put simply as the ability to work from anywhere.
- *Potential to reduce business risks*: Through outsourcing the service infrastructure to the cloud, the company also shifts the business risks to the service provider, who most of the time is better equipped for managing these risks [87].
- *Improve IT agility*: Cloud computing has the potential to improve organizations' IT agility as it allows them to reduce the time between identifying the need for a new IT resource and delivering it [87]. Cloud computing also enables the organization to concentrate internal resources on strategic IT and business priorities [87].
- *Better backup and disaster recovery services*: Cloud computing results in improved backup and disaster recovery services since data may be stored in multiple locations at the same time, and this minimizes the risk of the data being lost [88]. Also, cloud service providers have built-in robust modern disaster recovery systems.

14.2.8.5.2 Benefits of Cloud Computing in Business
Cloud computing supports interactive and user-friendly web applications. Different people have their own perspectives. Cloud computing may be considered a virtualized set of computer resources, dynamic development, and software deployment. Cloud computing leverages competitive advantage and provides improved IT capabilities. The benefits of CC in business are discussed next.

14.2.8.5.2.1 Flexibility Business flexibility is increased tremendously by the adoption of cloud computing, which provides flexibility to the working arrangements of employees within or outside their workplace. Employees who are on business trip can access the data as long as they have an internet connection through any kind of device. Every employee can get the updated version of the platforms and services.

14.2.8.5.2.2 Cost Reduction The main reason that organizations are adopting cloud computing in their businesses is that it helps in cost reduction. Cloud service providers provide "in-house" provision of these services with a significant cost reduction of up to 45.5%. Only the services actually used and accessed are paid for by the organization. Cloud computing resources can easily be installed and deployed, especially SaaS. The technical support provided by the providers reduces the burden on IT staffs, thereby reducing staff training costs and manpower requirements.

14.2.8.5.2.3 Automatic Software/Hardware Upgrades When an IT solution has been implemented in organizations, maintenance costs for hardware and software are a huge burden. Upon adoption of cloud computing, these problems are solved as the organization can now shift its capital expenses to operating expenses. This also helps to participate in technological advances, build a better relationship, provide standardized and low-cost services to customers, and increase profit. Because there are no capital investments for immediate access, it results in a faster time to market.

14.2.8.5.2.4 Agility In response to the customer's fast changing needs, organizations need the capability to stay competitive. Cloud computing is available 24/7 due to the availability of the internet around the clock. This enables organizations to deliver the services in the shortest possible time, and thus it can be used as a competitive tool for rapid development.

14.2.8.5.3 Summary
The top six benefits of cloud computing are as follows:

1. *Achieve economies of scale*: This results in an increase of volume output or productivity with fewer resources (computing and human).
2. *Reduce capital expenditure*: The move from CapEx to OpEx reduces capital expenditure (CapExto) on the pay-as-you-go operational expenditure (OpEx) model, based on demand/utility computing. This helps reduce capital expenditure (CapEx) on hardware and software licenses.
3. *Improve access*: Through omnichannel access, information can be accessed anytime, from any-where, and from any device.
4. *Implement agile development at low cost*: This is about the design, development, and rollout of new solutions and services using agile methodologies on cloud-based shared development operations.
5. *Leverage the global workforce*: One can roll out cloud computing services 24/7 through various data centers worldwide to ensure that services are available close to the end users.
6. *Gain access to advanced capabilities*: The latest advances in software (such as AI, blockchain, data mining) are available off the shelf as cloud services, enabling an organization to gain the benefits of these capabilities with minimal investment.

Presentation services in CC is about collaborating on presentations in CC concerning web-based or phone-based presentation applications. Common examples of presentation services are AuthorStream.com, SlideBoom.com, and SlideShare.net. The advantages of the presentation services in CC include

- the ability for users to cowork from multiple locations,
- no need to carry around presentation files,
- cost-effective – free or nominal fee!
- the ability to upload existing presentation files for sharing,
- support for formats like PPT, PDF, or ODP,
- prevention of editing existing files.

However, the cons include the fact that

- network access is critical,
- one doesn't always have the same range of features,
- existing presentations may have compatibility issues.

Similarly, database services are available in CC, and the common examples are

- Dabbledb.com [89],
- Teamdesk.net,
- Trackvia.com,
- Baseportal.com,
- Springbase.com,
- Viravis.com,
- Infodome.com,
- Creator.zoho.com,
- Quickbase.intuit.com.

14.2.8.6 Six Main Characteristics of Cloud Computing and How They Are Leveraged

Highlighted next are the six essential characteristics of cloud computing derived from the NIST framework.

1. On-demand Self-service

On-demand services means that users have access to cloud resources at anytime. With on-demand self-service, the consumer is enabled to provision computing capabilities on the cloud without necessarily requiring human interaction with the cloud provider [66]. Typically, once an organization has been provisioned for cloud services, minimal human interaction with the cloud service provider is necessary as most of the cloud solutions are designed to be self-service. With on-demand self-service, a consumer can access computing capabilities as needed automatically.

On-demand self services involves computer services such as email, applications, and network or server service, which are provided without requiring human interaction with each service provider. The most popular cloud service providers include Amazon Web Services (AWS), Microsoft, Google, IBM, and Salesforce.com. Gartner describes this characteristic as service based. On-demand self-service simply requires two things:

i. Availability of services 24/7
ii. Ability to modify the service received by the client organization without contacting the hosting provider

In summary, on-demand self-service is likely to be the least adopted of the characteristics in an enterprise-class hosted application.

2.   Broad Network Access

With broad network access, the computing resources on the web can be accessed from any standard device from anywhere and at anytime. Broad network access provides the ability to access cloud services through multiple devices, i.e., heterogeneous thin or thick client platforms (e.g., workstations, laptops, mobile phones, and tablets) [66]. In fact, these days this also speaks to access using any Internet-capable device. In broad network access, users can have access to cloud computing from anywhere and at any time using less disk device.

Broad network access is when any network-based appliance can access the hosted application from devices that can include but are not limited to

- laptops,
- desktops,
- smartphones,
- tablet devices.

Broad network access is more a function of the application being hosted.

3.   Resource Pooling

In shared resource pooling, users can share cloud resources. In a multitenant model, the cloud provider's pooled resources are used to service multiple consumers. Virtual and physical resources are dynamically assigned to satisfy consumer demand. The resources may be geographically spaced; hence in some instances physical location of the data may become an issue. The pooled resources include storage, memory, processing, and network bandwidth [66, 68]. Resource pooling means

- location independence,
- pooled provider resources serving multiple clients.

With regard to resource pooling, security needs are likely to be the bottleneck.

4.   Rapid Elasticity

Rapid elasticity is the provision of cloud services rapidly and elastically to quickly scale out and rapidly released to quickly scale in. Cloud services to the consumer can be rapidly scaled up or down according to demand, often in an automated process. To the consumer, the resources available often appear to be limitless and can be appropriated in any measure at any time [66, 68]. This is essential in project-intense areas such as mining where team sizes are dynamic and also ensures optimal licensing and may result in significant cost savings.

With rapid elasticity, computing resources may be accessed elastically by customers. Rapid elasticity means the ability to quickly scale into/out of service. In rapid elasticity, it is very easy to scale the resources up or down at anytime. Rapid elasticity is one of the key characteristics if your application experiences spikes in usage.

5.   Measured Service

Measured service is when cloud computing resources usage is controlled by the cloud service provider and can be measured and monitored. Cloud computing services use a metering capability that enables one to control and optimize resource use.

With measured service, cloud computing systems are used as adaptive systems. They automatically balance loads and optimize the use of resources. Transparency in bills can be achieved by permitting the customer to monitor and control resource usage. Cloud service providers make use of metering capability for automatic control and optimization of resources such as storage, processing, peak usage, bandwidth, and active user counts. Metering enables monitoring, controlling, and reporting, thereby providing transparency to both the cloud provider and the consumer [66, 68]. Measured service is about control and optimizing services based on metering. Metered usage goes hand in hand with elasticity.

6. Multitenacity

Multitenancy is the need for policy-driven enforcement, segmentation, isolation, governance, service levels, and chargeback/billing models for different consumer constituencies.

In conclusion, the characteristics of cloud computing are leveraged through the following:

- Massive scale
- Homogeneity
- Virtualization
- Resilient computing
- Low-cost software
- Geographic distribution
- Service orientation
- Advanced security technologies

14.2.8.7 Some Advantages of Network Function Virtualization

Network function virtualization (NFV) is a new paradigm for designing and operating telecommunication networks. Traditionally, these networks rely on dedicated hardware-based network equipment and their functions to provide communication services. However, this reliance is becoming increasingly inflexible and inefficient, especially in dealing with traffic bursts, for example, during large crowd events. NFV strives to overcome current limitations by (1) implementing network functions in software and (2) deploying them in a virtualized environment. The resulting virtualized network functions (VNFs) require a virtual infrastructure that is flexible, scalable, and fault tolerant.

The growing maturity of container-based virtualization and the introduction of production-grade container platforms promotes containers as a candidate for the implementation of NFV infrastructure (NFVI). Containers offer a simplified method of packaging and deploying applications and services.

14.2.8.7.1 Virtualization
Virtualization is basically making a virtual image, or "version," of something usable on multiple machines at the same time. This is a way of managing workload by transforming traditional computing to make it more scalable, efficient, and economical. Virtualization can be applied to hardware-level virtualization, operating system virtualization, and server virtualization. In virtualization, the costs of hardware are reduced and energy is saved when the resources and services are separated from the underlying physical delivery environment. A primary driver for virtualization is consolidation of servers in order to improve the efficiency and potential cost savings.

Virtualization entails

- underutilization of resources,
- division of resources,
- maintenance required, e.g., controlling job flow.

The benefits of virtualization include the lower costs and extended life of the technology, which has made it a popular option with small to medium-sized businesses. The physical infrastructure owned by the service provider is shared among many users using virtualization in order to increase the resource utilization. Virtualization facilitates efficient resource utilization and increased return on investment (ROI), which results in low capital expenditures (CapEx) and operational expenditures (OpEx).

With virtualization, one can attain better utilization rates of the resources of the service providers, increase ROI for both the service providers and the consumers, and promote green IT by reducing energy wastage. Virtualization technology has a drawback as follows:A single point of failure in the virtualization software could affect the performance of the entire system. Virtualization in general has tremendous advantages, as follows:

- Where the physical hardware is unavailable, the operating systems can still be run.
- Creating new machines, backup machines, etc. is easier.
- "Clean" installs of operating systems and software can be performed for software testing.
- More machines can be emulated than are physically available.
- Lightly loaded systems can timeshare one host.
- Debugging problems is easier (suspend and resume the problem machine).
- Migration of virtual machines is facilitated.
- Run legacy systems!

14.2.8.7.2 Why Virtualization?

With the help of virtualization, we can increase the use of resources available in order to gain more benefits. We should virtualize for the following reasons:

- Isolation among users
- Resource sharing
- Dynamic resources
- Aggregation of resources

The standard approach in virtualization is to implement on virtual machines (VMs) to benefit from

- abstraction of a physical host machine,
- allowing a hypervisor to manage and emulatethe instructions of VMs, VMWare, Xen, etc.,
- providing infrastructure API,i.e. plug-ins, to hardware/support structures.

Virtual machines make use of VM technology, allowing multiple virtual machines to run on a single physical machine. The virtual infrastructure managers (VIMs) is the key technology that entails

- the operating system of the cloud,
- the allocating resources in a cloud,
- aggregating resources from multiple sources,
- acting as the infrastructure sharing software, cloud operating system, and virtual infrastructure engines.

The key features of a virtual infrastructure (VI) manager are as follows:

- Virtualization support
  - Backbone
  - CPU, memory, storage
  - Sizing and resizing

- Self-service, on-demand resource provisioning
  - Directly obtaining services from the cloud
    - Creation of servers
    - Tailoring software
    - Configurations
    - Security policies
  - Elimination of having to go through a system administrator
- Multiple back-end hypervisors
  - Drawbacks of virtualization models
  - Uniform management of virtualization
- Storage virtualization
  - Abstracting logical storage from physical storage
  - Creation of an independent virtual disk
  - Storage area networks (SAN)
    - Fiber channel, iSCSI, NFS
- Interface to public clouds
  - Overloading requiring borrowing
  - VIMs obtaining resources from external sources using VIMs during spikes
- Dynamic resource allocation
  - Having to allocate and deallocate resources as needed
  - Difficulty in calculating demand prediction
  - Moving loads around to reduce overheating
  - Monitoring resource utilization and reallocating accordingly
- Virtual clusters
  - Holistically managing interconnected groups of virtual machines
- High availability and data recovery
  - Need for little downtime

14.2.8.7.3 Multicore Technology

In multicore technology, two or more CPUs work together on the same chip as a single integrated circuit (IC). These single ICs are called a *die*. Multicore technology can be used to speed up the processing in a multitenant cloud environment. Multicore architecture has become the recent trend of high-performance processors, and various theoretical and case study results illustrate that multicore architecture is scalable with the number of cores.

Most software vendors raise the compliant that their application is not supported in a virtual state or will not be supported if the end user decides to virtualize them. To accommodate the needs of the industry and operating environment and to create a more efficient infrastructure – the virtualization process has been modified as a powerful platform, such that process virtualization largely revolves around one piece of very important software. This is called a *hypervisor*. Thus a VM must host an OS kernel.

14.2.8.8 Virtualization and Containerization Compared and Contrasted

Virtualization allows the running of multiple operating systems on a single physical system while sharing the underlying hardware resources. Virtualization entails abstraction and encapsulation. Clouds rely heavily on virtualization, whereas grids do not rely as much on virtualization as clouds do. In virtualization, a hypervisor is a piece of computer software that creates and runs virtual machines.

Instead of installing the operating system as well as all the necessary software in a virtual machine, the docker images can be easily built with a Dockerfile since the hardware resources, such as CPU and memory, are returned to the operating system immediately. Therefore, many new applications are programmed into containers. Cgroups allow system administrators to allocate resources such as CPU, memory, network, or any combination of them, to the run containers, as illustrated in Figure 14.9.

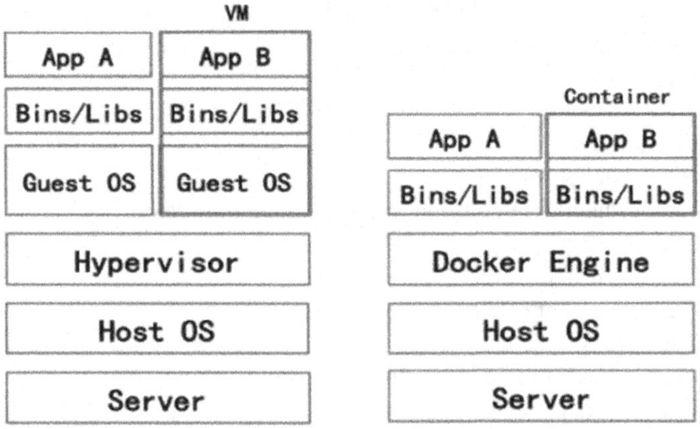

FIGURE 14.9   Architecture comparison of virtual machine vs. container.

Containers implement operating-system-level virtualization; i.e., each host runs multiple isolated Linux system containers, which are more lightweight than VMs. Portable, interoperable, distributed applications that are deployed in heterogeneous cloud environments require lightweight packaged containers. Container-based technologies provide poor isolation performance for the components such as memory, disk, and network (except the CPU). Application management is intended to guarantee that workloads are delivering an appropriate level of performance to end users.

14.2.8.8.1 Virtualization
Virtualization is the optimum way to efficiently enhance resource utilization. It refers to the act of creating a virtual (i.e., similar to actual) variations of the system. Physical hardware is managed with the help of software and converted into the logical resource that is in a shared pool or that can be used by the privileged user. This service is known as infrastructure-as-a-service. Virtualization is base of any public and private cloud development. Most public cloud providers such as Amazon EC2, Google Compute Engine, and Microsoft Azure leverage virtualization technologies to power their public cloud infrastructure [90]. The core component of virtualization is the hypervisor.

14.2.8.8.2 B. Hypervisor
A hypervisor is software that provides isolation for virtual machines running on top of physical hosts. This thin layer of software that typically provides the capabilities to virtual partitioning and that runs directly on hardwareprovides a potential for virtual partitioning and is responsible for running multiple kernels on top of the physical host. This feature makes the application and process isolation very expensive, so there will be a big impact if computer resources can be used more efficiently. The most popular hypervisors today are VMware, KVM, Xen, and HyperV.

As the generic term in the Linux universe, the container offers more opportunity for confusion. Basically, a container is nothing more than a virtual file system that is isolated, with some Linux kernel features such as namespaces and process groups, from the main physical system. A container framework offers an environment as close to the desired one as we can ask of a VM but without the overhead that comes with running on another kernel and simulating all the hardware. Due to lightweight nature of containers, more of them can run per host than virtual machines can. Unlike containers, virtual machines require emulation layers (either software or hardware), which consume more resources and add additional overhead.

Containers are different from virtualization in a number of aspects.

1. *Simple*: Easy sharing of a hardware resources clean command line interface, simple REST API
2. *Fast*: Rapid provisioning, instant guest boot, and no virtualization overhead as fast as bare metal

3. *Secure*: Secure by default, combining all available kernel security feature with AppArmor, user namespaces, SECCOMP

4. *Scalable*: Capability to broadcast the quality-of-service from a single container on a developer laptop to a container host in a data center, including remote image services with extensible storage and networking

5. *Control groups (cgroups):-* A kernel-provided mechanism for administration, grouping, and tracking through a virtual file system

   • *Namespaces*: A namespace (NS) that produces a confined instance of the global resource, basically used for the implementation of containers; a lightweight virtualization that gives the illusion that the processes are the only ones running in the system

   • *AppArmor*: A mandatory access control (MAC) mechanism, a security system that keeps applications from turning interrupted and that uses an LSM kernel for improving the programs restricted to certain resources

Openstack has increased significantly in private cloud areas. Modern organizations consider it a feasible solution for private clouds, because the Openstack environment provides effective resource management with high scalability. A high number of different virtual machines are currently on the market; however, due to the huge growth in popularity, Openstack has been chosen for this particular study.

Docker containers share the operating system and important resources, such as depending libraries, drivers, or binaries, with its host, and therefore they occupy less physical resources.

14.3 RESEARCH METHODOLOGY

14.3.1 Presentation of the Methodology

The pragmatism paradigm was used in this research, and this is intricately related to the mixed methods research (MMR).

14.3.1.1 Research Approach and Philosophy

14.3.1.1.1 Research Approach
The researcher adopts a qualitative approach in the form of a focus group discussion on research. The analysis is done to establish differences in data analytics models for cyber security without the necessity of quantifying the analysis [91].

14.3.1.1.2 Research Philosophy
The researcher adopts a postmodern philosophy to guide the research. First, the researcher notes that the definition, scope, and measurement of cyber security differs among countries and across nations [6]. Further, the postmodern view is consistent with descriptive research designs that seek to interpret situations or models in their particular contexts [92].

14.3.1.2 Research Design and Methods

14.3.1.2.1 Research Design
The researcher adopts a descriptive research design since the intention is to systematically describe the facts and characteristics of big data analytics models for cyber security. The purpose of the study is essentially an in-depth description of the models [91].

14.3.1.2.2 Research Methods

A *case study* research method was adopted in this study. In this respect, each data analytics model for cyber security is taken as a separate case to be investigated in its own separate context [92]. Prior research has tended to use case studies in relation to the study of cyber security [6]. However, the researcher develops a control case that accounts for an ideal data analytics model for cyber security for comparative purposes.

14.3.2 Population and Sampling

14.3.2.1 Population

The research population for the purpose of this study consists of all data analytics models for cyber security that have been proposed and developed in literature, journals, conference proceedings, and working papers. This is consistent with previous research that involves a systematic review of literature [3].

14.3.2.2 Sample

The researcher identified two data analytics models or frameworks from a review of literature and the sample size of 8. Eight participants in total were interviewed. However, while this may be limited data, it is sufficient for the present needs of this study. Research in the future may review more journals to identify more data analytics models that can be applied to cyber security.

14.3.3 Sources and Types of Data

The researcher uses secondary data in order to investigate the application of data analytics models in cyber security.

14.3.4 Model for Analysis

In analyzing the different data analytics models for cyber security, the researcher makes reference to the characteristics of an ideal data analytics model for cyber security. In constructing an ideal model, the researcher integrates various literature sources. The basic framework for a big data analytics model for cyber security consists of three major components:big data, analytics, and insights [4]. However, a fourth component may be identified – prediction (or predictive analytics) [3]). This is depicted in Figure 14.10.

14.3.4.1 Big Data

The first component in the big data analytics framework for cyber security is the availability of big data about cyber security. Traditional sources of big data are systems logs and vulnerability scans [4]. However, sources of big data about cyber security have extended to include computer-based data, mobile-based data, physical data of users, human resources data, credentials, one-time passwords, digital certificates, biometrics, and social media data [21]. Some authors identify the sources of big data about cyber security as business mail, access control systems, CRM systems and human resources systems, a number of pullers in linked data networks, intranet/internet and IIoT/IoT, collectors and aggregators in social media networks, and external news tapes [52]. Big data about cyber security should be imported from multiple sources to ensure effectiveness in detection and prediction of possible threats [5]. Further, some authors specify the characteristics of security data as consisting of a heterogeneous format, with diverse semantics and correlations across data sources, and classify them into categories, for example, nonsemantic data, semantic data, and security knowledge data [5].

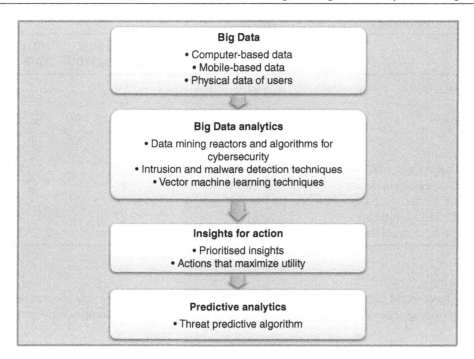

Big Data
- Computer-based data
- Mobile-based data
- Physical data of users

Big Data analytics
- Data mining reactors and algorithms for cybersecurity
- Intrusion and malware detection techniques
- Vector machine learning techniques

Insights for action
- Prioritised insights
- Actions that maximize utility

Predictive analytics
- Threat predictive algorithm

FIGURE 14.10   Big data analytics model for cyber security.

14.3.4.2 Big Data Analytics

To address the concerns of cyber security of big data, more robust big data analytics models [93] for cyber security have been developed in data mining techniques [94] and machine learning [4]. Big data analytics [95] employs data mining reactors [96] and algorithms [97], intrusion and malware detection techniques, and vector machine learning techniques for cyber security [4, 42]. However, it has been observed that adversarial programs have tended to modify their behavior by adapting to the reactors and algorithms designed to detect them [4]. Further, intrusion detection systems are faced with challenges such as unbounded patterns, data nonstationarity, uneven time lags, individuality, high false alarm rates, and collusion attacks [3]. This necessitates a multilayered and multidimensional approach to big data analytics for cyber security [2, 5]. In other words, an effective big data analytics model for cyber security must be able to detect intrusions and malware at every layer in the cyber security framework.

14.3.4.3 Insights for Action

Big data analytics for cyber security should be able provide prioritized and actionable insights to cyber security personnel, for example, setting up effective network defenders that are able to detect flaws in the network and be able to trace the source of threats or attacks [4]. Alternatively, cyber security personnel may update existing network defenders in light of new prioritized insights about the cyber security system [4]. The goal of analysts should be to maximize utility derived from the cyber security system.

14.3.4.4 Predictive Analytics

Predictive analytics refers to the application of a big data analytics model for cyber security to derive, from current cybersecurity data, the likelihood of a cyber security event occurring in future [3]. In essence, a data analytics model for cyber security should be able to integrate these components if it is to be effective

TABLE 14.3 Comparative Detection Accuracy Rate (%)

CLASSIFIER	DETECTION ACCURACY (%)	TIME TAKEN TO BUILD THE MODEL (SECONDS)	FALSE ALARM RATE (%)
Decision trees (J48)	81.05	**	**
Naïve Bayes	76.56	**	**
Random forest	80.67	**	**
SVM	69.52	**	**
AdaBoost	90.31	**	3.38
Mutlinomal naïve Bayes + N2B	38.89	0.72	27.8
Multinomal naïve Bayes updateable + N2B	38.94	1.2	27.9
Discriminative multinomial Bayes + PCA	94.84	118.36	4.4
Discriminative multinomial Bayes + RP	81.47	2.27	12.85
Discriminative multinomial Bayes + N2B	96.5	1.11	3.0

in its major functions of gathering big data about cyber security, in analyzing big data about cyber security threats, in providing actionable insights, and in predicting likely future cyber security incidents.

14.3.5 Validity and Reliability

The researcher solicited comments from peers on the emerging findings and also feedback to clarify the biases and assumptions of the researcher to ensure the internal validity of the study [92]. The researcher also needs to demonstrate reliability or consistency in research findings by explaining in detail the assumptions and theories underlying the study [92].

14.3.6 Summary of Research Methodology

In Chapter 3, the researcher developed appropriate methodology for investigating the ideal data analytics models for cyber security.

14.3.7 Possible Outcomes

The expected accuracy rate for the research should be according to Table 14.3, which shows the international benchmark.

14.4 ANALYSIS AND RESEARCH OUTCOMES

14.4.1 Overview

Figure 14.11 shows the landscape for intrusion detection. Service provision by each specific piece of equipment with a known IP address determines the network traffic behavior. To avoid denial of service attacks, the knowledge representation model can be established using sensibility analysis [8]. Figure 14.12

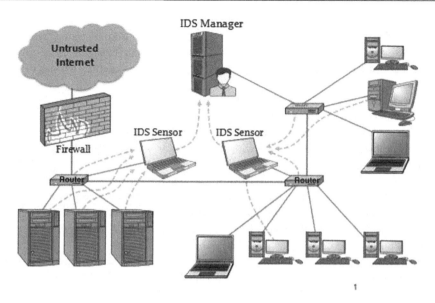

FIGURE 14.11   Landscape for intrusion detection.

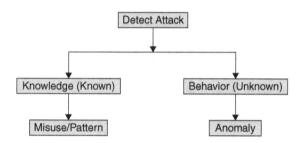

FIGURE 14.12   Analysis of attack.

Source: [38].

details the simple rules for the analysis of attack. The occurrence of an unusual behavior on the network triggers an alarm on the IDS in anomaly-based intrusion detection.

In [99], the use of machine learning (ML), neural network, and fuzzy logic to detect attacks on private networks on the different artificial intelligence (AI) techniques was highlighted. It is not technically feasible to develop a perfect sophisticated intrusion detection system, since the majority of IDSs are signature based.

The IDS is divided into either a host IDS (HIDS) or as a network IDS (NIDS). Analysis of the network traffic can be handled by a NIDS that distinguishes unlicensed, illegitimate, and anomalous behavior on the network. Packets traversing through the network should generally be captured by the IDS using network taps or span port in order to detect and flag any suspicious activity [99]. Anomalous behavior or malicious activity on the specific device can be effectively detected by a device-specific IDS.

The ever increasing Internet of Things (IoT) creates a huge challenge to achieving absolute cyber security. The IDS is characterised by network observation, analysis, and identification of possible behavior anomalies or unauthorized access to the network, with some protecting the computer network in the event of an intrusion. However, existing methods have several limitations and problems of the existing methods [100].

Provision of a reliable way of protecting the network system or of trusting an existing IDS is a greater challenge. In cases of specific weaknesses and limitations, administrators would be required to regularly update the protection mechanisms, which further challenges the detection system. The vulnerability of networks and susceptibility to cyber attacks is exacerbated by the use of wireless technology [100].

The gross inadequacies of classical security measures have been overtly exposed. Therefore, effective solutions for a dynamic and adaptive network defence mechanism should be determined. Neural networks can provide better solutions for the representative sets of training data [100]. [100] argues for the use of ML classification problems that are solvable with supervised or semisupervised learning models for the majority of the IDS. However, the one major limitation of the work done by [100] is on the informational structure in cyber security for the analysis of the strategies and the solutions of the players.

Autonomous robotic vehicles attract cyber attacks, which prevent them from accomplishing the intrusion prevention mission. Knowledge-based and vehicle-specific methods have limitations in detection, which is applicable to only specific known attacks [41]. The attack vectors of the attack scenarios used by [41] is shown on Figure 14.13.

In this experiment, the system is allowed to undertake several missions by the robotic vehicle, which diverts the robotic vehicle test bed. The practical experimental setup for the attack vectors used is shown on Figure 14.14.

Table 14.4 shows a comparison of the data mining techniques that can be used in intrusion detection.

Intrusion attack classification requires optimization and enhancement of the efficiency of data mining techniques. The pros and cons of each algorithm using the NSL-KDD dataset are shown on Table 14.5.

An intrusion detection system determines whether an intrusion has occurred and so monitors computer systems and networks, and the IDS raises an alert when necessary [102]. However, [102] addressed the problems of anomaly based signature (ABS), which reduces false positives by allowing a user to interact with the detection engine and by raising classified alerts. The advantages and disadvantages of ABSs and SBSs are summarized in Table 14.6.

An IDS must keep track of all the data, networking components, and devices involved. Additional requirements must be met when developing a cloud-based intrusion detection system due to its complexity and integrated services.

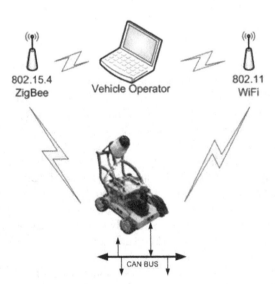

FIGURE 14.13 Attack vectors of the attack scenarios.

Source: [41].

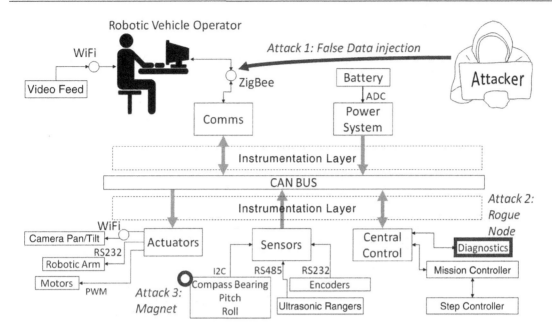

FIGURE 14.14   Attack vectors robotics experimental setup.
Source: [41].

TABLE 14.4   Advantages and Disadvantages of Data Mining Techniques

TECHNIQUE	ADVANTAGES	DISADVANTAGES
Genetic algorithm	• Finding a solution for any optimization problem • Handling multiple solution spaces	• Complexity to propose a problem space • Complexity to select the optimal parameters • Need to have local searching technique for effective functioning
Artificial neural network	• Adapts its structure during training without the need to program it.	• Not accurate results with test data as with training data
Naive Bayes classifier	• Very simple structure • Easy to update	• Not effective when there is high dependency between features
Decision tree	• Easy to understand • Easy to implement	• Works effectively only with attributes having discrete values • Very sensitive to training sets, irrelevant features, and noise.
K-mean	• Very easy to understand. • Very simple to implement in solving clustering problems.	• Number of clusters is not automatically calculated. • High dependency on initial centroids

Source: [101].

TABLE 14.5   Performance of Support Vector Machines, Artificial Neural Network, K-nearest Neighbor, Naïve Bayes, and Decision Tree Algorithms

PARAMETER	SVM	ANN	KNN	NB	DT
Correctly classifies instances	24519	24123	25051	22570	25081
Incorrectly classified instances	673	1069	141	2622	111
Kappa statistic	0.9462	0.9136	0.9888	0.7906	0.9911
Mean absolute error	0.0267	0.0545	0.0056	0.1034	0.0064
Root mean squared error	0.1634	0.197	0.0748	0.3152	0.0651
Relative absolute error	5.3676%	11.107%	1.1333%	20.781%	1.2854%

TABLE 14.6   Advantages and Disadvantages of ABSs and SBSs Models

DETECTION MODEL	ADVANTAGES	DISADVANTAGES
Signature-based	Low false positive rate	Cannot detect new attacks
	Does not require training	Requires continuous updates
	Classified alerts	Tuning could be a thorny task
Anomaly-based	Can detect new attacks	Prone to raise false positives
	Self-learning	Black-box approach
		Unclassified alerts
		Requires initial training

Source: [102].

14.4.2 Support Vector Machine

A support vector machine is a classification of the artificial intelligence and machine learning algorithm with a set containing points of two types in an X-dimensional place. A support vector machine generates a $(X - 1)$ dimensional hyperplane for separating these points into two or more groups using either linear kernel or nonlinear kernel functions [17]. Kernel functions provide a method for polynomial, radial, and multilayer perception classifiers such as classification of bank performance into four clusters of *strong*, *satisfactory*, *moderate*, and *poor performance*. The class of bank performance is defined by the function

$$\text{Performance class} = f\left(\vec{x}.\vec{w}\right) = f\left(\sum\nolimits_{j} x_j w_j\right)$$

where $\vec{x}$ is the input vector to the support vector classifier, $\vec{w}$ is the real vector of weights, and f is the function that translates the dot product of the input and real vector of weights into desired classes of bank performance. $\vec{w}$ is learned from the labeled training data set.

14.4.3 KNN Algorithm

The KNN algorithm is a nonparametric supervised machine learning technique that endeavors to classify a data point from given categories with the support of the training dataset [17]. Table 14.10 shows its performance. Predictions are performed for a new object (y) by searching through the whole training dataset for the K most similar instances or neighbors. The algorithm does this by calculating the Euclidean distance as follows:

$$d(x,y) = \sqrt{\sum_{i=1}^{m}\left(x_i - y_i\right)^2},$$

where $d(x, y)$ is the distance measure for finding the similarity between new observations and training cases and then finding the K-closest instance to the new instance. Variables are standardized before calculating the distance since they are measured in different units. Standardization is performed by the following function:

$$X_s = \frac{X - \text{Mean}}{s.d.},$$

where X_s is the standardized value, X is the instance measure, and mean and s.d. are the mean and standard deviation of instances. Lower values of K are sensitive to outliers, and higher values are more resilient to outliers and more voters are considered to decide the prediction.

14.4.4 Multilinear Discriminant Analysis (LDA)

The linear discriminant analysis is a dimensionality reduction technique. Dimensionality reduction is the technique of reducing the number of random variables under consideration through finding a set of principal variables [17], also known as course of dimensionality. The LDA calculates the separability between n classes also known as between-class variance. Let D_b be the distance between n classes.

$$D_b = \sum_{i=1}^{g} N_i\left(\overline{x}_i - \overline{x}\right)\left(\overline{x}_i - \overline{x}\right)^T.$$

where $\overline{x}$ the overall is mean, and $\overline{x}_i$ and N_i are the sample mean and sizes of the respective classes. The within-class variance is then calculated, which is the distance between the mean and the sample of every class.et S_y. be the within-class variance.

$$S_y = \sum_{t=1}^{g}\left(N_i - 1\right)S_i = \sum_{t=1}^{g}\sum_{j}^{N_i}\left(X_{i,j} - \overline{X}_i\right)\left(X_{i,j} - \overline{X}_i\right)^2.$$

The final procedure is to then construct the lower dimensional space for maximization of the seperability between classes and the minimization of within-class variance. Let P be the lower dimensional space.

$$P = \arg_p.\max\frac{\left|P^T D_b P\right|}{\left|P^T S_y P\right|}.$$

The LDA estimates the probability that a new instance belongs to every class. Bayes theorem is used to estimate the probabilities. For instance, if the output of the class is a, and the input is b. Then

$$P\left(Y = x \mid B = b\right) = \left(P \mid a * fa(b)\right) / \sum\left(P \mid a * f \mid (b)\right)$$

where $P|a$ is the prior probability of each class as obseed ithe training dataset, $f(b)$ is the estimated probability of b belonging to the class, and $f(b)$ uses the Gaussian distribution function to determine whether b belongs to that particular class.

14.4.5 Random Forest Classifier

The random forest classifier is an ensemble algorithm used for both classification and regression problems. It creates a set of decision trees from a randomly selected subset of the training set [17]. It then makes a decision by aggregating the votes from individual decision trees to decide the final class of an instance in the classification problem [103]. The tree with the higher error rates is given a low weight in comparison to the other trees, increasing the impact of trees with low error rates.

14.4.6 Variable Importance

Variable importance was implemented using the Boruta algorithm to improve model efficiency. The Boruta algorithm endeavors to intern all the key, interesting features existing in the dataset with respect to an outcome variable. Figure 14.15 shows that net profit is the most significant feature, followed by ROA, total assets, ROE, and other variables.

The next procedure was fitting these variables into our algorithms and hence evaluating their performance using the metrics discussed in the models section. The Boruta algorithm also clusters banks on important variables, as shown in Figure 14.16, for effective risk management and analysis.

14.4.7 Model Results

Before we discuss the results of our models, it is imperative to discuss the distribution of our dataset. We classify bank performance into four classes: strongly, satisfactorily, moderately, and poor performing banks. A strongly performing bank is the one with incredible CAMELS indicators. Its profitability indicators are high, and the management quality is top of the class and less sensitive to market movements with a high quality asset base. A satisfactory bank is the one with acceptable but not outstanding performance.

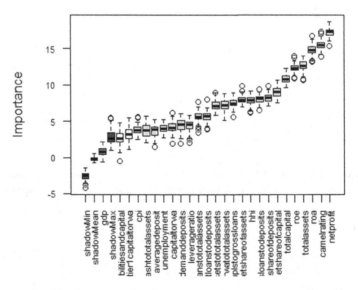

FIGURE 14.15 Boruta algorithm important features.

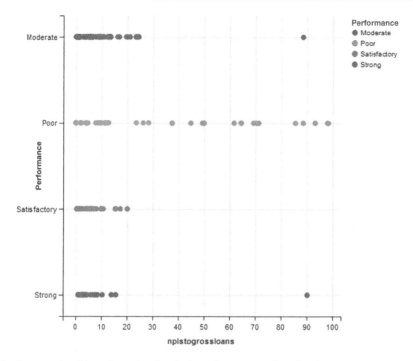

FIGURE 14.16   Boruta algorithm clustering banks based on nonperforming loans.

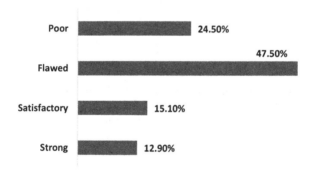

FIGURE 14.17   Distribution of the big dataset.

The CAMELS indicators are quite fine for such banks. Moderate performance is the one characterized by fundamental weakness or imperfections. A poorly performing bank is the one whose performance is below standard expectations or defective and hence can be categorized as an already failed bank. Our dataset is comprised of thousands of records from banking institutions returns. The distribution of performance classes is shown in Figure 14.17. We can see that strong banks comprise 12.9%, satisfactory banks 15.1%, moderate banks 47.5%, and poor banks 24.5%. The figure visualizes the effectiveness of Boruta algorithm in determining the most important variables that determine the condition of a bank.

14.4.8 Classification and Regression Trees (CART)

Table 14.7 shows the performance results of our CART algorithm in predicting bank failure on the training set. The algorithm's level of accuracy on the training dataset was 82.8%. The best tune, or complexity,

TABLE 14.7   CART Model Performance

COMPLEXITY PARAMETER	ACCURACY	KAPPA	ACCURACY SD	KAPPA SD
0.06849315	0.8275092	0.7519499	0.04976459	0.07072572
0.15753425	0.7783150	0.6683229	0.07720896	0.14039942
0.42465753	0.5222344	0.1148591	0.08183351	0.18732422

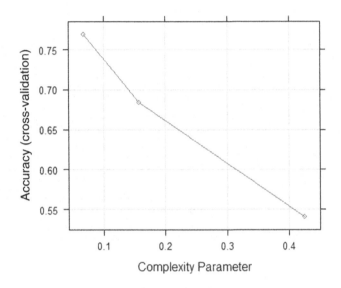

FIGURE 14.18   CART accuracy curve.

parameter of our optimal model was 0.068. The Kappa statistic was 75%, envisaging that our classifier was effective, as also shown with the Kappa SD of 0.07 in the classification of bank categories. On the test dataset, the algorithm achieved an accuracy level of 92.5% and a kappa of 88.72%. The algorithm only misclassified two instances as moderate and one as satisfactory.

The accuracy of the CART model based on the complexity parameters of different test runs is shown on Figure 14.18. The complexity, or the best tune, parameter of 0.068 optimized the model performance.

14.4.9 Support Vector Machine

The accuracy level of the SVM model on the training dataset was 79.1% in predicting bank solvency, as shown in Table 14.8. The best tune sigma and cost values of our highly performing model were 0.05 and 1, as shown on Figure 14.19. The Kappa statistic and the Kappa SD were 67.9% and 0.13, respectively. On the test dataset, the algorithm achieved an accuracy level of 92.5% and a kappa of 88.54%. The algorithm only misclassified three instances as moderate in comparison to the CART algorithm.

14.4.10 Linear Discriminant Algorithm

On the training dataset, the LDA achieved an accuracy level of 80% as in Table 14.9. The Kappa statistic and the Kappa SD were 70% and 0.16, respectively. On the test dataset, the algorithm achieved an accuracy level of 90% and a kappa of 84.64%. The algorithm only misclassified four instances as moderate whose performance is poor in comparison to the CART algorithm.

TABLE 14.8 Support Vector Machine Performance

SIGMA	C	ACCURACY	KAPPA	ACCURACY SD	KAPPA SD
0.050398	0.25	0.783223	0.678536	0.095598	0.140312
0.050398	0.50	0.776007	0.661354	0.087866	0.132552
0.050398	1.00	0.791391	0.678694	0.080339	0.126466

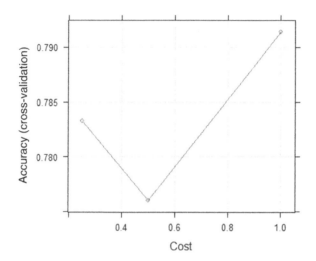

FIGURE 14.19 SVM accuracy curve.

TABLE 14.9 Linear Discriminant Algorithm Performance

ACCURACY	KAPPA	ACCURACYSD	KAPPASD
0.8042399	0.7038131	0.1016816	0.159307

TABLE 14.10 KNN Algorithm Performance

K	ACCURACY	KAPPA	ACCURACY SD	KAPPA SD
5	0.5988645	0.3698931	0.1280376	0.2158109
7	0.6268864	0.4072928	0.1564920	0.2703504
9	0.6621978	0.4715556	0.1747903	0.2881390

14.4.11 K-Nearest Neighbor

The level of accuracy on the training dataset was 66.2%. The best tune parameter for our model was $k = 9$, or 9 neighbors as shown on the accuracy curve in Figure 14.20. The Kappa statistic and the Kappa SD were 47.2% and 0.17, respectively. On the test dataset, the algorithm achieved an accuracy level of 67.5% and a kappa of 49%. The algorithm was not highly effective in classifying bank performance in comparison to other algorithms.

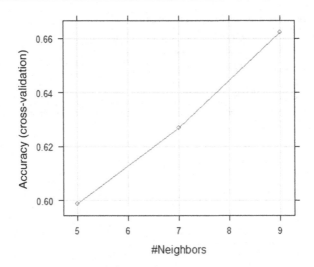

FIGURE 14.20 KNN confusion accuracy graph.

TABLE 14.11 Random Forest Performance

MTRY	ACCURACY	KAPPA	ACCURACY SD	KAPPA SD
2	0.8272527	0.7421420	0.10396454	0.15420079
14	0.8554212	0.7829891	0.06069716	0.09303130
16	0.8482784	0.7718935	0.06455248	0.09881991

14.4.12 Random Forest

On the training set, the accuracy of our random forest was 85.5%, as designated in Table 14.11. The best tune parameter for our model was the mtry of 14, which is the number of randomly selected predictors in constructing trees, as shown on Figure 14.21. The Kappa statistic and the Kappa SD were 78.3% and 0.09, respectively. On the test dataset, the algorithm achieved an accuracy level of 96% and a Kappa of 96%. The algorithm was highly effective in classifying bank performance in comparison to all algorithms.

14.4.13 Challenges and Future Direction

As the number of banking activities increase, data submission to the Reserve Bank continues to grow exponentially. This challenging situation, in combination with advances in machine learning (ML) and artificial intelligence (AI), presents unlimited opportunities to apply neural-network-based deep learning (DL) approaches to predict Zimbabwean Bank's solvency. Future work will focus on identifying more features that could possibly lead to poor bank performance and incorporate these in our models to develop a robust early warning supervisory tool based on big data analytics, machine learning, and artificial intelligence.

The researcher analyses the two models that have been proposed in the literature with reference to an ideal data analytics model for cyber security presented in Chapter 3.

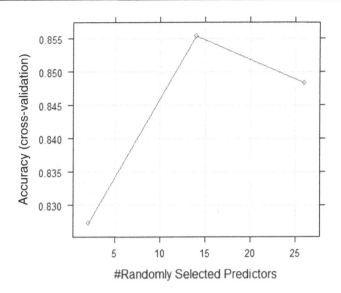

FIGURE 14.21   Random forest accuracy graph.

14.4.13.1 Model 1: Experimental/Prototype Model

In the first case, the researcher makes reference to the model presented in [52], which, although developed in the context of the public sector, can be applied to private sector organizations. Table 14.12 summarizes the main characteristics of the experimental model. (The reader is referred to the prototype model also demonstrated in [52].)

Software and Hardware Complex (SHC): Warning–2016

The proposed model, it is to be noted, was demonstrated to be effective in integrating big data analytics with cyber security in a cost-effective way [52].

14.4.13.2 Model 2: Cloud Computing/Outsourcing

The second model involves an organization outsourcing its data to a cloud computing service provider. Cloud computing service providers usually have advanced big data analytics models, with advanced detection and prediction algorithms and better state-of-the-art cyber security technologies and better protocols because they specialize in data and networks. However, it is to be noted that cloud computing service providers are neither exempt nor immune from cyber threats and attacks [21].

14.4.13.3 Application of Big Data Analytics Models in Cyber Security

There is overwhelming evidence to support the assertion that big data analytics need to be applied to cyber security, complete with many infallible proofs that such application is not only necessary in recent times but a means to survival [21, 52]. The researcher demonstrated, by identifying the characteristics of an effective data analytics model, the ideal model, that it is possible to evaluate different models. In the third hypothesis, the researcher postulated that there is an appropriate big data analytics model for cyber security for every institution. While the review of literature showed that institutions and countries adopt different big data analytics models for cyber security, the researcher also demonstrated that, beside the unique requirements that these models share, major common characteristics, for example reactors and

TABLE 14.12   Experimental Big Data Analytics Model for Cyber Security

MODEL ATTRIBUTES	DESCRIPTION
HBase working on HDFS (Hadoop Distributed File System)	• HBase, a nonrelational database, facilitates analytical and predictive operations • Enables users to assess cyber threats and the dependability of critical infrastructure
Analytical data processing module	• Processes large amounts of data, interacts with standard configurations servers and is implemented at C language • Special interactive tools (based on JavaScript/CSS/DHTML) and libraries (for example, jQuery) developed to work with content of the proper provision of cyber security
Special interactive tools and libraries	• Interactive tools based on JavaScript/CSS/DHTML • Libraries, for example jQuery developed to work with content • Designed to ensure the proper provision of cyber security
Data store for example (MySQL)	• Percona Server with the ExtraDB engine • DB servers are integrated into a multimaster cluster using the Galera Cluster.
Task queues and data caching	• Redis
Database servers balancer	• Haproxy
Web server	• nginx, involved PHP-FPM with APC enabled
HTTP requests balancer	• DNS (multiple A-records)
Development of special client applications running Apple iOS	• Programming languages used: Objective C, C++, Apple iOS SDK based on Cocoa Touch, CoreData, and UIKit
Development of applications running Android OS	• Google SDK
Software development for the web platform	• PHP and JavaScript
Speed of the service and protection from DoS attacks	• CloudFare (through the use of CDN)

Source: [52].

detection algorithms, are usually present in every model but differ in terms of complexity. Further, given the models presented in this chapter, it is worthy of note that many small organizations usually adopt Model 2, whereas very large organizations and sensitive public sector organizations adopt Model 1. This may also explain why the models used may differ even though the framework used in designing a data analytics model for cyber security in a cloud computing services provider may share similar characteristics with that developed by an institution on its own.

14.4.13.4 Summary of Analysis

In this chapter, the researcher presented two models for adopting data analytics models to cyber security. The first experimental or prototype model involves the design and implementation of a prototype by an institution, and the second model involves the service provided by cloud computing companies. The researcher also demonstrated how this study addressed the hypotheses postulated. In the information era we are currently living in, voluminous varieties of high-velocity data are being produced daily, and within them lay intrinsic details and patterns of hidden knowledge that needs to be extracted and utilized.

By applying such analytics to big data, valuable information can be extracted and exploited to enhance decision making and to support informed decisions. Thus the support of big data analytics to decision making was depicted.

14.5 CONCLUSION

Machine learning algorithms as part of artificial intelligence can be clustered into supervised, unsupervised, semisupervised, and reinforcement learning algorithms. The main characteristic of ML is the automatic data analysis of large data sets and the production of models for the general relationships found among data.

Big data analytics not only is about the size of data but also hinges on the volume, variety, and velocity of the data. Volume denotes big data as massive; velocity denotes the high speed of big data; variety denotes the diversity of big data; veracity denotes the degrees of trustworthiness in big data; vocabulary denotes the conformity of big data to different schema, models, and ontologies; and value denotes the cost and worth of big data. Big data [104] has necessitated the development of big data mining tools and techniques widely referred to as big data analytics. Big data analytics refers to a combination of well-known tools and techniques, for example in machine learning and and data mining, that are capable of leveraging useful data usually hidden in big data and creating an interface in the form of linear and visual analytics.

The information that is evaluated in big data analytics includes a mixer of unstructured and semistructured data, for instance, social media content, mobile phone records, web server logs, and internet click stream data. Big data analytics makes use of analytic techniques such as data mining, machine learning, artificial learning, statistics, and natural language processing. Big data came into existence when the traditional relational database systems were not able to handle the unstructured data generated by organizations, social media, or any other data generating sources.

Passive data sources can include computer-based data, for example geographical IP location, computer security health certificates, keyboard typing and clickstream patterns, or WAP data. Data over networks may be secured through the use of antivirus software, firewall, encryption, secure protocols, etc. However, hackers can always devise innovative ways of breaking into network systems. An intrusion detection and prevention system is placed inside the network to detect possible network intrusions and, where possible, prevent cyber attacks.

REFERENCES

[1]. Berman, D.S., Buczak, A.L., Chavis, J.S., and Corbett, C.L. (2019). "Survey of Deep Learning Methods for Cyber Security," *Information* 2019, 10, p. 122. DOI:10.3390/info10040122

[2] Sarker, I. H., Kayes, A. S. M., Badsha, S., Alqahtani, H., Watters, P., & Ng, A. (2020). Cybersecurity data science: an overview from machine learning perspective. *Journal of Big Data.* https://doi.org/10.1186/s40 537-020-00318-5

[3] Gheyas, I. A. & Abdallah, A. E. (2016). Detection and prediction of insider threats to cyber security: A systematic Literature Review and Meta-Analysis. *Big Data Analytics*, 1, p. 6.

[4] Kantarcioglu, M & Xi B (2016). *Adversarial Data Mining: Big data meets cybersecurity*, CCS, 16 October 24–28, 2016, Vienna, Austria.

[5] Lei, G., Liu, C., Li, Y., Chen, D., Guo, Y., Zhu, J. (2019). Robust Design Optimization of a High-Temperature Superconducting Linear Synchronous Motor Based on Taguchi Method. *IEEE Transactions on Applied Superconductivity*, 29(2), pp. 1–6.

[6] Min, K S, Chai S W, & Han M (2015). An international comparative study on cyber security strategy. *International Journal of Security and Its Applications*, 9(2), pp. 13–20.

[7] Lee, J. W., & Xuan, Y. (2019). Effects of technology and innovation management and total factor productivity on the economic growth of China. *Journal of Asian Finance, Economics and Business*, 6(2), pp. 63–73. https://doi.org/10.13106/jafeb.2019. vol6.no2.63.

[8] Bringas, P.B., and Santos, I. (2010). *Bayesian Networks for Network Intrusion Detection, Bayesian Network*, Ahmed Rebai (Ed.), ISBN: 978-953-307-124-4, InTech, Available from: www.intechopen.com/books/bayesian-network/bayesian-networks-for-network-intrusion-detection

[9] Umamaheswari, K., and Sujatha, S. (2017). Impregnable Defence Architecture using Dynamic Correlation-based Graded Intrusion Detection System for Cloud. *Defence Science Journal*, 67(6), pp. 645–653. DOI: 10.14429/dsj.67.11118.

[10] Nielsen, R. (2015). *CS651 Computer Systems Security Foundations 3d Imagination Cyber Security Management Plan*, Technical Report January 2015, Los Alamos National Laboratory, USA.

[11] Stallings, W. (2015). Operating System Stability. Accessed on 27th March, 2019. www.unf.edu/public/cop4 610/ree/Notes/PPT/PPT8E/CH15-OS8e.pdf

[12] Truong, T.C; Diep, Q.B.; & Zelinka, I. (2020). Artificial Intelligence in the Cyber Domain: Offense and Defense. *Symmetry* 2020, 12, 410.

[13] Proko, E., Hyso, A., and Gjylapi, D. (2018). Machine Learning Algorithms in Cybersecurity, www.CEURS-WS.org/Vol-2280/paper-32.pdf

[14] Pentakalos, O. (2019). Introduction to machine learning. *CMG IMPACT 2019*. https://doi.org/10.4018/978-1-7998-0414-7.ch003

[15] Russom, P (2011) Big Data Analytics. In: *TDWI Best Practices Report*, pp. 1–40.

[16] Hammond, K. (2015). *Practical Artificial Intelligence For Dummies®, Narrative Science Edition*. Hoboken, New Jersey: Wiley.

[17] Bloice, M. & Holzinger, A. (2018). *A Tutorial on Machine Learning and Data Science Tools with Python*. Graz, Austria: s.n.

[18] Cuzzocrea, A., Song, I., Davis, K.C. (2011). Analytics over Large-Scale Multidimensional Data: The Big Data Revolution! In: *Proceedings of the ACM International Workshop on Data Warehousing and OLAP*, pp. 101–104.

[19] Moorthy, M., Baby, R. & Senthamaraiselvi, S. (2014). An Analysis for Big Data and its Technologies. *International Journal of Computer Science Engineering and Technology(IJCSET)*, 4(12), pp. 413–415.

[20] Menzes, F.S.D., Liska, G.R., Cirillo, M.A. and Vivanco, M.J.F. (2016) Data Classification with Binary Response through the Boosting Algorithm and Logistic Regression. *Expert Systems with Applications*, 69, pp. 62–73. https://doi.org/10.1016/j.eswa.2016.08.014

[21] Mazumdar, S & Wang J (2018). Big Data and Cyber security: A visual Analytics perspective, in S. Parkinson et al (Eds), *Guide to Vulnerability Analysis for Computer Networks and Systems*.

[22] Economist Intelligence Unit: The Deciding Factor: Big Data & Decision Making. In: Elgendy, N.: *Big Data Analytics in Support of the Decision Making Process*. MSc Thesis, German University in Cairo, p. 164 (2013).

[23] Herodotou, H., Lim, H., Luo, G., Borisov, N., Dong, L., Cetin, F.B., Babu, S.: Starfish: A Self-tuning System for Big Data Analytics. In: *Proceedings of the Conference on Innovative Data Systems Research*, pp. 261–272 (2011).

[24] EMC (2012): Data Science and Big Data Analytics. In: EMC Education Services, pp. 1–508.

[25] Kubick, W.R.: Big Data, Information and Meaning. In: Clinical Trial Insights, pp. 26–28 (2012).

[26] Wilson, B. M. R., Khazaei, B., & Hirsch, L. (2015, November). Enablers and barriers of cloud adoption among Small and Medium Enterprises in Tamil Nadu. In: *2015 IEEE International Conference on Cloud Computing in Emerging Markets (CCEM)* (pp. 140–145). IEEE.

[27] Hashem, I. A. T., Yaqoob, I., Anuar, N. B., Mokhtar, S., Gani, A., & Ullah Khan, S. (2015). The rise of "big data" on cloud computing: Review and open research issues. In *Information Systems*. https://doi.org/10.1016/j.is.2014.07.006

[28] Hadi, J. (2015) "Big Data and Five V'S Characteristics," *International Journal of Advances in Electronics and Computer Science* (2), pp. 2393–2835.[29]. Serrat, O.: Social Network Analysis. Knowledge Network Solutions 28, 1–4 (2009).

[29] Siti Nurul Mahfuzah, M., Sazilah, S., & Norasiken, B. (2017). An Analysis of Gamification Elements in Online Learning To Enhance Learning Engagement. *6th International Conference on Computing & Informatics*.

[30] Pu, C. and Kitsuregawa, M., 2019, Technical Report No. GIT-CERCS-13-09; Georgia Institute of Technology, CERCS.

[31] Shen, Z., Wei, J., Sundaresan, N., Ma, K.L.: Visual Analysis of Massive Web Session Data. In: Large Data Analysis and Visualization (LDAV), pp. 65–72 (2012).

[32] Asur, S., Huberman, B.A. (2010). Predicting the Future with Social Media. *ACM International Conference on Web Intelligence and Intelligent Agent Technology*, 1, pp. 492–499.

[33] Van Der Valk, T., Gibers, G (2010): The Use of Social Network Analysis in Innovation Studies: Mapping Actors and Technologies. *Innovation: Management, Policy & Practice* 12(1), 5–17.

[34] Zeng, D., Hsinchun, C., Lusch, R., Li, S.H.: Social Media Analytics and Intelligence. *IEEE Intelligent Systems* 25(6), 13–16 (2010).

[35] Bolzoni, D. (2009). Revisiting Anomaly-based Network Intrusion Detection Systems, Ph.D Thesis, University of Twente, The Netherlands, ISBN: 978-90-365-2853-5, ISSN: 1381-3617. DOI:10.3990/1.9789036528535

[36] Gercke, M. (2012). "Cybercrime Understanding Cybercrime," *Understanding cybercrime: phenomena, challenges and legal response.*

[37] Karimpour, J., Lotfi, S., and Siahmarzkooh, A.T. (2016). Intrusion detection in network flows based on an optimized clustering criterion, Turkish Journal of Electrical Engineering & Computer Sciences, Accepted/Published Online: 17.07.2016, http://journals.tubitak.gov.tr/elektrik

[38] Murugan, S., and Rajan, M.S. (2014). Detecting Anomaly IDS in Network using Bayesian Network, IOSR Journal of Computer Engineering (IOSR-JCE), e-ISSN: 2278-0661, p- ISSN: 2278-8727, Volume 16, Issue 1, Ver. III (Jan. 2014), PP 01-07, www.iosrjournals.org

[39] National Institute of Standards and Technology (2018). Framework for Improving Critical Infrastructure Cybersecurity Version 1.1.

[40] Fernando, J. I., & Dawson, L. L. (2009). The health information system security threat lifecycle: An informatics theory. *International Journal of Medical Informatics.* https://doi.org/10.1016/j.ijmedinf.2009.08.006

[41] Bezemskij, A., Loukas, G., Gan, D., and Anthony, R.J. (2017). Detecting cyber-physical threats in an autonomous robotic vehicle using Bayesian Networks, 2017 IEEE International Conference on Internet of Things (iThings) and IEEE Green Computing and Communications (GreenCom) and IEEE Cyber, Physical and Social Computing (CPSCom) and IEEE Smart Data (SmartData), 21–23 June 2017, IEEE, United Kingdom, https://ieeexplore.ieee.org/document/8276737

[42] Gercke, M. (2012). "Cybercrime Understanding Cybercrime," *Understanding cybercrime: phenomena, challenges and legal response.*

[43] National Institute of Standards and Technology (2018). Framework for Improving Critical Infrastructure Cybersecurity Version 1.1.

[44] Analysis, F., Cybersecurity, F., Development, S., & Nemayire, T. (2019). *A Study on National Cybersecurity Strategies A Study on National Cybersecurity Strategies.*

[45] CENTER for Cyber and Information Security (https://ccis.no/cyber-security-versus-information-security/

[46] Op-Ed column, Gary Marcus and Ernest Davis (2019). How to Build Artificial Intelligence We Can Trust, www.nytimes.com/2019/09/06/opinion/ai-explainability.html

[47] Cox, R. & Wang, G. (2014). Predicting the US bank failure: A discriminant analysis. *Economic Analysis and Policy*, Issue 44.2, pp. 201–211.

[48] Fernando, J. I., & Dawson, L. L. (2009). The health information system security threat lifecycle: An informatics theory. *International Journal of Medical Informatics.* https://doi.org/10.1016/j.ijmedinf.2009.08.006

[49] Lakshami, R.V. (2019), Machine Learning for Cyber Security using Big Data Analytics. *Journal of Artificial Intelligence, Machine Learning and Soft Computing*, 4(2), pp. 1–8. http://doi.org/10.5281/zenodo.3362228

[50] Aljebreen, M.J. (2018, p.18), Towards Intelligent Intrusion Detection Systems for Cloud Comouting, Ph.D. Dissertation, Florida Institute of Technology, 2018.

[51] Yang, C., Yu, M., Hu, F., Jiang, Y., & Li, Y. (2017). Utilizing Cloud Computing to address big geospatial data challenges. *Computers, Environment and Urban Systems.* https://doi.org/10.1016/j.compenvurbsys.2016.10.010

[52] Petrenko, S A & Makovechuk K A (2020). Big Data Technologies for Cybersecurity.

[53] Snowdon, D. A., Sargent, M., Williams, C. M., Maloney, S., Caspers, K., & Taylor, N. F. (2019). Effective clinical supervision of allied health professionals: A mixed methods study. *BMC Health Services Research.* https://doi.org/10.1186/s12913-019-4873-8

[54] Bou-Harb, E., & Celeda, P. (2018). *Survey of Attack Projection, Prediction, and Forecasting in Cyber Security.* September. https://doi.org/10.1109/COMST.2018.2871866

[55] Wang, H., Zheng, Z., Xie, S., Dai, H. N., & Chen, X. (2018). Blockchain challenges and opportunities: a survey. *International Journal of Web and Grid Services.* https://doi.org/10.1504/ijwgs.2018.10016848

[56]   Zhang, L., Wu, X., Skibniewski, M. J., Zhong, J., & Lu, Y. (2014). Bayesian-network-based safety risk analysis in construction projects. *Reliability Engineering and System Safety*. https://doi.org/10.1016/j.ress.2014.06.006

[57]   Laney, D., and Beyer, M.A. (2012), "The importance of 'big data': A definition," Russell, S., & Norvig, P. (2010). *Artificial Intelligence: A Modern Approach* (3rd edition). Prentice Hall.

[58]   Editing, S., Cnf, D. N. F., Paul M. Muchinsky, Hechavarría, Rodney; López, G., Paul M. Muchinsky, Drift, T. H., 研究開発戦略センター国立研究開発法人科学技術振興機構, Basyarudin, Unavailable, O. H., Overview, C. W., Overview, S. S. E., Overview, T., Overview, S. S. E., Graff, G., Birkenstein, C., Walshaw, M., Walshaw, M., Saurin, R., Van Yperen, N. W., … Malone, S. A. (2016). Qjarterly. *Computers in Human Behavior*. https://doi.org/10.1017/CBO9781107415324.004

[59]   Iafrate, F. (2015), From Big Data to Smart Data, ISBN: 978-1-848-21755-3 March, 2015, Wiley-ISTE, 190 Pages.

[60]   Pence, H. E. (2014). What is Big Data and Why is it important? *Journal of Educational Technology Systems*, *43*(2), pp. 159–171. https://doi.org/10.2190/ET.43.2.d

[61]   Thomas, E. M., Temko, A., Marnane, W. P., Boylan, G. B., & Lightbody, G. (2013). Discriminative and generative classification techniques applied to automated neonatal seizure detection. *IEEE Journal of Biomedical and Health Informatics*. https://doi.org/10.1109/JBHI.2012.2237035

[62]   Pavan Vadapalli (2020). "AI vs Human Intelligence: Difference Between AI & Human Intelligence," 15th September, 2020, www.upgrad.com/blog/ai-vs-human-intelligence/

[63]   Dezzain.com website (2021), www.dezzain.com/

[64]   Cunningham, Lawrence A. (2008). The SEC's Global Accounting Vision: A Realistic Appraisal of a Quixotic Quest. North Carolina Law Review, Vol. 87, 2008, GWU Legal Studies Research Paper No. 401, GWU Law School Public Law Research Paper No. 401, Available at SSRN: https://ssrn.com/abstract=1118377

[65]   Buyya, R., Yeo, C. S., & Venugopal, S. (2008). Market-oriented cloud computing: Vision, hype, and reality for delivering IT services as computing utilities. *Proceedings – 10th IEEE International Conference on High Performance Computing and Communications, HPCC 2008*. https://doi.org/10.1109/HPCC.2008.172

[66]   Mell, P. M., & Grance, T. (2011). *The NIST definition of cloud computing*. https://doi.org/10.6028/NIST.SP.800-145

[67]   Xin, Y., Kong, L., Liu, Z., Chen, Y., Li, Y., Zhu, H., Gao, M., Hou, H., & Wang, C. (2018). Machine Learning and Deep Learning Methods for Cybersecurity. *IEEE Access*, 6, pp. 35365–35381. https://doi.org/10.1109/ACCESS.2018.2836950

[68]   Hassan, H. (2017). Organisational factors affecting cloud computing adoption in small and medium enterprises (SMEs) in service sector. *Procedia Computer Science*. https://doi.org/10.1016/j.procs.2017.11.126

[69]   He, Y., Lee, R., Huai, Y., Shao, Z., Jain, N., Zhang, X., Xu, Z.: RCFile: A Fast and Space-efficient Data Placement Structure in MapReduce-based Warehouse Systems. In: *IEEE International Conference on Data Engineering (ICDE)*, pp. 1199–1208 (2011).

[70]   Lee, R., Luo, T., Huai, Y., Wang, F., He, Y., Zhang, X.: Ysmart: Yet Another SQL-to-MapReduce Translator. In: *IEEE International Conference on Distributed Computing Systems (ICDCS)*, pp. 25–36 (2011).

[71]   Burt, D., Nicholas, P., Sullivan, K., & Scoles, T. (2013). Cybersecurity Risk Paradox. *Microsoft SIR*.

[72]   Marzantowicz (2015), Corporate Social Responsibility of TSL sector: attitude analysis in the light of research, "Logistyka" 2014, No. 5, pp. 1773–1785.

[73]   Pai & Aithal (2017). The basis of social responsibility in management, Poltext, Warszawa.

[74]   Sen and Tiwari (2017). Port sustainability and stakeholder management in supply chains: A framework on resource dependence theory, *The Asian Journal of Shipping and Logistics*, No. 28 (3): 301–319.

[75]   Tashkandi, A. N., & Al-Jabri, I. M. (2015). Cloud computing adoption by higher education institutions in Saudi Arabia: An exploratory study. *Cluster Computing*. https://doi.org/10.1007/s10586-015-0490-4

[76]   Fehling, C., Leymann, F., Retter, R., Schupeck, W., & Arbitter, P. (2014). Cloud Computing Patterns. In *Cloud Computing Patterns*. https://doi.org/10.1007/978-3-7091-1568-8

[77]   Sether, A. (2016), Cloud Computing Benefits (2016).

[78]   Handa, A., Sharma, A., & Shukla, S. K. (2019). Machine learning in cybersecurity: A review. In *Wiley Interdisciplinary Reviews: Data Mining and Knowledge Discovery*. https://doi.org/10.1002/widm.1306

[79]   KPMG (2018), Clarity on Cybersecurity. Driving growth with confidence.

[80]   Gillward & Moyo (2013). Green performance criteria for sustainable ports in Asia, International Journal of Physical Distribution & Logistics Management, No. 43(5): p. 5.

[81]   Greengard, S. (2016). Cybersecurity gets smart. In *Communications of the ACM*. https://doi.org/10.1145/ 2898969

[82]   Rivard, Raymond, and Verreault (2006). Resource-based view and competitive strategy: An integrated model of the contribution of information technology to firm performance, DOI:10.1016/j.jsis.2005.06.003, Corpus ID: 206514952

[83]   Kobielus, J. (2018). Deploying Big Data Analytics Applica- tions to the Cloud: Roadmap for Success. Cloud Standards Customer Council.

[84]   Furht (2010). Information orientation, competitive advantage, and firm performance: a resource-based view. *European Journal of Business Research*, 12(1), 95–106.

[85]   Lee, J. (2017). *HACKING INTO CHINA'S CYBERSECURITY LAW,* In: *IEEE International Conference on Distributed Computing Systems* (2017).

[86]   Zhang, Q., Cheng, L., & Boutaba, R. (2010). Cloud computing: state-of-the-art and research challenges. *Journal of Internet Services and Applications*, 1(1), pp.7–18.

[87]   Oliveira, T., Thomas, M., & Espadanal, M. (2014). Assessing the determinants of cloud computing adoption: An analysis of the manufacturing and services sectors. *Information and Management*. https:// doi.org/10.1016/j.im.2014.03.006

[88]   Hsu, T. C., Yang, H., Chung, Y. C., & Hsu, C. H. (2018). A Creative IoT agriculture platform for cloud fog computing. *Sustainable Computing: Informatics and Systems*. https://doi.org/10.1016/j.sus com.2018.10.006

[89]   Bou-Harb, E., & Celeda, P. (2018). Survey of Attack Projection, Prediction, and Forecasting in Cyber Security. September. https://doi.org/10.1109/COMST.2018.2871866

[90]   Berman, D.S., Buczak, A.L., Chavis, J.S., and Corbett, C.L. (2019). "Survey of Deep Learning Methods for Cyber Security," *Information* 2019, 10, p. 122. DOI:10.3390/info10040122

[91]   Kumar, R. (2011). *Research Methodology: A step by step guide for beginners*. 3rd ed. London: Sage.

[92]   Merrian, S.B. & Tisdell E.J. (2016). *Qualitative Research: A guide to design and implementation*, 4th Edition. Jossey-Bass, A Wiley Brand.

[93]   TechAmerica (2012): Demystifying Big Data: A Practical Guide to Transforming the Business of Government. In: TechAmerica Reports, pp. 1–40 (2012).

[94]   Sanchez, D., Martin-Bautista, M.J., Blanco, I., Torre, C.: Text Knowledge Mining: An Alternative to Text Data Mining. In: *IEEE International Conference on Data Mining Workshops*, pp. 664–672 (2008).

[95]   Zhang, L., Stoffel, A., Behrisch, M., Mittelstadt, S., Schreck, T., Pompl, R., Weber, S., Last, H., Keim, D.: Visual Analytics for the Big Data Era – A Comparative Review of State-of-the-Art Commercial Systems. In: *IEEE Conference on Visual Analytics Science and Technology (VAST)*, pp. 173–182 (2012).

[96]   Song, Z., Kusiak, A. (2009). Optimizing Product Configurations with a Data Mining Approach. *International Journal of Production Research* 47(7), 1733–1751.

[97]   Adams, M.N. (2010). Perspectives on Data Mining. *International Journal of Market Research*, 52(1), pp. 11–19.

[98]   Fu, J., Zhu, E. Zhuang, J. Fu, J. Baranowski, A. Ford and J. Shen (2016) "A Framework-Based Approach to Utility Big Data Analytics," in *IEEE Transactions on Power Systems*, 31(3), pp. 2455–2462.

[99]   Napanda, K., Shah, H., and Kurup, L. (2015). Artificial Intelligence Techniques for Network Intrusion Detection, International Journal of Engineering Research & Technology (IJERT), ISSN: 2278-0181, IJERTV4IS110283 www.ijert.org, Vol. 4 Issue 11, November-2015.

[100] Stefanova, Z.S. (2018). "Machine Learning Methods for Network Intrusion Detection and Intrusion Prevention Systems," Graduate Theses and Dissertations, 2018, https://scholarcommons.usf.edu/etd/7367

[101] Almutairi, A. (2016). Improving intrusion detection systems using data mining techniques, Ph.D Thesis, Loughborough University, 2016.

[102] Bolzoni, D. (2009). Revisiting Anomaly-based Network Intrusion Detection Systems, Ph.D Thesis, University of Twente, The Netherlands, ISBN: 978-90-365-2853-5, ISSN: 1381-3617. DOI:10.3990/ 1.9789036528535

[103] Mouthami, K., Devi, K.N., Bhaskaran, V.M.: Sentiment Analysis and Classification Based on Textual Reviews. In: *International Conference on Information Communication and Embedded Systems (ICICES)*, pp. 271–276 (2013).

[104] Manyika, J., Chui, M., Brown, B., Bughin, J., Dobbs, R., Roxburgh, C., Byers, A.H.: Big Data: The Next Frontier for Innovation, Competition, and Productivity. In: McKinsey Global Institute Reports, pp. 1–156 (2011).

Using ML and DL Algorithms for Intrusion Detection in the Industrial Internet of Things

15

Nicole do Vale Dalarmelina,[1] Pallavi Arora,[2]
Baljeet Kaur,[3] Rodolfo Ipolito Meneguette,[1]
and Marcio Andrey Teixeira[4]

[1]*Instituto de Ciências Matemáticas e de Computação (ICMC), University of São Paulo, São Carlos, SP, Brasil*

[2]*IK Gujral Punjab Technical University, Kapurthala, Punjab, India*

[3]*Guru Nanak Dev Engineering College, Ludhiana, Punjab, India*

[4]*Federal Institute of Education, Science, and Technology of Sao Paulo, Catanduva, SP, Brazil*

Contents

DOI: 10.1201/9781003187158-18

15.1 INTRODUCTION

The internet has turned into a worldwide essential resource for humanity, taking place not only on smartphones and computers but also in TVs, refrigerators, and even automobiles. This advance is named the Internet of Things (IoT). According to Nick [1], in 2021 there were 46 billion internet-connected devices. Running alongside these developments, industrial devices' technology and connectivity are also increasing and giving rise to a new technology, the Industrial Internet of Things (IIoT), which is the IoT concept applied to industrially dedicated network applications.

IIoT technologies are being heavily used in critical environments, such as manufacture and energy management, water supply control, and many others. The system that operates this kind of environment is called an industry control system (ICS). Once the industrial machines and devices are connected to the internet, it becomes a potential target for cyber attacks, being at the mercy of data exposure, theft, modification, or destruction [2].

To trick modern security mechanisms, cyber attacks are constantly moving forward. The aftermaths are countless and can reach out to ordinary people as well as large companies and turn into catastrophic scenarios. To try to hold back this cybernetic epidemic, a multitude of intrusion detection systems (IDS) are being developed using machine learning (ml) and deep learning (dl) algorithms to detect attack patterns in network traffic. To detect these patterns, the ML and DL algorithms can be trained using network traffic datasets. Despite the several datasets available to train ML and DL models, is important to identify the best features in the dataset so that the training can be made properly. In this chapter, we discuss the use of ML and DL algorithms in practical applications to develop efficient IDSs for industrial networks considering the best features of each attack present in the datasets used.

15.2 IDS APPLICATIONS

There are various IDS types, each one intended for a different environment. Among the existing types, the characteristics of host-based and network-based IDS are described in Figure 15.1.

Cybernetic threats can reach environments, networks, and hosts. Host intrusion detection systems (HIDS) are designed to focus on computers and operational systems, inspecting the host in order to detect malicious activities in workstations to guarantee the security inside each station. On the other hand, network intrusion detection systems (NIDS) are developed to detect malicious activities on a network level. This kind of IDSs can analyze the information flow in the network, monitoring the traffic, opened and closed doors, requests and packages, and services transitions. There are many ways to detect intrusion in an IDS that can be classified as misuse-based or signature-based; network traffic behavior; hybrid.

The misuse-based method is used to recognize exclusive patterns of unauthorized behaviors. This approach scored a minimal number of false-positives [3] as it's capable of achieving high levels of accuracy in identifying previously classified intrusions. In this approach, ML/DL algorithms can be trained to recognize new threat patterns identifying new attack signatures.

The network traffic behavior approach is worried about identifying anomalous events and behaviors considering the system's usual behavior, not only the attack signature. Each event is analyzed and classified as normal or attack, according to a trained model. This approach can reach high levels of false-positives because every unknown activity is classified as an attack.

Hybrid detection aims to merge the two other methods, using the best of each approach, in order to intensify detection, being able to identify known and unknown attacks.

Several types of cyber attacks are described in the literature [4]. However, in our study, we are considering two types of attacks: the reconnaissance attack and the denial of service (DoS) attack. To

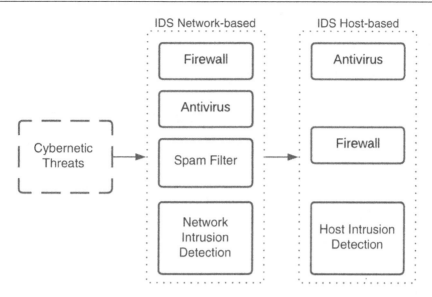

IDS Network-based   IDS Host-based

FIGURE 15.1 Cyber security traditional systems.
Source: Adapted from Teixeira et al. [3].

understand how they behave, these attacks are described here, and more information about other types of attacks can be found in [5–6].

- *Reconnaissance attack*: Usually, reconnaissance is the first step of an attack in order to scan the network searching for relevant information about the target, such as open doors, devices connected, and possible vulnerabilities. Commonly this kind of attack offers no harm to the target [5]; however, it is used to learn the best ways to attack the target, once all the activity has been watched.
- *Denial of Service (DoS)*: The goal of this attack is to make the target services unavailable, denying legitimate users access to the system. This kind of cyber attack was used for many years against websites and is able to cut off an entire organization from the network [6]. One variation of this attack is called distributed denial of service (DdoS) and was designed to control multiple computers so that they can "obey" a master computer. In this way, all infected computers must make a certain request to a certain service simultaneously.

Usually, to generate the dataset of an industrial network traffic, the data of the sensors is collected over a long period on what can result in a large-scale dataset. Use of a dataset of this dimension can result in very time-consuming training and detection [6]. In order to reduce this time required and to make the best use of ML algorithms for the IDS, experiments to choose the features with the higher scores are showed. In other words, these experiments help in choosing the features with higher correlations with the "target." In our case, the "target" means to verify whether a network flow is normal (0) or under attack (1) – using the random forest classifier and the Pearson correlation coefficient.

15.2.1 Random Forest Classifier

The Random forest classifier [7] is a ML algorithm available in the "skit-learn" library from Python [8]. This algorithm uses the concept of training simple predictions models – called weak learners – to generate a much more complex model – called a strong learner. In this way, the random forest algorithm trains a lot

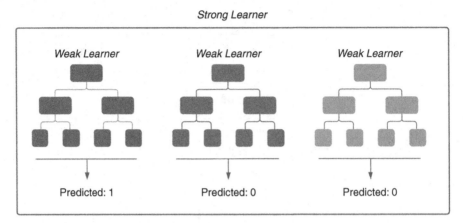

FIGURE 15.2   Random forest decision tree.

of decision trees so that each one can adapt to different slices of the dataset; the outcome of this training is the ability to classify all variations of the data in the dataset.

Figure 15.2 illustrates how the random forest classifier works. The weak learners predict the results, and the strong learner, which is the combination of the weak learners, calculates the most likely value. In the example in Figure 15.2, two decision trees predicted 0, and one predicted 1, so the strong learner predicted 0 because it was the value that appears the most among the weak learners.

15.2.2 Pearson Correlation Coefficient

The Pearson correlation coefficient calculates the linear statistical relationship between two variables, aiming to point out how they are associated with each other. The equation for this calculation r is:

$$r = \frac{1}{n-1} \Sigma \left(\frac{x_i - \bar{X}}{S_x} \right) \left(\frac{y_i - \bar{Y}}{S_y} \right),\tag{15.2.1}$$

where n is the number of samples, x is the first element, and is second. $\bar{X}$ and $\bar{Y}$ represent the samples average of x and y. S is the standard deviation, S_x being the standard deviation of the x element and S_y the standard deviation of the y element. The Pearson correlation formula can be calculated, resulting in a real number between -1 and 1.

A negative result – where $r < 0$ – indicates that as the value of one of the elements increases, the value of the other decreases. In a positive result – where $r > 0$ – the value of one of the elements increases as the value of the other element increases. Finally, a correlation resulting in 0 does not mean that the elements have no correlation, only that they do not have a linear correlation with each other.

15.2.3 Related Works

Intending to detect known and unknown cyber threats, several IDS have been designed using diverse techniques and technologies, as shown in [9–13]. Apruzzese et al. [14] analyzed technologies for cyber security issues, such as malware analysis, spam detection, and intrusion detection in order to verify whether these technologies supplied the identification and prevention of these issues. According to the authors, the studied approaches showed vulnerability to attacks and needed constant retraining.

TABLE 15.1   Comparison Between Our Proposal Application and the Others Existing in the Literature

WORK	FEATURE SELECTION	IDS	IIOT	ML	DATASET BALANCING	ENSEMBLE LEARNING
[6]	✓	✗	✓	✓	✓	✗
[9]	✓	✓	✗	✗	✗	✗
[10]	✓	✓	✗	✓	✗	✗
[11]	✓	✓	✓	✗	✗	✗
[12]	✓	✓	✗	✓	✗	✗
[13]	✗	✓	✓	✓	✗	✗
Our approach	✓	✓	✓	✓	✓	✓

Aiming the development of an efficient IDS at industrial networks, first the most important features in the training dataset need to be found. The next step is to apply the chosen algorithm to design the detection system, and this can be done using ML and DL algorithms.

Table 15.1 illustrates the main differences between the existing approaches and the practical application that will be described in section 15.4. It is possible to observe that this practical approach addresses issues such as feature selection, which is not presented in [13], and dataset balancing that is taken into account only in [6]. Still, the development of an IDS system [9–13] for IIoT environments, is only studied in [6, 11, 13], using ML algorithms – also seen in [6, 10, 12, 13] – and taking into account the imbalanced dataset problem [6], and ensemble learning, which is not found in any other of these works.

15.3 USE OF ML AND DL ALGORITHMS IN IIOT APPLICATIONS

Machine learning and deep learning are subareas of artificial intelligence, where systems can make decisions based on their own experiences and known data. The algorithm is trained with datasets that carry data already classified or unclassified, depending on whether the learning is supervised or unsupervised. In supervised learning, the training data is submitted into a previous classification and labeling, where all the input and output are known. To develop a supervised model, techniques like regression can be used. On the other hand, in unsupervised learning, the algorithm has to find the best correlation between the uncategorized information and classify the data according to the patterns found. To design an unsupervised model, clustering strategies can be used, where the data is analyzed and classified depending on resemblance.

Intending to improve IIOT applications, so the performance and security could achieve better results, ML and DL algorithms are being used. An example of the use of ML in IIOT environments is presented in [13], where ML algorithms are used to detect reconnaissance attacks in a water storage tank, where the treatment and distribution of water are carried out. Some of the ML algorithms used were forest, logistic regression, and KNN; the authors even highlight the efficiency of ML models to detect attacks in real time and affirm that the test scenario provides a good understanding of the effects and consequences of attacks in real SCADA (supervisory control and data acquisition) environments.

To achieve high levels of performance using ML and DL for training models, the used datasets must be classified correctly and balanced. Real-world IIoT systems may have a significant number of attacks and intrusions [6]. However, this number is small compared to the total amount of data that is not attacked (normal data). For this reason, IIOT datasets usually present an uneven proportion of normal and samples under attack.

According to Melo [15], a dataset with 50% of a unique class can be already considered as an imbalanced dataset. Zolanvari et al. [16] presented a study about the problems of imbalanced datasets during the training of a model to IDSs. The training of a ML model with this kind of dataset can be turned into a big issue in the industry environment, since it can result in lots of false alarms, classifying an attack as a "normal flow" for not being able to classify it as a malicious activity due to the low number of attack samples.

Designed to meet the needs of the industry environment, an IDS must be developed with ML and DL algorithms trained with balanced and well classified datasets. Beyond that, the dataset must contain only pertinent information, so the time and computational resources spent could be minimal.

15.4 PRACTICAL APPLICATION OF ML ALGORITHMS IN IIOT

Several research projects about IDSs development can be found, as shown in the previous section, using different technologies. However, there is still a gap in the intrusion detection in IIOT networks, which needs a better security system, since each data manipulation or steal could result in terrible consequences. In this section, the best approach to develop an IDS for industrial environments using ML algorithms is discussed. To this end, the best features to make a satisfactory training for the IDS are analyzed, as well as several tests with prediction algorithms in order to choose the best algorithm to design the model. The dataset used on this experiment was developed by Teixeira et al., and all details about the dataset can be find in [17].

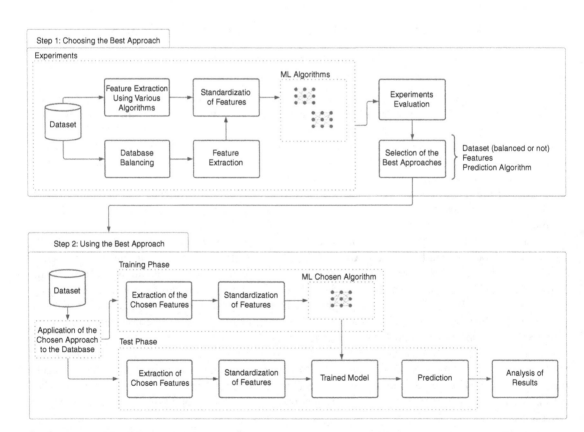

FIGURE 15.3 Our IDS approach.

To develop our IDS, we will use two different datasets, one with a different type of cyber attack, and the goals are divided into two steps. Step 1 involves analysis and choosing the best approach, and step 2 is using the best approach.

In step 1, the Pearson correlation coefficient and the random forest are used to pick only the features with the higher scores. After features selection, prediction tests are made with the datasets containing only the best features, using different prediction algorithms.

To implement this step, three different scenarios are used:

1. (RF) dataset with only the features selected with the random forest algorithm
2. (PC) dataset with only the features selected with the Pearson correlation coefficient
3. (AF) dataset with all the features

The same experiments will be made with these datasets using different prediction algorithms to find the best algorithm for this purpose.

Afterward, the datasets are balanced using the under sampling method, and the same experiments are made to analyze whether there is any on the training with an imbalanced dataset.

At the end of the step 1, all the experiments made are evaluated to pick up the best approach, which must satisfy the following terms: balanced or imbalanced dataset according to the best performance; specific features of each cyber attack or general features as attested by the best performance on the tests; the prediction algorithm with the best rate at the evaluation.

Ins step 2, the best approach is put into practice using ensemble learning to develop the IDS to detect malicious activities in industrial networks. In order to implement this approach, the features are extracted – using the features selection approach from the first step – and standardized, so that the dataset can be divided into two parts: 80% of the dataset for training and 20% for testing. At the testing point, the developed IDS makes its predictions, which are finally evaluated.

To measure the performance of our approach, metrics from the confusion matrix are used, as shown in Table 15.2.

The confusion matrix, illustrated in Table 15.2, can be read in the context as follows:

- *Input* represents the network flow input that will be analyzed by the model; the input value can be 0 – normal flow – or 1 – flow under attack.
- *True-negative* represents the number of normal flows correctly predicted as normal.
- *False-positive* represents the number of normal flows incorrectly predicted as under attack.
- *False-negative* represents the number of under attack flows incorrectly predicted as normal.
- *True-positive* represents the number of under attack flows correctly predicted as under attack.

The metrics from the confusion matrix used are accuracy, false alarm rate (FAR), undetected rate (UR), Matthews correlation coefficient (MCC), and sensitivity.

Accuracy is the rate of predictions made correctly considering the total number of predictions and can be represented as shown in equation (15.4.1).

TABLE 15.2   Confusion Matrix

		PREDICTION	
		CLASSIFIED AS NORMAL	CLASSIFIED AS ATTACK
Input	**Normal**	True-negative (TN)	False-positive (FP)
	Attack	False-negative (FN)	True-positive (TP)

$$\text{Accuracy} = \frac{TP + TN}{TP + TN + FP + FN} \times 100. \tag{15.4.2}$$

FAR is the percentage of normal traffic incorrectly predicted as an attack.

$$\text{FAR} = \frac{FP}{FP + TN} \times 100. \tag{15.4.3}$$

URis the fraction of traffic under attack incorrectly predicted as normal.

$$\text{UR} = \frac{FN}{FN + TP} \times 100. \tag{15.4.4}$$

The MCC measures the prediction quality, presenting the correlation between the predictions and the real classification of the flow.

$$\text{MCC} = \frac{TP\,TN - FP\ FN}{\sqrt{(TP + FP)\,(TP + FN)\,(TN + FP)\,(TN + FN)}} \times 100. \tag{15.4.5}$$

The effectiveness in predicting flows under attack correctly is measured using sensitivity.

$$\text{Sensitivity} = \frac{TP}{TP + FN} \times 100. \tag{15.4.6}$$

15.4.1 Results

For the development of this practical application, two different datasets were used – available in [13] – with different cyber attacks:

- **A**: distributed denial of service (DdoS)
- **B**: reconnaissance attack

To find the best algorithm to choose the most important dataset features, two different techniques are used: the random forest algorithm and the Pearson correlation coefficient. These techniques are used in three different scenarios:

- **RF**: dataset with only the features selected with the random forest algorithm
- **PC**: dataset with only the features selected with the Pearson correlation coefficient
- **AF**: dataset with all the features

In the first scenario (**RF**), the random forest [18] was used to find the features with the best scores in the dataset **A**. As a result, the best features found were *DstBytes*, *DstLoad*, and *DstRate*, described in the Table 15.3, where the description and score are shown.

Still in the scenario **RF**, the best features from the dataset **B** were found, with a score over 0.15: *Ploss*, *Dport*, *SrcPkts*, and *TotPkts*. The features are described in Table 15.4.

For the scenario **P**, the Pearson correlation coefficient was calculated so that the features achieving the higher scores – over 0.55 – of correlations with the target – which identify the flow as 0 (normal) or 1 (under attack) – could be identified. Using the dataset **A**, the best features were: *DstLoss*, *Ploss*, *Mean*, and *SrcLoss*. The description and scores of these features are illustrated in the Table 15.5.

The same calculation was performed over dataset **B**, and the best features were *DstLoss*, *Dport*, *SrcLoss*, and *Ploss*. The best features and scores in this experiment are described at the Table 15.6.

TABLE 15.3 Best Features Selected with the Random Forest Algorithm from Dataset A

FEATURES	SCORES	DESCRIPTION
Destination bytes (DstBytes)	0.272996	Destination/source bytes count
Destination load (DstLoad)	0.182034	Destination bits per second
Destination rate (DstRate)	0.181972	Destination packets per second

TABLE 15.4 Best Features Selected with the Random Forest Algorithm from Dataset B

FEATURES	SCORES	DESCRIPTION
Total percent loss (Ploss)	0.309460	Percent packets retransmitted/dropped
Destination port (Dport)	0.183496	Destination port number
Source packets (SrcPkts)	0.181972	Source/destination packet count
Total packets (TotPkts)	0.180653	Total transaction packet count

TABLE 15.5 Best Features Selected with the Pearson Correlation Coefficient from Dataset A

FEATURES	SCORES	DESCRIPTION
Destination loss (DstLoss)	0.949714	Destination packets retransmitted/dropped
Total percent Loss (Ploss)	0.818986	Percent packets retransmitted/dropped
Mean flow (Mean)	0.578331	Average duration of active flows
Soure loss (SrcLoss)	0.571413	Source packets retransmitted/dropped

TABLE 15.6 Best Features Selected with the Pearson Correlation Coefficient from Dataset B

FEATURES	SCORES	DESCRIPTION
Destination loss (DstLoss)	0.905914	Destination packets retransmitted/dropped
Destination port (Dport)	0.847983	Destination port number
Source loss (SrcLoss)	0.729565	Source packets retransmitted/dropped
Total percent loss (Ploss)	0.671127	Percent packets retransmitted/dropped

After finding the best features of each dataset according to each technique, the datasets were standardized using the Standard Scaler, from *sklearn* [4] and were partitioned in 80% for training and 20% for test, so that the prediction experiments could be done. The logistic regression class from *sklearn* [4] was used to develop the prediction model using the features chosen before.

The same training was performed using both datasets in the three different scenarios (**RF, PC,** and **AF**). In terms of learning time, as shown in Figure 15.3, it was observed that the dataset **A** in the scenario **RF** consumed more time than the other, precisely 424.97 seconds, and that, in the **PC** scenario, the same dataset spent 12.48 seconds. If considering only the learning time consumed using dataset **A**, it is possible to observe that the **AF** scenario has the second higher level and that, using the dataset **B**, achieve the higher learning time.

The accuracyequation was used to calculate the accuracy of the training models, and the results are illustrated in Figure 15.4. In the figure, it is observed that the **AF** achieves the higher accuracy out of the

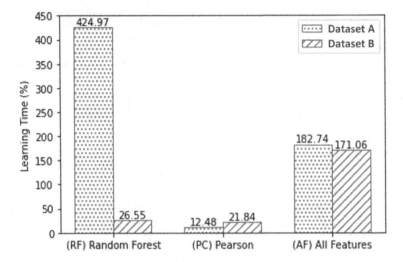

FIGURE 15.4   Learning time consumed using logistic regression.

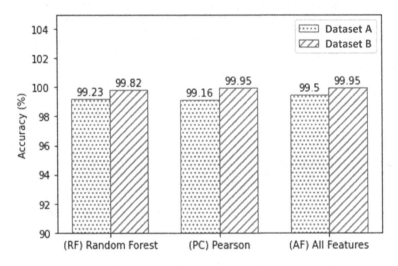

FIGURE 15.5   Accuracy.

three scenarios, 99.50% using the dataset **A** and 99.95% using dataset **B**. This can be explained because all of the features were used in this scenario, so the accuracy could be more precise than the others. Still, the accuracy achieved using the dataset **B** in the **PC** was also 99.95%, showing a great performance.

Besides the high levels of accuracy achieved in all tests so far – over 99% – it's necessary to consider other metrics to evaluate the training model performance with imbalanced datasets. For this purpose, the false alarm rate (FAR) was used to calculate the percentage of normal traffic classified wrongly as traffic under attack. Applying the FARequation, presented in the previous section of this chapter, satisfying results of under 1% were achieved.

Analyzing Figure 15.5, the training using dataset **A** predicted wrongly more normal flows than the training using dataset **B**. While the lower rate using dataset **B** was 0.01%, the lower rate using dataset **A** was 0.51%, a rate higher than the highest rate using dataset **B**.

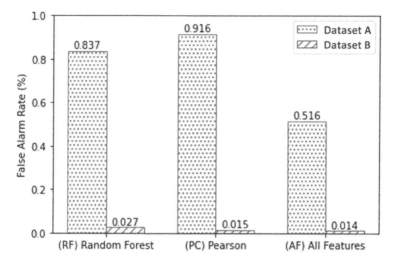

FIGURE 15.6   FAR.

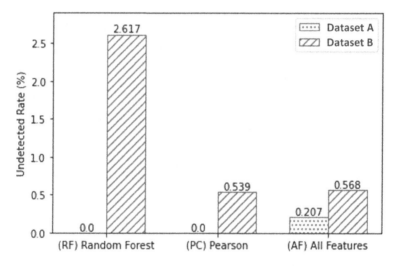

FIGURE 15.7   UR.

To represent the rate of traffic under attack predicted as normal flow during the experiments, the URequation was applied. Considering Figure 15.6, it can be noticed that, besides the low rate, the experiments using dataset **B** achieve a high UR, which can be a big problem in the industrial environment in that an unnoticed attack could cause a lot of damage.

A very important metric used to evaluate performances of IDSs trained using imbalanced datasets is the Matthews correlation coefficient (MCC), once it measures the classification quality, showing the correlation between the real values and the prediction values. To calculate the MCC, the equation MCC was used, and the results are shown in Figure 15.7.

Analyzing Figure 15.7, it's possible to observe that all the experiments so far were satisfying for this approach's objectives, presenting 99.57% at the **PC** and **AF** and 98.35% at the **RF** using the dataset **B**. The experiments using dataset **A** presented lower results; however, the lowest rate was at the **RF**, with 95.39%, which is still a high rate achieved.

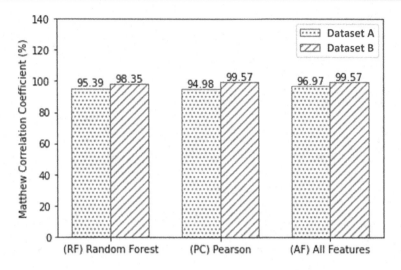

FIGURE 15.8   MCC.

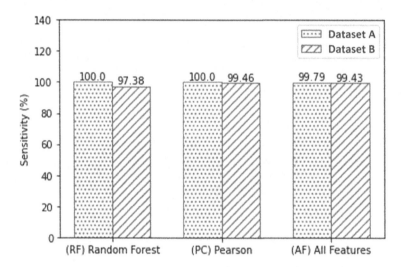

FIGURE 15.9   Sensitivity.

Finally, the sensitivity equation was employed to measure the efficiency of the models to predict anomalous activities correctly. As seen in Figure 15.8, the models showed high sensitivity in intrusion detection in network traffic being trained using imbalanced datasets. The experiments using dataset **A** presented the best results using this metric, achieving the highest rate – 100% – at **RF** and **PC** scenarios, as using dataset **B**, the rates were 97.38% and 99.46% respectively (see Figure 15.9).

Considering the results described in this section, of the experiments made with the logistic regression algorithm, it can be observed that the **PC** achieved the best rates out of the three presented in most of the calculations. However, it's also necessary to calculate the metrics using balanced datasets in order to develop the best approach to an IDS for industrial networks.

15.5 CONCLUSION

In this chapter, we presented the methods and metrics used to analyze the best approach to develop an IDS for industrial networks using ML algorithms. Using datasets present in the literature, several experiments were conducted aiming to choose the best approach to develop an efficient IDS for industrial networks. To achieve this goal, the performance of the training model was evaluated using the logistic regression algorithm in three different scenarios, considering the learning time duration and other metrics from the confusion matrix, such as accuracy, FAR, UR, MCC, and sensitivity.

For upcoming steps, experiments using other prediction algorithms will be done, such as experiments with balanced datasets to analyze whether balanced and imbalanced datasets can lead to different results in training prediction models, so that the best approaches can be used to develop the IDS for industrial networks using ML and ensemble learning algorithms. Moreover, we will conduct the experiments considering a dataset with more types of attacks.

REFERENCES

[1]   Nick G. How Many IoT Devices Are There in 2021? [All You Need To Know]. [S.l.], 2021. [Online]. https://techjury.net/blog/how-many-iot-devices-are-there/#gref

[2]   ISO/IEC 27000:2018(en). Information technology – Security techniques – Information security management systems – Overview and vocabulary. [S.l.], 2018. [Online]. www.iso.org/obp/ui/#iso:std:iso-iec:27000:ed-5:v1:en

[3]   Brown, D. J.; Suckow, B.; Wang, T. A survey of intrusion detection systems. Department of Computer Science, University of California, San Diego, 2002.

[4]   Biju, Jibi Mariam, Neethu Gopal, and Anju J. Prakash. "Cyber attacks and its different types." International Research Journal of Engineering and Technology 6.3 (2019): 4849–4852.

[5]   Burton, J.; Dubrawsky, I.; Osipov, V.; Tate Baumrucker, C.; Sweeney, M. (Ed.). Cisco Security Professional's Guide to Secure Intrusion Detection Systems. Burlington: Syngress, 2003. P. 1–38. ISBN 978-1-932266-69-6. Available in: www.sciencedirect.com/science/article/pii/B9781932266696500215.

[6]   Pajouh, H. H.; Javidan, R.; Khayami, R.; Dehghantanha, A.; Choo, K.-K. R. A two-layer dimension reduction and two-tier classification model for anomaly-based intrusion detection in IOT backbone networks. IEEE Transactions on Emerging Topics in Computing, v. 7, n. 2, p. 314–323, 2019.

[7]   License, B. (Ed.). RandomForestClassifier. [S.l.], 2020. Available in: https://scikit-learn.org/stable/modules/generated/sklearn.ensemble.RandomForestClassifier.html.

[8]   Foundation, P. S. Python. [S.l.], 2021. Available in: www.python.org

[9]   Ambusaidi, M. A.; He, X.; Nanda, P.; Tan, Z. Building an intrusion detection system using a filter-based feature selection algorithm. IEEE Transactions on Computers, v. 65, n. 10, p. 2986–2998, 2016.

[10]   Sarker I.H.; Abushark, Y. A. F. K. A. IntruDTree: A Machine Learning Based Cyber Security Intrusion Detection Model. [S.l.], 2020.

[11]   Keshk, M.; Moustafa, N.; Sitnikova, E.; Creech, G. Privacy preservation intrusion detection technique for SCADA systems. In: 2017 Military Communications and Information Systems Conference (MilCIS). [S.l.: s.n.], 2017. P. 1–6.

[12]   Zargar, G. R.; Baghaie, T. et al. Category-based intrusion detection using PCA. Journal of Information Security, Scientific Research Publishing, v. 3, n. 04, p. 259, 2012.

[13]   Teixeira, M. A.; Salman, T.; Zolanvari, M.; Jain, R.; Meskin, N.; Samaka, M. Scada system testbed for cybersecurity research using machine learning approach. Future Internet, v. 10, n. 8, 2018. ISSN 1999-5903. Available in: www.mdpi.com/1999-5903/10/8/76

[14] Apruzzese, G.; Colajanni, M.; Ferretti, L.; Guido, A.; Marchetti, M. On the effectiveness of machine and deep learning for cyber security. In: 2018 10th International Conference on Cyber Conflict (CyCon). [S.l.: s.n.], 2018. P. 371–390.

[15] Melo, C. Como lidar com dados desbalanceados? [S.l.], 2019.

[16] Zolanvari, M.; Teixeira, M.; Jain, R. Effect of imbalanced datasets on security of industrial iot using machine learning. 12 2019.

[17] Teixeira, Marcio Andrey and Zolanvari, Maede and Khan, Khaled M. Flow-based intrusion detection algorithm for supervisory control and data acquisition systems: A real-time approach. 2021.

[18] Pedregosa, F. and Varoquaux, G. and Gramfort, A. and Michel, V. and Thirion, B. and Grisel, O. and Blondel, M. and Prettenhofer, P. and Weiss, R. and Dubourg, V. and Vanderplas, J. and Passos, A. and Cournapeau, D. and Brucher, M. and Perrot, M. and Duchesnay, E., Journal of Machine Learning Research. Scikit-learn: Machine Learning in {P}ython. 2011.

PART IV

Applications

On Detecting Interest Flooding Attacks in Named Data Networking (NDN)–based IoT Searches

16

Hengshuo Liang, Lauren Burgess, Weixian Liao, Qianlong Wang, and Wei Yu

*Department of Computer and Information Sciences,
Towson University, Maryland, USA*

Contents

DOI: 10.1201/9781003187158-20

16.1 INTRODUCTION

The Internet of Things (IoT) has drawn much attention in recent years [1, 2]. The volume of IoT is increased by the millisecond and will expand to 50 billion by 2025 [3]. To better manage IoT devices and such massive data traffic, the IoT search engine (IoTSE) is designed to meet those requirements [4]. Generally speaking, the IoTSE system consists of three components, as shown in Figure 16.1:

1. *Users*: There are two types of users: human users and machine users (e.g., asmart vehicle).
2. *IoT searching engine (IoTSE)*: This has three logical layers:
   a. *Network communication layer*: This sets network communication protocols determined based on the type of network that the devices and users are on, such as a wired network or wireless network.
   b. *Data service layer*: This is the core layer of IoTSE, where the search engine is deployed. At this layer, the requests from the users are processed based on the request's needs in data storage. It is then forwarded to the upper layer for any advanced requirements or sent back to the users.
   c. *Application layer*: This layer analyzes the requested data to extract insightful patterns and generate further decisions.
3. *Data source*: IoT devices generate data and send the data back to the data storage of IoTSE. In general, the user posts a request to IoTSE at the beginning. If the requested data exists in data storage, the IoTSE then returns it to the user. Otherwise, IoTSE requests data from IoT devices and sends it back to the user.

The most popular IoTSE is the web-crawler-based system that uses the TCP/IP network structure to transmit data. This type of website-based configuration raises some issues:

1. *Domain name system (DNS) issue*: As the number of connected IoT devices grows, so does the number of required IP assignments. The IP management is needed in both public and private networks, which significantly increases demands for network address translation (NAT) and DNS management.
2. *Network transferring issue*: Since TCP/IP is an end-to-end communication protocol, if a packet loss occurs during the process, the end user can only discover this issue after the time-out, and the end-user may not resend the request again until then.
3. *Vulnerability to IP-based flooding attack*: If the adversaries can spoof the IP address of interest, they can quickly launch a flooding attack to pollute the whole network by continuously transmitting junk data. Furthermore, the adversaries can even send crafted malicious data to a specific target to cause critical damage in the system.

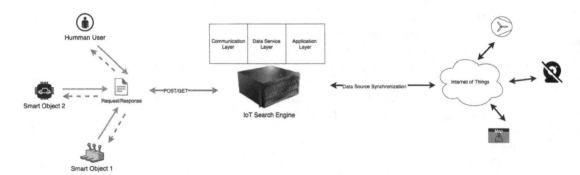

FIGURE 16.1   Internet of Things search engine (IoTSE) structure.

To mitigate performance degradation caused by attacks, machine learning (ML) provides promising solutions. Even for high volumes of data traffic, ML can extract the key structures and features by analyzing internal relationships between data features [5]. This differs from traditional methods that use the model to distinguish between data points as much as possible and use subjective guessing by simple mathematics models. The ML methods provide in-depth analysis and understanding of the relationship between data. Thus it can provide effective enhancement for analyzing network traffic based on the magnitude of the big data collected [6].

On the other hand, in recent years, a promising network structure, called named data network (NDN), has been developed for replacing the current TCP/IP structure. Generally speaking, NDN is a user-driven, pull-requested communication architecture. Unlike traditional TCP/IP-based networks, all the object retrieval processes in NDN are based on data names instead of IP addresses. Naming in NDN is under a hierarchical scheme, where the request is defined in a tree manner with a backslash sign (/), such as "/ USA/MD/Towson University/CS Department/Lab1." Due to this human-readable data name, users can name the data freely without any location constraints. When users send a request to the network, it sends to all possible links. Each node records incoming and outgoing network interfaces. Once the request hits a data producer with the same namespace, the data producer sends the packet in the reverse way. In the whole process, there is no receiver address or sender address. Thus the adversary cannot launch any attack based on the address of the target. Even though NDN can stop all IP-based attacks, it can still be affected by namespace-spoofing-based attacks [7]. Therefore, in this study, we focus on detecting flooding attacks based on NDN-embedded IoTSE.

In this study, we make the following three contributions:

1. We summarize three different interest flooding attack methods based on the ratios between the number of attacking nodes and the data producers.
2. We create our dataset for further study of ML-based detection, using two different scenarios: small scale (one to one and many to one) and large scale (many to many). Using ns3, we design and create the attacking scenarios with different settings and collect row traffic data that we then prepossess.
3. We use the collected dataset to train several ML-based detection schemes and compare their performance. Our experimental results show that that our investigated attacks can be easily recognized by ML-based detection schemes.

The remainder of this chapter is organized as follows. In section 16.2, we briefly introduce NDN, IoTSE, and ML. In section 16.3, we present the framework design, our scenarios, and ML models used in our study. In section 16.4, we present performance evaluation and discuss the evaluation results. In section 16.5, we discuss some remaining issues for future research. In section 16.6, we review the existing research efforts that are relevant to our study. Finally, we conclude the chapter in section 16.7.

16.2 PRELIMINARIES

In this section, we first provide a brief introduction to NDN, IoTSE, and machine learning.

16.2.1 Named Data Networking (NDN)

NDN refers to one development of the future Internet paradigm designed by naming data [8–9]. Nowadays, the Internet architecture that utilizes Transmission Control Protocol/Internet Protocol (TCP/IP) typically is application specific, in which the content names of one application do not affect the content names of another application unless the applications need to exchange data [8–9]. Nonetheless, the current point-to-point driven network has some limitations and weaknesses. From a data-driven perspective, NDN tends to address these by utilizing two types of packets: interest and data packets [10–11].

The consumer applications send these interest packets for named data, which the routers then forward based on the names [9, 11]. Once these interest packets reach the data source, known as a producer, a proxy, or the in-network cache, the interest returns the data packet following the path in reverse [9–11]. NDN utilizes key components to send interest toward data producers: (1) the *forwarding information base* (*FIB*) contains a list of name-prefixes, finds the longest match based routing protocols, and can have multiple output interfaces for each prefix; (2) the *content store* (*CS*) caches the returning data packets, which must match exactly to satisfy any future interest with the same name; (3) the *pending interest table* (*PIT*) records any outgoing, incoming data interfaces, and any interest packets that are not completed; and (iv) the *forwarding strategy* allows NDN to use in-network states per namespace and determines when and where to transfer the packets [9–11].

16.2.2 Internet of Things Search Engine (IoTSE)

A massive number of IoT devices are continuously deployed around the world. The heterogeneous network of devices make sharing the gathered data difficult due to the differences in the standard formats between the vastly different devices that could belong to different organizations [4, 11–12]. IoTSE aims to find a way to efficiently share and build intelligence within the IoT systems by developing a universal format for users, accelerating the progress of data sharing among the vastly different devices [4, 11, 4, 13].

In the current networks utilizing TCP/IP, IoTSE has the challenge of the number of IP addresses used, causing it to be harder to scale [4, 11, 13]. Nonetheless, with the use of NDN, which focuses on naming, the goal would be to create a scalable naming service that can be used by IoT systems [4]. Similar to current web-based search engines, IoTSE provides functionalities, including data collection using web-crawler-based schemes, and can index and organize that collected data while storing and cataloging the data into the different subsystems for future queries [4, 11].

To make searching on IoTSE as efficient as possible, utilizing artificial intelligence (AI) techniques, such as ML and DL, would be beneficial as they are capable of reducing the workloads and determining which queries are looking for similar data [11]. Our research investigates whether ML and DL are capable of learning and detecting attacks against IoTSE in an NDN network.

16.2.3 Machine Learning (ML)

Artificial Intelligence (AI) is prevalent nowadays. For instance, it has been involved in smart homes using certain smart devices, e.g., NEST thermostats and ring cameras. Furthermore, AI-enabled smart home devices, such as Google Home, Amazon Alexa, and Apple's Siri, are already capable of handling speech recognition. [14–16]. These advancements are increasing our attention in seeing how far AI can go. One of the more noticeable developments that came out from AI, using sensor fusion and computer vision, has been self-driving cars [16]. Computer vision is the way to extract meaningful features from video and images while completing tasks such as object localization, image classification, and image segmentation [17].

AI has also been used in the detection and prevention of intrusions into the network by using ML and DL techniques [14, 15, 18]. As a category that falls under AI, ML is associated with human intelligence and has the ability to learn and analyze based on the use of computational algorithms [14, 15, 19]. ML allows the computational algorithms to recognize the various patterns to train the machine on decisions and recommendations [19]. Thus, after sufficient amounts of training, the algorithm can take an input of data and predict the output [15, 19]. ML works on mathematical optimization, and, when problem-solving, it works to break the problem apart into multiple subproblems to obtain the final result [15].

16.3 MACHINE LEARNING ASSISTED FOR NDN-BASED IFA DETECTION IN IOTSE

This section introduces our proposed attack model, attack scale, attack scenarios, and machine learning models.

16.3.1 Attack Model

The interest flooding attack (IFA) is a typical variation of the denial of service (DoS) attack in the NDN architecture. The adversary in IFA could generate interest requests to overuse the PIT in the NDN network within a limited short period. Therefore, IFA can block the NDN router from receiving other interest requests from all legitimate users or drop the current unsatisfied legitimate interest request in PIT due to time-out.

IFA can be classified into two categories: non collusive and collusive [7]. In *noncollusive* IFA, the adversary pretends to be a legitimate user spreading malicious requests for exhausting the computation resource in IoTSE. In *collusive IFA*, the adversary spreads malicious requests to the adversary's IoTSE through the whole network. The adversary's IoTSE responds to those requests and sends data back. In that IFA, the adversary targets the entire network by overusing the network resource. It causes network traffic congestion and blocks other legitimate users. In this chapter, we assume that the adversary does not know the profile of IoTSE, including the naming scheme of the interest packet. Therefore, the adversary can only generate nonexistent interest packets to overuse the network (noncollusive IFA). In the nonexistent interest packet, the adversary generates an interest packet with random data name without knowledge about the profile in IoTSE, and it is only used for overusing network resources in NDN entities. So this nonexistent interest-packet-based IFA is a mutation of the noncollusive IFA.

Figure 16.2 shows the nonexistent interest packet's attack model. As seen in the figure, this model targets IoTSE. The adversary sends a series of nonexistent interest requests to the neighboring NDN routers. Once those routers receive requests, they generate PIT entities. Within a short time, those PITs are fully occupied by the requests generated from the adversary and cannot accept any incoming interest requests. Those routers forward the requests to IoTSE. Therefore, it causes traffic congestion in the whole network and cannot serve the legitimate request because of the delay.

Figure 16.3 demonstrates the nonexistent interest packets attack workflow as follows:

1. The adversary generates nth interest with a fake name (/. /. /Nowhere/ ././Fake user (n))
2. These interests are sent into the PIT of an NDN router.
3. Since those are nonexistent requests, those routers cannot satisfy from the cache content store. Those requests remain in the PIT until the time-out of those requests.
4. Therefore, no data is sent back from IoTSE to satisfy those requests.
5. Simultaneously, the NDN router cannot receive any interest from legitimate users since there is no place in PIT for them.
6. As a result, the existing data in IoTSE cannot be sent back to the user since there is no request from the legitimate user that can be sent to IoTSE due to the overuse of PIT by the adversary. Therefore, the request from the legitimate consumer is denied.

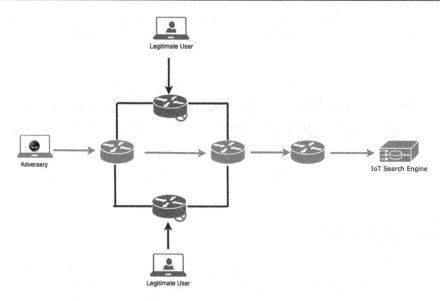

FIGURE 16.2   Attack model with nonexistent interest packets.

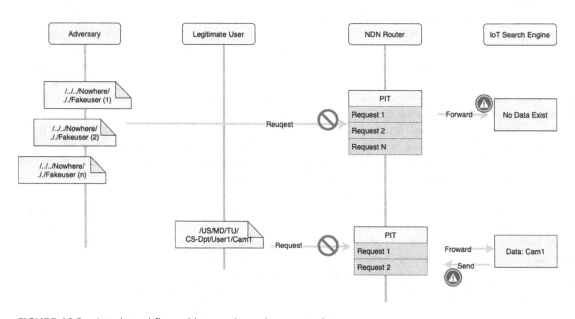

FIGURE 16.3   Attack workflow with nonexistent interest packets.

16.3.2 Attack Scale

To quantify the scope of the attacks, we categorize them into three different scenes according to the ratio between the number of adversary users and the number of legitimate users in IoTSE:

- *One to one*: In this attacking scene, we deploy an IoTSE with limited legitimated users. These legitimate users send a request to the IoTSE and wait for a response, while a single adversary user frequently sends fake data requests to cause overutilization of the resources in the network.

- *Many to one*: In this attacking scene, we add more adversarial users by following the first scenario. While legitimate users are still sending requests to the IoTSE, there are now multiple adversarial users who send fake interest requests to flood the network.
- *Many to many*: For this attack scene, we build multiple networks with an IoTSE endpoint connected by a backbone network. Although several legitimate users are sending interest requests and awaiting responses from the IoTSE, we set different quantities of adversary groups, which cross-send fake interest requests to poison the network.

We consider the first two scenes as the small-scale attack, while the latter scene is the large-scale attack.

16.3.3 Attack Scenarios

We now discuss how to design the attacking cases with the previous attacking scenes in terms of attack scale. To better quantify attacking performance, we need to design attacking cases from four aspects:

1. *Number of adversary users*: For each attack scene, the total number of users determines the primary user group of the network. It should be noticed that the larger user group, no matter the adversary or the legitimate, usually leads to higher chance of utilizing all of the network resources. For these experiments, the goal is for the adversary users to hinder the network performance for the legitimate user.
- *Location of adversary users*: The distribution and location of the adversary users are based on the total number of legitimate users. The distance between the adversary and the IoTSE determines the data transmission delay. For instance, if the adversary is closer to the IoTSE, the lower latency is in sending fake interest requests, causing a more negative impact on the performance of legitimate users.
- *Active time of adversary user*: This refers to how long it takes for the adversary to flood the network. It depends on the active amount of time when these users send the fake interest packets. For this research, the ideal case is to have each adversary user always be active to produce maximum damage to the network.
- *Flooding rate of adversary user*: The number of packets sent by the adversary user every second determines the bandwidth usage. The faster the adversary user sends the fake interest packet, the more hindrance is placed on bandwidth usage, and the more floods the network prevents the legitimate users from getting responses from the IoTSE.

Based on these aspects, we design two attacking scenarios:

1. *Small-scale attacking scenario*: In this scenario, we create a single tree topology, similar to the one shown in Figure 16.4. In the figure, one IoTSE is shown in blue, four gateways are shown in gray, and two different leaf nodes are shown in red and green. The red group is considered the adversary group, whereas the green is designated as legitimate users. The adversary users are randomly active with a 50% probability for every second during the attacking period; links in this topology use 10 Mbps bandwidth and 10 ms delay.
2. *Large-scale attacking scenario*: In this scenario, we use Rocketfuel internet server provider (ISP) topology [20]. We utilize AT&T in the United States as our ISP topology due to the large size, including 296 leaf nodes, 108 gateways, and 221 backbone nodes. For this scenario, we randomly select 70 gateways, shown in gray, and backbone nodes, shown in black, with one node being the IoTSE. Out of the 296 leaf nodes, we randomly set 140 leaf nodes as the adversary group, shown in red, leaving the remaining nodes as legitimate users, shown in green, which creates the topology shown in Figure 16.5.

FIGURE 16.4   Simple tree topology.

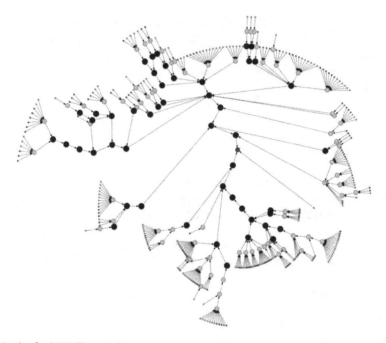

FIGURE 16.5   Rocketfuel ISP–like topology.

To create severe damage, we utilize the link configurations shown in Table 16.1. We set the end user with less bandwidth and a higher delay, while the other links obtain the opposite. For this case, the adversary user can easily cause congestion in the link from the gateway to leaf nodes and hinder the performance of legitimate users. All 140 adversary nodes are randomly activated during the attack period with a 50% probability for every second.

16.3.4 Machine Learning (ML) Detection Models

The following section discusses the six machine learning models, two models under sector vector classifiers, used in our research.

TABLE 16.1   Rocketfuel ISP Topology Link Table

LINK NAME	BANDWIDTH (MBPS)		DELAY (MS)	
	MIN	MAX	MIN	MAX
Backbone to backbone	40	100	5	10
Backbone to gateway	10	20	5	10
Gateway to gateway	10	20	5	10
Gateway to leaf node	1	3	10	70

1. *Decision tree classifier [21]*: Decision trees are a supervised learning method used for classification and regression. Decision trees learn from data to approximate a sine curve with a set of if-then-else decision rules. The deeper the tree, the more complex the decision rules, and the fitter the model.
2. *Support vector machine classifier with radial kernel [22] and with linear kernel [23]*: The objective of the support vector machine algorithm is to find a hyperplane in N-dimensional space (N as the number of features) that distinctly classifies the data points. The radial kernel is for the nonlinear regression problem, and the linear kernel is for a linear problem when the feature space can be easily split by a single straight plane.
3. *K-nearest neighbors classifier [24]*: K-nearest neighbors (KNN) algorithm uses similarity of feature to predict the values of new data. Then KNN computes the distance between new data and all train data, choosing K numbers of shortest distance as the prediction result.
4. *Random forest classifier [25]*: A random forest consists many individual decision trees that run as an ensemble. Each tree in the random forest represents one class prediction, and the class with the most votes become our model's result.
5. *Extremely randomized trees classifier [26]*: Extremely randomized trees classifier (extra trees classifier) is a type of ensemble learning algorithm that aggregates the results of multiple decorrelated decision trees collected in a forest to output its result. Extra trees classifier is very similar to a random forest classifier and only differs from it in constructing the decision trees in the forest. Unlike random forest, which uses a greedy algorithm to select an optimal split point, the different trees algorithm selects a split point at random.
6. *Logistic regression with LASSO [27]*: Logistic regression is a statistical approach that uses a logistic function to classify the input into a binary variable. However, this model always faces overfitting issue. For that, as a method of regularization, LASSO helps the model improve the issue to enhance the interpretability of the resulting statistical model.

16.4 PERFORMANCE EVALUATION

In the following section, we first introduce how to prepare the IFA dataset, set up the experiments, and evaluate the performance of IFA and the result of IFA performance. We then introduce how to prepossess the dataset collected. Lastly, we introduce how to evaluate the performance of machine learning and then show the result.

16.4.1 Methodology

Based on the attacking scenarios in section 16.3, we generate IFA network traffic. As mentioned in section 16.3, we have defined two attacking scenarios, small scale and large scale. We use ns3 [28] as our main simulation platform and NDNsim [29] as the running protocol in ns3.

For the configuration of both attacking scenarios, we set up the following: (1) The simulation is fixed to 900 seconds. (2) All the legitimate users only send the request at a maximized data rate but do not bypass network congestion. This variable is calculated on the network topology configuration, including numbers of users, locations of IoTSE, and the link table in both scenarios. (3) We launch the attack from 300 to 600 seconds. For all the adversary users, those are randomly active to attack with a 50% probability for each second, with a fixed data generating rate of 1,000 packets per second.

To evaluate attacking performance, we consider the following performance metric to qualify the impact of attacks: the *packet successful ratio*, which refers to successful packet reception ratio of legitimated users. During the whole simulation time, we keep tracking all network activities in all users (including IoTSE), including received or forward packet numbers received or forward request numbers and the timestamp of each node behavior. We can easily calculate the successful packet reception of all legitimate users for each second based on those. If this rate drops, the attack is successful. Ideally, if the rate becomes zero, the attack damage is maximized.

We use the statistic-based machine learning models described in section 16.3.4 via sklearn [30]. We also randomly split the dataset into 20% test dataset and 80% training dataset. In addition, we define the following performance metrics to measure the detection results of ML based models: (1) *Highest training accuracy*: During the training, we keep recording the highest training accuracy and use that configuration for further testing. (ii) *Average testing accuracy*: During the test, we record each test result and compute the average of actual results to measure the overall accuracy. (iii) *Average precision and average recall in testing*: During the test session, we record each time test with precision and recall, computing an average result. (iv) *f1-score*: Since this is a biclassifying problem, the F1 score would be a better metric to determine performance, based on the previous two metrics in testing.

16.4.2 IFA Performance

We now show the performance of IFA. Then based on those effective datasets, we introduce how to prepossess the dataset.

16.4.2.1 Simple Tree Topology (Small Scale)

In Figure 16.6, at time 0 second, all legitimate users start to send the requests to IoTSE. Since the routing table is empty initially, it needs more time to build the link and keep that helpful information in its interface table. For that, the packet successful ratio increases from 90% to 99% during the 0–20 seconds. Then the ratio is stable until 300 seconds because those transmission paths and interfaces to be used are in the routing table. From 300 to 600 seconds, it is the attacking period. From the figure, we can see that the packet successful ratio is dropping immediately to around 70%. During this period, the ratio is under 20%, even close to zero, like 320–450 seconds. Since only two nodes are adversary nodes in four nodes, there is only a 50% chance to be active at every time. After 600 seconds, the ratio recovers to 99% and then works on stable performance before the attack. In this scenario, IFA is working as well as we expected.

16.4.2.2 Rocketfuel ISP like Topology (Large Scale)

Figure 16.7 shows that, at the beginning, all the nodes are building routing tables just as in the previous one. The packet success ratio is stable until the 300 seconds, and since there are 140 adversary nodes out of 296 nodes, the packet successful ratio is meager, almost close to zero during the attack period. This proves that IFA is working well on this large scale.

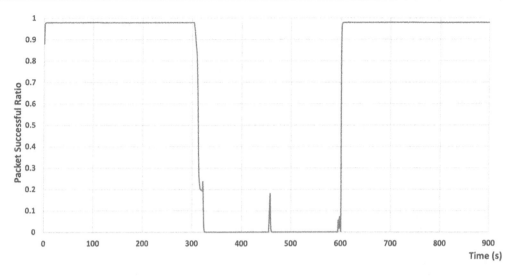

FIGURE 16.6   Packets successful ratio of simple tree topology (small scale).

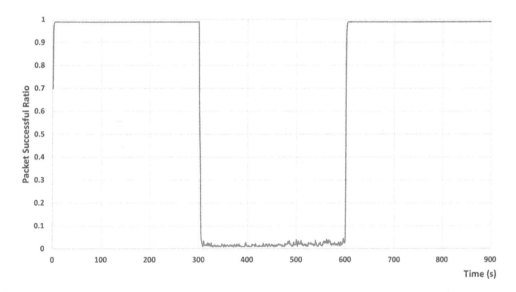

FIGURE 16.7   Packets successful ratio of Rocketfuel-ISP-like topology (large scale).

16.4.3 Data Processing for Detection

Based on what we collected in previous IFA data generation. We summarize those network traffic records as follows: (1) *Simple tree topology network traffic*: In Table 16.2 we have large amounts of network traffic, around 229,000, for all the nodes (mixed of legitimated users, adversary users, and IoTSE). We find that related network traffic to legitimate and adversary users was about 20% of total records; the rest were middle pipeline traffic. (2) *Rocketful-ISP-like topology network traffic*: In Table 16.3, we have over 12 million records but find that, similarly to the previous result, the directly related network traffic of legitimate and adversary users are about 15% of total records.

TABLE 16.2   Simple Tree Topology Network Traffic Records

NODE TYPE	# OF NETWORK FLOW RECORD
Middle gate record	189,655
legitimated user record	23,374
Adversary record	15,574
Total	228,603

TABLE 16.3   Rocketful-ISP-like Topology Network Traffic Records

NODE TYPE	# OF NETWORK FLOW RECORD
Middle gate record	10,630,045
Legitimated user record	1,005,082
Attacking record	611,895
Total	12,247,022

TABLE 16.4   Reduced Simple Tree Topology Network Traffic Records

NODE TYPE	# OF NETWORK FLOW RECORD
Legitimated user request	1,799
Adversary user request	1196
Total	2,995

TABLE 16.5   Reduced Rocketful-ISP-like Topology Network Traffic Records

NODE TYPE	# OF NETWORK FLOW RECORD
Legitimated user request	140,245
Adversary user request	83,885
Total	224,130

After summarizing those two datasets, we decided to reduce dataset density and dimensions for the following reasons: (1) The scope of our chapter is to determine whether the request is from legitimated or adversary users. Thus we only consider the requests directly coming from users, not in the middle pipeline. We keep only records directly coming from adversary nodes and legitimated users. Because we only want to analyze requests, we deleted all nonrequest data types. (2) Since we only run one protocol (NDN), we only keep out-forwarding activities of network interfaces in the pending interest table. (3) Because adversaries send fake interest requests to all data producers, the namespace does not matter the adversary to overload network. Thus we skip analyzing different kinds of namespace in legitimated users.

Due to those reasons, the datasets are briefed into Table 16.4 and Table 16.5. There are around 3,000 records for the small scale, including 1,800 records about legitimate users and 1,200 records about adversary users. There are over 224,000 records in total for the large scale, including 140,000 records about legitimate users and 83,000 records about adversary users. Each record has two features: the number of sent packets and the label of attacks. Finally, we transform this complicated task into a simplified task to build a binary classifier.

16.4.4 Detection Results

16.4.4.1 ML Detection Performance in Simple Tree Topology

Table 16.6 shows that each of these eight models has a similar result with the testing dataset. The decision tree classifier has the best train accuracy. We also find some of them have the same result in certain metrics like recall in K-nearest neighbors classifier recall in random forest classifier. This is due to the limitation of volume for the dataset, which is only around 3,000. Nevertheless, it is still observed that the adopted eight models can effectively recognize this NDN network traffic and detect illegal requests.

16.4.4.2 ML Detection in Rocketful ISP Topology

Table 16.7 shows that the nonlinear models (the first five models) have the better test accuracy than those linear models (the last three models). These five nonlinear models have much better training accuracy than those three linear models. This time, the volume of the dataset is 100 times the previous one; those three linear models are not capable of learning such an amount of dataset well. If we compare with the result in simple tree topology, we also find out those nonlinear models are more capable and adoptive to a large dataset than those linear models.

We also found some interesting results in both scenarios. For example, the support vector machine classifier with a linear kernel in simple tree topology has 100% recall in Table 16.6. That is because the model predicts all legitimate network traffic as legitimate behaviors during the test (he false-negative case is 0). So it makes perfect recall. A similar result is found in the logistic regression with LASSO in Table 16.6. We also found another alternative result (higher precision) in Table 16.7. In this case, that model recognizes all malicious network traffic as malicious behaviors during the test (the false-positive case decreased to 0). So it makes the precision extraordinarily high.

TABLE 16.6   ML Detection Performance in Simple Tree Topology

MODEL	TRAIN DATASET	TEST DATASET			
	HIGHEST ACCURACY	PRECISION	RECALL	AVERAGE ACCURACY	F1 SCORE
Decision tree classifier	**0.824738183**	**0.729420732**	**0.988636364**	**0.847245409**	**0.839473684**
Support vector machine classifier with radial kernel	0.821038475	0.717505635	0.986570248	0.837646077	0.830795998
K-nearest neighbors classifier	0.719165424	0.729420732	0.988636364	0.847245409	0.839473684
Random forest classifier	0.824738183	0.729420732	0.988636364	0.847245409	0.839473684
Extremely randomized trees classifier	0.827899589	0.729420732	0.988636364	0.847245409	0.839473684
Support vector machine classifier with linear kernel	0.670561362	0.677871148	1	0.808013356	0.808013356
Logistic regression with LASSO	0.69240877	0.698916968	1	0.825959933	0.822779431

TABLE 16.7   ML Detection Performance in Rocketful-ISP-like Topology

	TRAIN DATASET	TEST DATASET			
MODEL	HIGHEST ACCURACY	PRECISION	RECALL	AVERAGE ACCURACY	F1 SCORE
Decision tree classifier	**0.999135667**	**0.999582189**	**0.998717834**	**0.99936413**	**0.999149825**
Support vector machine classifier with radial kernel	0.999046706	0.999254321	0.998941468	0.999325086	0.99909787
K-nearest neighbors classifier	0.999016291	0.99928386	0.998568745	0.999196796	0.998926174
Random forest classifier	0.998987084	0.99952258	0.998822196	0.999380864	0.999172266
Extremely randomized trees classifier	0.998986978	0.999522652	0.998971285	0.999436642	0.999246892
Support vector machine classifier with linear kernel	0.662324974	0.999909834	0.496004413	0.811425575	0.999246892
Logistic regression with LASSO	0.652869166	1	0.478680264	0.8049609	0.647442555

16.5 DISCUSSION

There are some potential future directions for IFA in NDN-based IoTSE, concerning the improvement of detection, launching more comprehensive IFA, and IFA mitigation.

Improvement of detection: In this chapter, we assume the adversary does not know the profile of the IoTSE, including the naming scheme of the interest request. Thus the adversary can only target PIT in the NDN router by sending massive, faked interest requests quickly. We consider a packets rate as a detection feature and leverage ML-based schemes to carry out anomaly detection. Since legitimate users have a static packet sampling rate, we can easily identify abnormal behaviors. Nonetheless, supposing the adversary has knowledge of part of setting the user's profile (e.g., request sampling rate), the adversary can use the same configuration, and the adversary's behavior will be indistinguishable from legitimated users. Therefore, the detection scheme based on packet rate will be ineffective. To deal with such an issue, we shall consider other schemes, e.g., setting more rules [8] to restrict the network behaviors, designing an attributed-based detection algorithm [31] to identify unusual traffic statistics in PIT entity, and using graph neutral network-based schemes [32] to detect unusual network traffic, as well as to investigate features that are fundamental and difficult for the adversary to manipulate [33].

Studying comprehensive IFA: In this chapter, we set the adversary with a fixed sampling rate with a 50% probability to be on-off for each second and assume that the adversary does not know the profile of IoTSE. This kind of attack can evolve into a much more comprehensive one. There are two potential ways to hinder the detection: (1) The adversary can send those nonexistent interest packets to the adversary's IoTSE. Then the adversary's IoTSE responds to those requests by sending data back. Since the target is to overuse the whole bandwidth of the network instead of the IoTSE, the detection based on packet rate as a feature in IoTSE would consider that malicious traffic as normal traffic of other IoTSE. (2) With the knowledge of the setting configuration for IoTSE and users, the adversary can launch satisfied interest packets with random selections or nonexistent interest requests with a known naming scheme; therefore, those malicious requests become indistinguishable from those of legitimate users. Thus it can cause more damage to the network and become more challenging to detect.

IFA mitigation: To better mitigate the IFA, we intend to design a countermeasure in the future study. There are some existing methods [34–37] to deal with IFA in NDN. Nonetheless, there are few limitations to those methods: (1) The efficacy of those algorithms highly depends on the shapes of topology. (2) Those designs are for single detection points, i.e., one single PIT table or one single interface. Thus we plan to investigate more comprehensive mitigation schemes in our future study. One is to consider different detection points to increase detection accuracy and reduce the detection sensitivity on the different topologies. The other is to leverage with deep learning-based schemes to deal with large-scale and complex IFA.

16.6 RELATED WORKS

In the following, we review various research efforts that are closely relevant to our study.

Recently, there have been a number of existing research efforts with ML to carry out tasks such as enhancement of IP-based network security [38–40]. For example, Doshi et al. [38] applied a low-cost ML-based algorithm into a home router to pervert DoS attack, which proves that it can be a high detection accuracy. He et al. [39] developed a statistics-based ML algorithm to mitigate DoS attacks in the cloud, which leads to a high detection accuracy and effective mitigation. Likewise, Santos et al. [40] used a ML algorithm in a SDN-based network to detect the DoS attack to secure network performance.

There are also some existing efforts on improving NDN security [41–42]. For example, Afanasyev et al. [41] developed four different schemes to mitigate flooding attacks in NDN for different levels of efficiency. Likewise, Khelifi et al. [42] leveraged blockchain in a NDN-based vehicle network to secure the cache in the vehicular environment and enhance the trust between cache stores and consumer vehicles. In addition, there are some research efforts on IoTSE [11, 43]. For example, Liang et al. [11] apply NDN in IoTSE and tested the performance of data transmission in different IoTSE scenarios, confirming that NDN can improve network performance. Likewise, Cheng et al. [43] designed a new neutral network structure with a feature map over the dataset to obtain better performance in image classification. Also, there are some research efforts for public safety and related applications in IoTs [44–45]. For example, Zhu et al. [44] designed an IoT-gateway to improve network reliability and network capacity utilization in large-scale communication. Likewise, Thirumalai et al. [45] developed an efficient public key secure scheme for cloud and IoT security. However, those existing works have heavy and complex network traffic to track in IP-based networks and cost more resources and time for deploying in machine learning models.

In addition, those methods are not suitable for the nature of IoTSE, which is a data-driven structure instead of location driven. In our study, we tackle the problem in two steps in the IoTSE scenario. First, we use NDN as our network communication protocol since NDN is a data-driven, pull-push model, the match for IoTSE. Without the address of the source and destination, the volume of network traffic records in NDN is reduced. Second, we simplify the dataset of network traffic for machine learning. Since the IoTSE has a static and particular task profile, the packet sending rate of all kinds of applications should be in a relatively stable range. Thus we can efficiently detect suspicious packets if the packet sending rate is out of the range in profile. We also reduce the size of dataset in order to reduce the cost of time and resources in training process.

16.7 FINAL REMARKS

This chapter has investigated three attacking scenes for interest flooding attacks (IFA) in NDN-based IoTSE. Using ns3 as the evaluation platform, we have designed two IFA scenarios with a nonexistent interest flooding attack strategy in small- and large-scale simulations, utilizing different link configurations. We first show the severity of the damage caused by IFA. Then, based on the collected dataset of the successful IFA, we process the dataset and reduce it into only the out-request and outer network interface in the PIT-related dataset. Finally, we have obtained a dataset with two features: packets and the label of attack. We compared the eight ML models to carry out attack detection and concluded that most ML models could effectively detect this kind of attack in a packet-rate-based dataset. However, we find that the linear-based ML model obtains worse detection accuracy when the dataset volume increases tremendously.

ACKNOWLEDGMENT

This material is based upon work supported by the Air Force Office of Scientific Research under award number FA9550-20-1-0418. Any opinions, findings, and conclusions or recommendations expressed in this material are those of the author(s) and do not necessarily reflect the views of the United States Air Force.

REFERENCES

[1] J. Lin, W. Yu, N. Zhang, X. Yang, H. Zhang, and W. Zhao, "A survey on internet of things: Architecture, enabling technologies, security and privacy, and applications," IEEE Internet of Things Journal, vol. 4, no. 5, pp. 1125–1142, 2017.

[2] H. Xu, W. Yu, D. Griffith, and N. Golmie, "A survey on industrial internet of things: A cyber-physical systems perspective," IEEE Access, vol. 6, pp. 78 238–78 259, 2018.

[3] J. Zhang and D. Tao, "Empowering things with intelligence: a survey of the progress, challenges, and opportunities in artificial intelligence of things," IEEE Internet of Things Journal, vol. 8, no. 10, pp. 7789–7817, 2020.

[4] F. Liang, C. Qian, W. G. Hatcher, and W. Yu, "Search engine for the internet of things: Lessons from web search, vision, and opportunities," IEEE Access, vol. 7, pp. 104 673–104 691, 2019.

[5] A. Abraham, F. Pedregosa, M. Eickenberg, P. Gervais, A. Mueller, J. Kossaifi, A. Gramfort, B. Thirion, and G. Varoquaux, "Machine learn- ing for neuroimaging with scikit-learn," Frontiers in neuroinformatics, vol. 8, p. 14, 2014.

[6] M. Mayhew, M. Atighetchi, A. Adler, and R. Greenstadt, "Use of machine learning in big data analytics for insider threat detection," in MILCOM 2015–2015 IEEE Military Communications Conference. IEEE, 2015, pp. 915–922.

[7] R.-T. Lee, Y.-B. Leau, Y. J. Park, and M. Anbar, "A survey of interest flooding attack in named-data networking: Taxonomy, performance and future research challenges," IETE Technical Review, pp. 1–19, 2021.

[8] A. Afanasyev, J. Burke, T. Refaei, L. Wang, B. Zhang, and L. Zhang, "A brief introduction to named data networking," in MILCOM 2018–2018 IEEE Military Communications Conference (MILCOM), 2018, pp. 1–6.

[9]   S. Shannigrahi, C. Fan, and C. Partridge, "What's in a name?: Naming big science data in named data networking." Proceedings of the 7th ACM Conference on Information-Centric Networking, pp. 12–23, 2020.

[10] L. Zhang, A. Afanasyev, J. Burke, V. Jacobson, K. Claffy, P. Crowley, C. Papadopoulos, L. Wang, and B. Zhang, "Named data networking," ACM SIGCOMM Computer Communication Review, vol. 44, no. 3, pp. 66–73, 2014.

[11] H. Liang, L. Burgess, W. Liao, C. Lu, and W. Yu, "Towards named data networking for internet of things search engine," 2021. [Online]. Available: to be published.

[12] F. Liang, W. Yu, D. An, Q. Yang, X. Fu, and W. Zhao, "A survey on big data market: Pricing, trading and protection," IEEE Access, vol. 6, pp. 15 132–15 154, 2018.

[13] W. G. Hatcher, C. Qian, W. Gao, F. Liang, K. Hua, and W. Yu, "Towards efficient and intelligent internet of things search engine," IEEE Access, 2021.

[14] H. Liang, L. Burgess, W. Liao, C. Lu, and W. Yu, "Deep learning assist iot search engine for disaster damage assessment," 2021. [Online]. Available: to be published.

[15] X. Yang, K. Lingshuang, L. Zhi, C. Yuling, L. Yanmiao, Z. Hongliang, G. Mingcheng, H. Haixia, and W. Chunhua, "Machine learning and deep learning methods for cybersecurity." IEEE Access, vol. 6, pp. 35 365–35 381, 2018.

[16] T. G. Dietterich, "Steps toward robust artificial intelligence." AI Magazine, vol. 38, no. 3, pp. 3–24, 2017.

[17] M. Capra, B. Bussolino, A. Marchisio, G. Masera, and M. Shafique, "Hardware and software optimizations for accelerating deep neural networks: Survey for current trends, challenges, and the road ahead," IEEE Access, vol. 8, pp. 225 134–225 180, 2020.

[18] W. G. Hatcher and W. Yu, "A survey of deep learning: Platforms, applications and emerging research trends," IEEE Access, vol. 6, pp. 24 411–24 432, 2018.

[19] J. M. Helm, A. M. Swiergosz, H. S. Haeberle, J. M. Karnuta, J. L. Schaffer, V. E. Krebs, A. I. Spitzer, and P. N. Ramkumar, "Machine learning and artificial intelligence: Definitions, applications, and future directions." Current Reviews in Musculoskeletal Medicine, vol. 13, no. 1, p. 69, 2020.

[20] N. Spring, R. Mahajan, and D. Wetherall, "Measuring isp topologies with rocketfuel," ACM SIGCOMM Computer Communication Review, vol. 32, no. 4, pp. 133–145, 2002.

[21] P. H. Swain and H. Hauska, "The decision tree classifier: Design and potential," IEEE Transactions on Geoscience Electronics, vol. 15, no. 3, pp. 142–147, 1977.

[22] B. Scholkopf, K.-K. Sung, C. J. Burges, F. Girosi, P. Niyogi, T. Poggio, and V. Vapnik, "Comparing support vector machines with gaussian kernels to radial basis function classifiers," IEEE transactions on Signal Processing, vol. 45, no. 11, pp. 2758–2765, 1997.

[23] C. Cortes and V. Vapnik, "Support vector machine," Machine learning, vol. 20, no. 3, pp. 273–297, 1995.

[24] L. E. Peterson, "K-nearest neighbor," Scholarpedia, vol. 4, no. 2, p. 1883, 2009.

[25] T. K. Ho, "Random decision forests," in Proceedings of 3rd international conference on document analysis and recognition, vol. 1. IEEE, 1995, pp. 278–282.

[26] P. Geurts, D. Ernst, and L. Wehenkel, "Extremely randomized trees," Machine learning, vol. 63, no. 1, pp. 3–42, 2006.

[27] L. Meier, S. Van De Geer, and P. Bu¨hlmann, "The group lasso for logistic regression," Journal of the Royal Statistical Society: Series B (Statistical Methodology), vol. 70, no. 1, pp. 53–71, 2008.

[28] G. F. Riley and T. R. Henderson, "The ns-3 network simulator," in Modeling and tools for network simulation. Springer, 2010, pp. 15–34.

[29] A. Afanasyev, I. Moiseenko, L. Zhang et al., "ndnsim: Ndn simulator for ns-3," 2012.

[30] F. Pedregosa, G. Varoquaux, A. Gramfort, V. Michel, B. Thirion, O. Grisel, M. Blondel, P. Prettenhofer, R. Weiss, V. Dubourg et al., "Scikit-learn: Machine learning in python," the Journal of machine Learning research, vol. 12, pp. 2825–2830, 2011.

[31] Y. Xin, Y. Li, W. Wang, W. Li, and X. Chen, "A novel interest flooding attacks detection and countermeasure scheme in ndn," in 2016 IEEE Global Communications Conference (GLOBECOM). IEEE, 2016, pp. 1–7.

[32] F. Scarselli, M. Gori, A. C. Tsoi, M. Hagenbuchner, and G. Monfar- dini, "The graph neural network model," IEEE transactions on neural networks, vol. 20, no. 1, pp. 61–80, 2008.

[33] H. Xu, W. Yu, X. Liu, D. Griffith, and N. Golmie, "On data in- tegrity attacks against industrial internet of things," in 2020 IEEE Intl Conf on Dependable, Autonomic and Secure Computing, Intl Conf on Pervasive Intelligence and Computing, Intl Conf on Cloud and Big Data Computing, Intl Conf on Cyber Science and Technology Congress (DASC/PiCom/CBDCom/CyberSciTech), 2020, pp. 21–28.

[34] H. Dai, Y. Wang, J. Fan, and B. Liu, "Mitigate ddos attacks in ndn by interest traceback," in 2013 IEEE Conference on Computer Communications Workshops (INFOCOM WKSHPS). IEEE, 2013, pp. 381–386.

[35] P. Gasti, G. Tsudik, E. Uzun, and L. Zhang, "Dos and ddos in named data networking," in 2013 22nd International Conference on Computer Communication and Networks (ICCCN). IEEE, 2013, pp. 1–7.

[36] T. Zhi, H. Luo, and Y. Liu, "A gini impurity-based interest flooding attack defence mechanism in ndn," IEEE Communications Letters, vol. 22, no. 3, pp. 538–541, 2018.

[37] A. Compagno, M. Conti, P. Gasti, and G. Tsudik, "Poseidon: Mitigating interest flooding ddos attacks in named data networking," in 38th annual IEEE conference on local computer networks. IEEE, 2013, pp. 630–638.

[38] R. Doshi, N. Apthorpe, and N. Feamster, "Machine learning ddos detection for consumer internet of things devices," in 2018 IEEE Security and Privacy Workshops (SPW). IEEE, 2018, pp. 29–35.

[39] Z. He, T. Zhang, and R. B. Lee, "Machine learning based ddos attack detection from source side in cloud," in 2017 IEEE 4th International Conference on Cyber Security and Cloud Computing (CSCloud). IEEE, 2017, pp. 114–120.

[40] R. Santos, D. Souza, W. Santo, A. Ribeiro, and E. Moreno, "Machine learning algorithms to detect ddos attacks in sdn," Concurrency and Computation: Practice and Experience, vol. 32, no. 16, p. e5402, 2020.

[41] A. Afanasyev, P. Mahadevan, I. Moiseenko, E. Uzun, and L. Zhang, "Interest flooding attack and countermeasures in named data networking," in 2013 IFIP Networking Conference, 2013, pp. 1–9.

[42] H. Khelifi, S. Luo, B. Nour, H. Moungla, and S. H. Ahmed, "Reputation- based blockchain for secure ndn caching in vehicular networks," in 2018 IEEE Conference on Standards for Communications and Networking (CSCN). IEEE, 2018, pp. 1–6.

[43] G. Cheng, Z. Li, J. Han, X. Yao, and L. Guo, "Exploring hierarchical convolutional features for hyperspectral image classification," IEEE Transactions on Geoscience and Remote Sensing, vol. 56, no. 11, pp. 6712–6722, 2018.

[44] Q. Zhu, R. Wang, Q. Chen, Y. Liu, and W. Qin, "Iot gateway: Bridging- wireless sensor networks into internet of things," in 2010 IEEE/IFIP International Conference on Embedded and Ubiquitous Computing. Ieee, 2010, pp. 347–352.

[45] C. Thirumalai, S. Mohan, and G. Srivastava, "An efficient public key secure scheme for cloud and iot security," Computer Communications, vol. 150, pp. 634–643, 2020.

Attack on Fraud Detection Systems in Online Banking Using Generative Adversarial Networks

17

Jerzy Surma and Krzysztof Jagiełło

Department of Computer Science and Digital Economy,
Warsaw School of Economics, Poland

Contents

DOI: 10.1201/9781003187158-21

17.1 INTRODUCTION

17.1.1 Problem of Fraud Detection in Banking

Millions of transactions are registered daily in electronic banking systems. Most of them are legal operations, but a small percentage are attempts at illegal activities. The *Black's Law Dictionary* [1] defines fraud as "a knowing misrepresentation of the truth or concealment of a material fact to induce another to act to his or her detriment." Fraud is usually a tort, but in some cases (especially where the conduct is willful), it may be a crime. The fraud causes significant losses for banks and heavyweight problems for users. Fraud detection describes a set of activities undertaken to identify financial fraud.

An essential element of fraud detection systems currently being developed is data mining, i.e., discovering significant and previously unknown patterns in large data sets [2]. Several dozen techniques are used in this context [3]. One of the simplest is logistic regression [4], and other widely used methods are the support vector machine (SVM) [5] and decision tree techniques. In recent years, neural networks have gained particular popularity [6]. Machine learning techniques can be categorized into supervised and unsupervised learning techniques. *Supervised* learning means it is clearly defined whether or not a specific transaction is fraudulent. Based on such data, a fraud detection model is created. On the other hand, *unsupervised* learning refers to a situation in which a model was created based on data without class assignment. In this way, transactions that are not necessarily fraudulent but significantly deviate from the pattern are identified [7].

In this chapter, we look at the use of the generative adversarial networks (GAN) approach that was introduced by [8] and is currently one of the most effective types of generative modeling. The use of neural networks allows one to obtain results unattainable with classical methods. However, they generally require much information to operate effectively and are also characterized by a prediction time that is often clearly longer than, for example, logistic regression [9].

17.1.2 Fraud Detection and Prevention System

The schematic flow of the process leading to the implementation of the fraud detection model can be broken down into three primary phases [2]:

1. *Data preprocessing*: problem description, identification and selection of data sources, data cleansing, data processing
2. *Data analysis and model creation*,
3. *Model management*: interpretation of the model results, model evaluation, implementation and maintenance of the model

The initial and fundamental element of the process is the identification of the business problem. Then potential data sources that can be used in the analysis are identified and selected by suitability in the context of the issue under consideration and availability. Once the data sources have been identified, a careful review and selection of variables take place. The next step is to clean the data and make the necessary transformations. These steps constitute the data preprocessing phase, which is generally the most time-consuming part of the process. After its completion, the proper analysis and creation of a fraud detection model may take place. The last phase relates to the model itself: its evaluation, interpretation of results, and – with satisfactory quality – implementation are performed. Figure 17.1 shows a diagram of the so-called fraud cycle in the banking fraud detection system (BFDS). It consists of four essential elements:

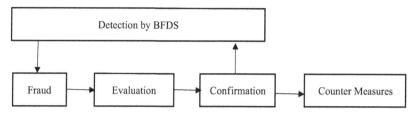

FIGURE 17.1 Fraud cycle.

Source: Authors' study based on Baesens et al. [10].

1. *Detection*: a prediction made on new transactions and assigning each of them a probability of being fraudulent
2. *Evaluation*: evaluation, often by an expert, of transactions identified as suspicious
3. *Confirmation*: assessment of the truthfulness of the suspicion that the transaction is fraudulent
4. *Countermeasures*: development of countermeasures in order to prevent similar fraud instances from occurring

These elements can create an effective system only when newly detected fraud is immediately added to the pool of historically detected ones. The BFDS should then be updated based on this enlarged learning set.

Some of the first commercially used systems were based on fuzzy logic. They could apply a specific fraud assessment policy to select the optimal threshold values [11]. The result of the system's operation was the likelihood of fraud. It is assumed that the quality of the assessment was comparable to that given by the expert. The credit fraud detection model introduced later [12] used the classification approach. Classification-based techniques have been systematically developed and have repeatedly proven their effectiveness in many business applications [13]. An alternative approach is to build regression models. Classification and regression may also be used complementarily for fraud detection [10].

17.2 EXPERIMENT DESCRIPTION

17.2.1 Research Goal

The fraud problem in electronic banking poses a significant threat to banks in terms of financial, reputational, and relational damage. Each BFDS vulnerability, mainly unidentified ones, can lead to significant losses. The development of technologies related to the generation of specific data distributions may make it possible to launch attacks of unprecedented specificity.

Possible attacks on BFDS systems, due to the availability of information about the model, are the so-called gray-box attacks. In this model, attackers have partial knowledge – e.g., they know its parameters or have obtained a data set used to build it [14]. According to Surma, adversarial attacks taxonomy [15]; this is an attack on the integrity of a machine learning system.

In our approach, a gray-box attack is presented, and the study aims to verify two hypotheses:

1. It is possible to imitate the characteristics of fraud and legal transactions.
2. Artificially generated transactions can be used to attack BFDS successfully.

The hypotheses are verified with the following assumptions:

- The hacker has complete knowledge about the learning set for BFDS creation.
- The hacker does not have direct access to BDFS.
- The hacker can have access to different bank accounts.

The attacker's scenario generally occurs in two phases of the preparation and execution of an attack. The attacker builds the so-called imitation fraud detection system (IFDS) in the first phase, which maps the real BFDS. It is based on a dataset that the attackers illegally obtained. In the second phase, IFDS enables quality testing of artificially generated transactions using GAN.

After such actions, the attacker can use artificially generated transactions to conduct an attack such as:

- fooling the fraud surveillance team by generating a false-negative error in the BFDS,[1]
- denial of service of transaction systems infrastructure.

The threats may be interdependent. In particular, attackers may trigger the occurrence of one of them in order to make the occurrence of another more probable. For example, to increase the likelihood that the fraud remains undetected, attackers may try to paralyze the work of the antifraud team in advance.

17.2.2 Empirical Data

The data set used for the analysis contains actual data collected by the systems supporting electronic banking in one of the biggest retail banks in Poland. The data comes from 2015 and 2016 and consists of 237,034 transactions, 163 of which were identified by the bank as fraudulent. The set consists of 122 variables. Table 17.1 shows some of them. The fraud variable, its dependent variable, takes the values 1 and 0 to mark the transaction as fraudulent or legal appropriately. The remaining 121 variables are the explanatory variables of the BFDS.

The set is characterized by a clearly unbalanced distribution of the target variable. The minority class constitutes only 0.07% of the majority class.

17.2.3 Attack Scenario

As already mentioned, the concept of generative adversarial networks is used to carry out the attack. The reason is the high quality of the generated data (in particular continuous data), which is crucial for the attack scenarios' success.

TABLE 17.1   List of Variables

VARIABLE NAME	DESCRIPTION
Fraud	Whether the transaction has been marked as fraudulent (dependent variable)
avg_amount	Average transaction amount
security_level	Authorization level
status	Status of the transaction
which_transaction	Which transaction is part of logging
local_time	Time of the transaction
service_time	Total duration of invoked services
no_of_errors	Number of errors during login

Source: Author's compilation.

The attack scenario steps are as follows:

1. The BFDS learning set is acquired (stolen) by attackers.
2. Access to accounts is gained, enabling transactions.[2]
3. Create an IFDS to emulate BFDS.
4. Create a model, generator, based on the GANs approach.
5. Use the generator from step 4 to generate a dataset imitating a specific distribution.
6. Test the data from step 5 on the classifier created in step 3 (IFDS).
7. With satisfactory results of step 6, use the generated data in actual banking transactions, i.e., attack BFDS.

The purpose of generating data that appears to be legitimate transactions is to produce a set of transactions that does not raise an alert by BFDS. This means:

- Forcing a false-negative classification for an attacker to carry out a fraudulent transaction
- Generating transactions that appear to be legal in significant numbers.

Assuming that the computational cost of processing each transaction is > 0, and the banking infrastructure is scaled to a limited extent, it is theoretically possible to carry out a DDoS attack on the BFDS infrastructure in this way. It is also possible to generate transactions that look like financial frauds (false-positive errors). Raising many false-positives, carried out at high intensity in the short term, can paralyze the functioning of the antifraud team. In the extreme case, multiple false-positive alerts can strongly create a lack of confidence in the BFDS.

The practical use of those scenarios is highly complicated and requires significant resources and coordination of transactions from multiple mule accounts.

17.3 GENERATOR AND DISCRIMINATION MODEL

17.3.1 Model Construction

17.3.1.1 Imitation Fraud Detection System Model

The imitation fraud detection system was created to reproduce the BFDS based on the learning set used for its construction without any additional information. It should clearly underline that the IFDS is only an approximation of the BFDS. The construction of the IFDS is based on the assumption that the BFDS was built according to best practices; therefore, if the currently built model is also created in this way, their convergence will be relatively high.

Selecting the appropriate algorithm used to build the IFDS is key to meeting the best practice assumption. It was made taking into account the following criteria: It is the binary classification; it is effective when there is a clear imbalance in the distribution of the target variable and has an acceptable prediction time. Based on [16] and [17], it was decided that the Extreme Gradient Boosting (XGBoost) algorithm would be the most appropriate in this context [18]. The selection of the algorithm configuration itself and the preparation and division of the dataset are essential. Using the same criteria based on [19] and a review of the preceding methods in the event of imbalance in the distribution of the target variable, it was decided to build a model

- on the entire data set using tenfold cross-validation,
- with hyperparameters selected using Bayesian optimization [19].

TABLE 17.2   Quality of IFDS Classification

ACCURACY	PRECISION	RECALL	F1 SCORE
0.99	0.69	0.66	0.68

Source: Author's compilation.

Table 17.2 shows the averaged results of the IFDS classification. The results were obtained by calculating each of the presented measures separately for each of the ten created models and then calculating their arithmetic mean. The classification accuracy = 0.99 of the IFDS is very good.

17.3.1.2 Generator Models

A total of two pairs of models (generator and discriminator) were created in order to generate attacks forcing IFDS to classify false-negative–model 1 and false-positive–model 2. The generator model is crucial because it produces data that looks like a specific type of transaction used to attack BFDS. For each GAN, the discriminant objective function is based on the truthfulness of the prediction.

The primary modification in relation to the original GAN concept introduces several changes based on the method of [20], the main element of which is implementing Wasserstein–GAN instead of the typical GAN. Additionally, mapping was applied at the generator learning stage after leaving the last layer but before calculating the objective function, which takes the closest available value for categorical variables. Both models use an actively updated learning coefficient thanks to stochastic optimization [21]. Due to the unequal difficulty of the generator and discriminator tasks, the ratio of their repetitions within one iteration was experimentally selected and set at 5:4, respectively. For each model, a limitation of the learning time to 200 epochs was introduced while adopting the criterion of an early end [22], which each time finished learning before reaching the limit. Additionally, based on the simplified concept described by Ham [23], for each pair of models, generator and discriminator, core learning was preceded by initial discriminator learning. The implementation was made in Python using the TensorFlow library.[3]

First, a model 1 imitating legal transactions was created (models 1, respectively). The most critical steps of building model 1 (see Figure 17.2) are

1. downloading all available legal transactions (236,871) and splitting them into 20 batches,
2. generating consecutive transactions (one at a time) within a batch,
3. the learning process based on the download of batches taken from step 2,
4. repeating steps 2 and 3 for subsequent epochs,
5. downloading the constructed generator model,
6. IFDS testing.

Second, a model 2 imitating fraud transactions was created using 163 fraud cases. The general approach is the same as presented for model 1.

17.3.2 Evaluation of Models

The evaluation of GANs-based models was twofold: internal and external. *Internal* evaluation is based on the mutual prediction of the generator and the discriminator (it is used only in step 3 of model learning). However, it is burdened with many potential disadvantages, the main of which is the possibility of falsely overestimating the statistics of one of the models due to the poor quality of the other. It should only be interpreted during the model construction phase and is purely technical. A much more reliable method is *external* evaluation, which is the IFDS testing (step 6).

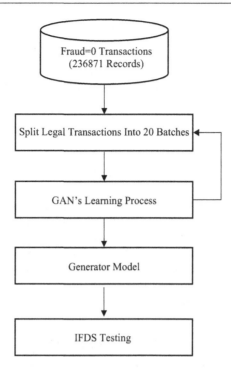

FIGURE 17.2   Construction of generator models imitating legal transactions (model 1).

Source: Author's compilation.

TABLE 17.3   Classification Accuracy of IFDS

	ACCURACY
Model 1	0,002
Model 2	0,831

Source: Authors' compilation.

Data generated by all generator models were IFDS classified with the same number (1,000) and with the same generator input vector. Classification accuracies for models 1 and 2 of IFDS are shown in Table 17.3. The purpose of the attack is to cause incorrect classification by IFDS, i.e., low classification accuracy determines the high effectiveness of the attack. As shown in Table 17.3, attacks aimed at provoking the false-negative classification (model 1) are incomparably more effective than those trying to force false-positive predictions (model 2).

Table 17.4 shows the difference between the proportion of a given class in a given set and the percentage of successful attacks on IFDS that enforce it. For legal transactions, it is small (the attack using model 1). On the other hand, the percentage of successful attacks (using model 2) forcing a false-positive classification was over 200 times greater than the representation of fraud in the data set.

The conducted experiment shows that it is possible to force a relatively frequent misclassification into IFDS using data generated using GAN. The percentage of successful extortions of false-negative classifications from IFDS is very high and comparable to the representation of legal transactions in the dataset. Model 1 reaches the percentage of successful false-negative classification extortions that justify the thesis that an effective attack on BFDS can be carried out. The accuracy of IFDS classification for attacks using model 1 is only 0.002, which means that the detection system is actually unable to detect an attack.

TABLE 17.4   Inside into IFDS Performance

	Fraud = 0 transactions	False-negative classification
Model 1	> 99%	> 99%
	Fraud = 1 transactions	**False-positive classification**
Model 2	< 1%	17 %

Source: Authors' compilation.

The forcing of a false-positive classification on IFDS was effective from several dozen to several hundred times more often than the representation of fraud in the dataset. Nevertheless, the accuracy of the IFDS classification of 0.831 for an attack using model 2 indicates that an attack scenario enforcing a false-positive classification is rather unlikely. The high quality of IFDS results with attacks forcing a false-positive classification may be primarily the result of a clearly greater representation of legal transactions than frauds in the IFDS learning set or a greater diversification of frauds compared to legal transactions.

17.4 FINAL CONCLUSIONS AND RECOMMENDATIONS

As shown, at least an attack on BFDS based on the false-negative error is possible to perform. The primary recommended method of mitigating the risk related to an attack on BFDS is to place special supervision over the data describing transactions and the models themselves related to fraud detection. In addition, it is worth considering the introduction of a policy resulting in faster blocking of subsequent queries of the model when suspecting an irregularity aimed at counteracting the presumption of BFDS activity by attackers. A more frequent update of models may also have a preventive effect on a coordinated attack, which, in addition to the possibility of replacing a model that attackers have somehow recognized, creates an element of uncertainty that is beneficial in terms of security.

At the stage of building BFDS, it is worth considering strengthening the relationship between the variables that can be modified by the attackers and those that are unavailable to them. Additionally, it is recommended to extend the consistency tests between the variables based on predefined rules and extend the number of BFDS explanatory variables related to the history of the client's activity. Actions that can be taken to mitigate the risk of an attack on BFDS using GAN have been proposed. It is recommended that the data describing transactions and models related to fraud detection be subject to special supervision. It is also recommended to introduce a policy resulting in faster blocking of subsequent BFDS activations when irregularities are suspected and more frequent updates.

DISCLAIMER. The technical details and the extensive model evaluation are presented in Krzysztof Jagiello master thesis [24] written under Professor Jerzy Surma's supervision.

NOTES

1   False negative is lack of a warning, while a system under attack is actually hacked.
2   For instance, by using mule accounts: https://en.wikipedia.org/wiki/Money_mule
3   www.tensorflow.org/

REFERENCES

[1] Garner, B. Black's law dictionary. Thomson West, St.. Paul., 2004.

[2] Han, J., Pei, J. i Kamber, M. Data mining: concepts and techniques. Elsevier, 2011.

[3] Albashrawi, M. i Lowell, M. Detecting financial fraud using data mining techniques: a decade review from 2004 to 2015. *Journal of Data Science, 14*(3), 553–569, 2016.

[4] Bhattacharyya, S., Jha, S., Tharakunnel, K. i Westland, J.C. Data mining for credit card fraud: A comparative study. *Decision Support Systems, 50*(3), 602–613, 2011.

[5] Suykens, J.A. i Vandewalle, J. Least squares support vector machine classifiers. *Neural processing letters, 9*(3), 293–300, 1999.

[6] Roy, A., Sun, J., Mahoney, R., Alonzi, L., Adams, S. i Beling, P. Deep learning de-tecting fraud in credit card transactions. *2018 Systems and Information Engineering Design Symposium (SIEDS)*, IEEE, 129–134, 2018.

[7] Bolton, R.J. i Hand, D.J. Statistical fraud detection: A review. *Statistical Science*, 235–249, 2002.

[8] Goodfellow, I., Pouget-Abadie, J., Mirza, M., Xu, B., Warde-Farley, D., Ozair, S., Courville, A. i Bengio, Y. Generative adversarial nets. In: *Advances in neural information processing systems*. MIT Press, 2014.

[9] Justus, D., Brennan, J., Bonner, S. i McGough, A.S. Predicting the computational cost of deep learning models. *2018 IEEE International Conference on Big Data*, IEEE, 3873–3882, 2018.

[10] Baesens, B., Van Vlasselaer, V. i Verbeke, W. Fraud analytics using descriptive, predictive, and social network techniques: a guide to data science for fraud detection. John Wiley & Sons, 2015.

[11] Von Altrock, C. Fuzzy logic and neuro fuzzy applications explained. Prentice-Hall, Inc., 1995.

[12] Groth, R. Data mining: a hands-on approach for business professionals. PTR Santa Clara: Prentice Hall, 1998.

[13] He, H., Wang, J., Graco, W. i Hawkins, S. Application of neural networks to detection of medical fraud. *Expert Systems with Applications, 13*(4), 329–336, 1997.

[14] Surma, J. Attack vectors on supervised machine learning systems in business applications. *Business Informatics, 3*(57), 65–72, 2020

[15] Surma, J. Hacking Machine Learning: Towards The Comprehensive Taxonomy of Attacks Against Machine Learning Systems, *ICIAI 2020: Proceedings of the 2020 the 4th International Conference on Innovation in Artificial Intelligence*, 1–4, 2020

[16] Niu, X., Wang, L. i Yang, X. A comparison study of credit card fraud detection: Supervised versus unsupervised, *arXiv preprint arXiv*:1904.10604, 2019

[17] Zhang, Y., Tong, J., Wang, Z. i Gao, F. Customer transaction fraud detection using xgboost model. *2020 International Conference on Computer Engineering and Application (ICCEA)*, IEEE, 554–558, 2020.

[18] Chen, T. i Guestrin, C. Xgboost: A scalable tree boosting system. *Proceedings of the 22nd ACM SIGKDD international conference on knowledge discovery and data mining*, 785–794, 2016.

[19] Wang, Y. i Ni, X.S. A xgboost risk model via feature selection and bayesian hyper-parameter optimization, *arXiv* preprint arXiv:1901.08433, 2019.

[20] Ba, H. Improving detection of credit card fraudulent transactions using generative adversarial networks, *arXiv preprint arXiv*:1907.03355, 2019.

[21] Kingma, D.P. i Ba, J. Adam: A method for stochastic optimization, *arXiv preprint arXiv*:1412.6980, 2014.

[22] Prechelt, L. Automatic early stopping using cross validation: quantifying the criteria. *Neural Networks, 11*(4), 761–767, 1998.

[23] Ham, H., Jun, T.J. i Kim, D. Unbalanced gans: Pre-training the generator of generative adversarial network using variational autoencoder, *arXiv preprint arXiv*:2002.02112, 2020.

[24] Jagiello, K. Model ataku na system detekcji nadużyć w bankowości elektronicznej z wykorzystaniem generatywnych sieci współzawodniczących. Master Thesis (supervisor: Jerzy Surma), Warsaw School of Economics, Poland, 2020

Artificial Intelligence-assisted Security Analysis of Smart Healthcare Systems

18

Nur Imtiazul Haque and Mohammad Ashiqur Rahman

*Department of Electrical and Computer Engineering,
Florida International University, USA*

Contents

DOI: 10.1201/9781003187158-22

18.1 INTRODUCTION

The high reliance on human involvement in conventional healthcare systems for consultation, patient monitoring, and treatment creates a significant challenge and feasibility issues for the healthcare sector in this pandemic situation, while COVID 19 has indiscriminately spread throughout the world. The widespread nature of the virus is introducing delayed and incorrect treatment resulting in serious health concerns and human mortality. Moreover, healthcare costs are also skyrocketing in parallel, making this basic human need overburdening for the mean population even in first world countries. For instance, the United States has experienced almost $3.81 trillion expenditure for the healthcare sector in 2019, which is projected to reach $6.2 trillion by 2028 [1]. Hence the contemporary healthcare sector is shifting toward adopting the internet of medical things (IoMT)–based smart healthcare system (SHS) to monitor and treat patients remotely with wireless body sensor devices (WBSDs) and implantable medical devices (IMDs) [2]. The WBSDs-provided sensor measurements are analyzed by an artificial intelligence (AI)–based controller for assessing patients' health status, which uses various supervised machine learning (ML) algorithms like decision tree (DT), logistic regression (LR), support vector machine (SVM), neural network (NN), etc. [3]. The patient statuses identified by the ML models allow the SHS controller to generate necessary control signals for actuating the IMDs to deliver automated treatment. Due to computational and device constraints, the WBSDs' measurements cannot be protected with computationally expensive cryptographic algorithms. Moreover, the data transfer among the sensor devices and the controller takes place in the open network.

The humongous attack surface in open network communication of WBSDs raises reliability issues on the sensor measurements. Recent reports and statistics suggest that the frequency of cyber attacks is increasing tremendously. For instance, Check Point Software reported a 45% increase in cyber attacks (626 average weekly attacks) at healthcare organizations since November 2020, which is significantly higher than other industry sectors [4]. Five years' worth of confidential patient records were stolen in Colorado through a ransomware attack in June 2020 [5]. Another report states that the University of Vermont Medical Center (UVMC) was losing $1.5 million daily due to a cyber attack [6]. Moreover, a security breach at Blackbaud cloud service provider exposed the information of almost 1 million patients of 46 hospitals and health systems. Hence a comprehensive anomaly detection model (ADM) with zero-day attack detection capability is required for next-generation safety-critical SHS. The effectiveness of using unsupervised ML models (e.g., density-based spatial clustering of applications with noise [DBSCAN], K-means) for SHS ADMs has been experimented with one of our ongoing works [7]. We propose an unsupervised machine learning (ML)–based ADM ensembling autoencoder (AE) and one-class SVM (OCSVM) for SHS abnormality detection. The effectiveness of the proposed ADM is evaluated in one of our recent works that developed ADM in the smart building domain [7]. The concept of the ensembled OCSVM and AE model is inspired by the fact that the OCSVM model can show significant performance in case-identifying malicious sensor measurements. However, the OCSVM model creates a lot of unnecessary alarms, while the opposite is true of AE. The proposed ensembled OCSVM-AE model can combine the benefit of both models and incurs significantly low false anomalous and benign sensor measurements.

In our proposed SHS design, both the patient status classification model (PSCM) and ADM use ML models, and the success of the models largely relies on their hyperparameters. Several contemporary research works attempt to get the best out of the ML models by obtaining optimal hyperparameters utilizing various optimization techniques. Various optimization algorithms can be used for ML model hyperparameter optimization. However, the ML models are trained with a massive number of samples, and hence computational efficacy is compulsory for feasible implementation. Therefore, we choose metaheuristic bio-inspired stochastic optimization algorithms for ML models' hyperparameter optimization due to their computational efficiency. The approaches mimic nature's strategy as a process of constrained optimization. We consider grey wolf optimization (GWO), whale optimization (WO), and firefly optimization (FO) algorithms for optimizing ML models' hyperparameters for both PSCM and ADM. The latter comes up with many more challenges due to the absence of abnormal samples in the case of zero-day attack detection. We propose a novel fitness function for ADMs' hyperparameter optimization as identified in one of our recently published works [7].

The SHS embedded with an ADM might still be vulnerable to cyber attacks. The robustness and resiliency analysis of the system is mandatory for ensuring the safety of patients' lives. Hence, we propose a formal attack analyzer leveraging ML models and formal methods to assess the predeployment vulnerability of the SHS. The proposed attack analyzer inspects the underlying PSCM and ADM's decisions by identifying the possible attacks that can be deployed by minimal alteration of sensor measurement values. Moreover, the analyzer verifies whether the attack goal is attainable based on the attackers' capability not. However, the formal constraint acquisition from the ML models is a challenging task. It becomes more difficult for the clustering-based ADMs since they impose plenty of constraints that are not solvable in feasible time. We develop a novel concave-hull-based boundary acquisition algorithm that can mimic the original clustering algorithms and create a lot less linear constraint. A smaller set of constraints make it feasible to solve them using formal methods like satisfiable modulo theories (SMT)–based solvers. We verify our attack analyzer's efficacy using the University of Queensland Vital Signs (UQVS) dataset and our synthetic dataset using several performance metrics [8].

18.2 SMART HEALTHCARE SYSTEM (SHS)

IoMT-enabled SHS is a next-generation safety-critical cyber-physical system (CPS) that has brought a revolution in the medical field by reducing treatment cost and consultation inaccuracy. The quality of human lives is improving significantly due to the introduction of personalized, accessible, and efficient SHS by incorporating modern healthcare devices and disease diagnosis tools [9]. However, the effectiveness of SHS in the case of patient status classification and abnormality detection depends on the amount of training data used to train the underlying ML models. Fortunately, a massive amount of healthcare and medication data exists, which can be a great source to understand intricate relationship patterns of various human vital signs. The extracted pattern can be utilized to classify patients' health conditions and determine the plausibility of the sensor measurements to assess the status of the adversarial alteration

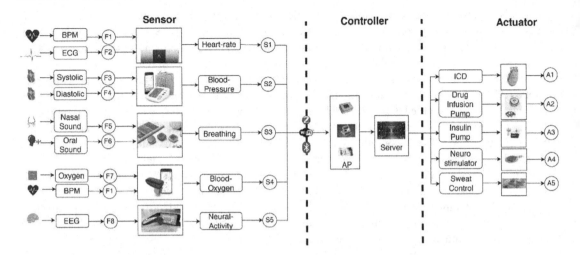

FIGURE 18.1   SHS architecture.

accurately and precisely [10]. This section gives an overview of the basic architecture of the ML-based SHSs, discussing some state-of-the-art ML algorithms that show remarkable performance for patient status classification. It also discusses potential attacks that can pose serious security issues in SHSs.

An SHS leverages WBSDs to obtain patients' vital data ceaselessly, to analyze and process it through an AI-driven controller to generate a control signal, and, depending on the generated signal, to trigger IMDs to deliver real-time healthcare service (medication/treatment). Figure 18.1 shows a simplistic representation of an SHS, where multimodal sensor fusion has been adopted for reliable control signal generation. WBSDs, as shown in the figure, collect vital signs from the patients and forward the sensor measurements to the controller. These sensor devices cannot be protected using strong encryption algorithms as they are computationally expensive and can quickly deteriorate the battery life of the devices, making them infeasible to be implemented in real-life applications [11]. Hence lightweight encryption mechanisms are installed in the device's firmware to encrypt the sensor measurements before sending them to the open communication network. The sensor measurements are delivered to the controller through various communication protocols, namely, Bluetooth, Zigbee, Wifi, etc. The controller uses ML algorithms to identify the criticality of the patients' status and detect the abnormality in the sensor measurements. Hence, the SHS controller comprises two different ML models: the PSCM and ADM. A simplistic SHS can be developed using a blood glucose monitoring sensor device, an ML-based controller trained with diabetes patients' data, and an implantable insulin pump. The blood glucose level can be constantly transmitted to the SHS controller through Dario's blood glucose monitoring device. Depending on the value of the measurement, the controller can decide on the need for insulin delivery. For example, when the measurement values are plausible (determined by the ADM) and PSCM has identified the patient in a state that needs immediate medication, the controller can activate the implanted insulin pump and inject necessary insulin.

18.2.1 Formal Modeling of SHS

We perform formal abstraction of the ML models of an SHS for analyzing its security. Formally modeling the PSCM and ADM requires acquiring constraints from them, which are fed into a first-order logic solver for threat analysis. Let P be the set of a patient's sensor measurements with the patient status/label, j. We aim to detect and analyze critical attack samples that can bypass both PSCM- and ADM-driven constraints. The entire write-up will follow the notations described in Table 18.1.

TABLE 18.1   Modeling Notations

MODEL TYPE	ML MODEL	SYMBOL	DESCRIPTION	TYPE
All	All	n_s	Total sensor count	Integer
		n_l	Total considered patient status	Integer
		a, b	Pair of sensor measurement	String
		j	A particular patient status	String
		S	Set of all sensor measurements	Set
		L	Set of all patient status	Set
PSCM	DT	Np^f	Set of nodes in the path f	Set
		Np	Set of all nodes	Set
		Pth	Set of all paths starting from root to all leaves	Set
		θ_{gi}	Model parameter (weight) associated with g-th patient status and i-th sensor measurement	Real
		θ_g	Set of all model parameter (weight) associated with g-th patient status	Set
		θ	Set of all model parameters (weight)	
		$intercept_g$	Model parameter (bias) for g-th patient status	Set
		$intercept$	Set of all model parameter (bias)	Real
		NL_i	Set of nodes in i-th Layer	Set
		NL	Set of all layers	Set
ADM	DBSCAN, K-Means	$C_k^{a,b,j}$	k-th cluster for the sensor pair (a,b) for a patient with status, j	Sets
		$C^{a,b,j}$	Set of all cluster for the sensor pair (a,b) for a patient with status, j	Cluster
		$Ls_l^{a,b,j}$	l-th line segment of the clusters for the sensor pair (a,b) and patient's status, j	Set
		$Ls^{a,b,j}$	Set of all line segments of the clusters for the sensor pair (a,b) and patient's status, j	Tuple
		C_{all}	Sets of all clusters	Sets
		Ls_{all}	Set of all line segments	Sets

18.2.2 Machine Learning (ML)–based Patient Status Classification Module (PSCM) in SHS

The CPSs use machine learning (ML) algorithms provided their control decisions are driven through organizational, personal, or societal influence, and SHS is one of its kind. The prevalence of healthcare sensor devices has accelerated the data collection process from various sources in the emerging healthcare sector. Analyzing these data using ML models can extricate complex patterns from the data distribution. The learned pattern can be utilized to classify the patients' health conditions and discover abnormalities in real-time sensor measurements [12]. This section is concentrated on discussing the ML models for PSCM of SHS. The PSCM mostly uses supervised ML algorithms being trained on the enormous data sample collected from state-of-the-art test beds and data centers. Among them, NN is a mentionable one that can precisely separate the dissimilar data distribution using nonlinear boundaries in high-dimensional feature space and conversely group the same data distribution [13]. Moreover, existing research shows that the DT algorithm exhibits notable performance in the patient status classification [14]. Other popular ML algorithms include LR, SVM, etc., which also can classify data for multiclass settings like NN and DT. A short overview of these ML models is presented here. These ML models are formally modeled based on our ongoing work [15].

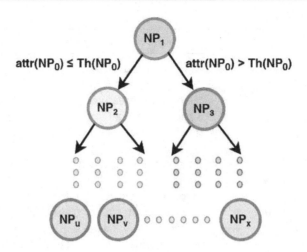

FIGURE 18.2   Flow diagram of DT model.

18.2.2.1 Decision Tree (DT)

The comprehensive, swift, straightforward, and versatile nature of the DT algorithm has made it a popular classification algorithm that draws out inference rules from a complex feature space [14]. The DT algorithm categorizes a sample through its tree-structured decision rules. The ultimate challenges of the decision tree algorithm are to determine the depth of the inference tree, select the best features at a particular level, and choose optimal splitting points for the selected features. In this work, we have adopted a CART (classification and regression trees)–based algorithm leveraging Gini index-based attribute selection metric. Figure 18.2 shows the tree structure of the DT model with formal notations.

The formal constraint acquisition from the DT model is quite simple and straightforward since the trained DT model outputs paths from source to destination using a hierarchical rule-based system. According to the inference rules from DT, we define a Boolean function inference (P, j) that returns true if the patient's sensor measurements are following the rules based on the DT-provided label, j.

18.2.2.2 Logistic Regression (LR)

LR is a classification problem-solving effective baseline algorithm that can accurately separate various classes using nonlinear boundaries. Although LR was primarily designed to focus on solving binary classification problems, the modified algorithm can be leveraged in multiclass classification settings. However, solving multiclass classification problems give rise to several challenges (e.g., selecting one-vs.-one or one-vs.-rest approach). Moreover, LR requires a solver to optimize the cost function to determine the optimal decision boundaries. In our security analyzer, we use a one-vs.-rest scheme for dealing with multiclass classification, and the limited-memory Broyden–Fletcher–Goldfarb–Shanno solver has been adopted to optimize the underlying cost function [16]. The LR model outcomes probability values based on the input features being processed by a softmax activation function. The inference function for the LR model can be represented as follows in equation (18.2.1).

$$\text{Inference}\left(P, j\right) \Leftrightarrow \operatorname*{argmax}_{g} \frac{\sum_{h=1}^{n_l} \exp\left(\left(\sum_{i=1}^{n_s} P_h \theta_{hi}\right) + \epsilon_h\right)}{\exp\left(\left(\sum_{i=1}^{n_s} P_g \theta_{gi}\right) + \epsilon_g\right)} = j. \tag{18.2.1}$$

18.2.2.3 Neural Network (NN)

NN is one of the most popular ML algorithms that facilitate nonlinear pattern extraction from underlying complex feature relationships of data [17]. The NN model is comprised of a network of nodes, where the arrangement and the number of the nodes depend on the data distribution. The architecture of NN is inspired by the working principle of the human brain, which can simultaneously perform several tasks while maintaining system performance. NN can observe a notable performance for solving multiclass classification problems. However, training such a network requires a lot of tuning, such as learning rate, batch size, number of hidden layers, etc.

An NN consists of an input layer, one output layer, and one or more hidden layers. The node inputs are calculated using the weight and bias values of connecting links along with the outputs of the connected previous layer nodes (with an exception for layer 1). The calculation is carried out as follows.

$$\forall_{m \in (1,N)} \text{input}\left(\text{NL}_{mn}\right) = \sum_{o=1}^{|\text{NL}_{m-1}|} \left(\text{output}\left(\text{NL}_{(m-1)o}\right) \times W_{on}^{m-1}\right) + b_{mn}. \tag{18.2.2}$$

The input and output of the layer 1 nodes are basically the input feature values or patient sensor measurements, in our case as in equation (18.2.3). A N-layer NN model is shown in Figure 18.3, where the last hidden layer is $\in = N - 1$.

$$\text{Input}\left(\text{NL}_1\right) = \text{Output}\left(\text{NL}_1\right) = \text{P}. \tag{18.2.3}$$

Some nonlinear transfer/activation functions (e.g., sigmoid, relu, tanh) are used for processing each node's input values, as demonstrated in equation (18.2.4).

$$\forall_{m \in (1,N)} \text{output}\left(NL_{mn}\right) = \text{activation}\left(\text{input}\left(NL_{mn}\right)\right). \tag{18.2.4}$$

The patient sensor measurements in consideration get a label, j, from the NN model, if and only if the softmax function output of the jth output nodes gets the maximum values compared to the other output nodes. The inference rules of NN for a patient vital sign, P is shown in equation (18.2.5).

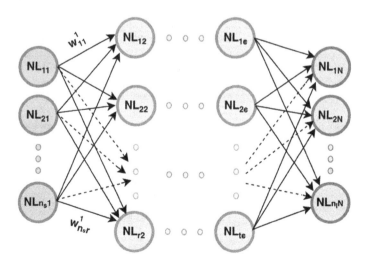

FIGURE 18.3   Flow diagram of NN model.

$$\text{Inference}\big(P, j\big) \Leftrightarrow \underset{g}{\text{argmax}} \, \frac{\exp(\text{input}\big(NL_{ng}\big)\}}{\sum_{q=1}^{n_l} \exp\big(\text{input}\big(NL_{nq}\big)\big)} = j. \tag{18.2.5}$$

18.2.3 Hyperparameter Optimization of PSCM in SHS

The choice of hyperparameters largely determines the performance of ML models. In this work, we optimize the hyperparameters of both ADMs and PSCMs. The hyperparameter optimization of ADMs is discussed in section 18.4.2.5. We have used metaheuristic BIC-based algorithms for hyperparameter optimization of SHS patient data classification anomaly detection (Figure 18.4). Three potential BIC algorithms are explored: whale optimization (WO), grey wolf optimization (GWO), and firefly optimization (FFO).

18.2.3.1 Whale Optimization (WO)

The WO algorithm, inspired by the foraging behavior of humpback whales, was proposed by Mirjalili et al. [18]. The humpback whales' hunting behavior is referred to as bubble-net feeding. They create a circle (upward spiral) or nine-shaped bubbles (double loops) to hunt small fishes close to the water surface. Several engineering and intricate real-life problems have been solved by leveraging the hunting traits of humpback whales. The overall process can be divided into the three-stage process, which is described as follows. We present the mathematical modeling of the three phases of the WO algorithm: prey encircling, bubble-net attacking, and prey searching.

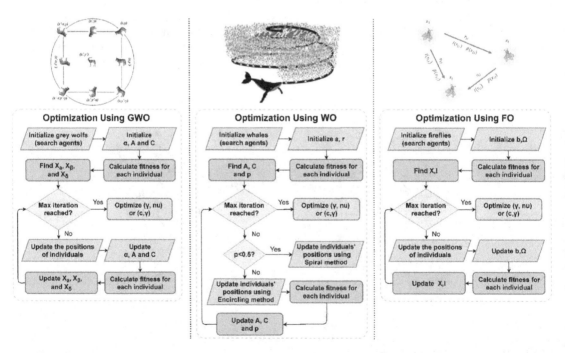

FIGURE 18.4 Schematic diagram of the BIC-based optimization algorithms of the ML-based PSCM's hyperparameter optimization.

18.2.3.1.1 Prey Encircling

The WO algorithm looks for the prey's location in the prey encircling phase. The process is performed by several agents, and the best path found by an agent is followed by the rest. The behavior can be modeled using equations (18.2.6) and (18.2.7).

$$\text{vec}D^{wh} = \left| \overrightarrow{C^{wh}} \overrightarrow{X^{wh}_{best}}(t) \right| - \overrightarrow{X^{wh}}(t). \tag{18.2.6}$$

$$\overrightarrow{X^{wh}}(t+1) = \overrightarrow{X^{wh}}(t+1) - \vec{A}.\vec{D}. \tag{18.2.7}$$

Here, $\overrightarrow{X^{wh}_{best}}(t)$ denotes the current best position found by an agent at time, t, while the position vector is indicated by $\overrightarrow{X^{wh}}(t)$. The following equations are used to determine the coefficient vectors $\vec{A}$ and $\vec{C}$.

$$\overrightarrow{A^{wh}} = 2\overrightarrow{A^{wh}}.\overrightarrow{r^{wh}} - \overrightarrow{A^{wh}}. \tag{18.2.8}$$

$$\overrightarrow{C^{wh}} = 2.\overrightarrow{r^{wh}} \tag{18.2.9}$$

In equation (18.2.8), $\overrightarrow{A^{wh}}$ is linearly changed from 2 to 0 over iterations using a random vector $\overrightarrow{r^{wh}}$ (i.e., $0 \le r^{wh} \le 1$). The overall search space proximal to an agent can be explored by tuning the $\overrightarrow{A^{wh}}$ and $\overrightarrow{C^{wh}}$ parameters.

18.2.3.1.2 Bubble-net Attacking

Two different types of mechanisms are there for bubble-net attacking, which is an exploitation phase.

1. *Shrinking encircling*: Follows equation (18.2.8), where the coefficient vector, $\vec{A}$ is altered randomly in the interval $[-a, a]$. The process allows a linear movement throughout iterations.
2. *Spiral updating position*: The agent uses a spiral updating equation resembling a movement similar to a helix shape, which can be represented as equation (18.2.10).

$$\overrightarrow{X^{wh}}(t+1) = \overrightarrow{D'^{wl}}.e^{bl}.\cos(2\pi l) + \overrightarrow{X^{wh}_{best}}(t). \tag{18.2.10}$$

Here, $\overrightarrow{D'^{wl}} = \left| \overrightarrow{X^{wh}_{best}}(t) - \overrightarrow{X^{wh}}(t) \right|$, which signifies ith agent's distance from the prey, b indicates a logarithmic spiral shape constant, and $-1 \le l \le 1$ denotes a random value.

These techniques are used simultaneously based on a probability value, p.

Prey searching: In prey searching, the agents vary the coefficient $\overrightarrow{A^{wh}}$ for exploring the search space. Making the value of $\vec{A}$ out of $[-1,1]$ interval, the exploration process ensures traversing the remote places from the agent of interest. The mathematical representation of exploration can be expressed as follows:

$$\overrightarrow{D^{wh}} = \left| \overrightarrow{C^{wh}}.\overrightarrow{X^{wh}_{rand}} *(t) - \overrightarrow{X^{wh}}(t) \right|, \tag{18.2.11}$$

$$\overrightarrow{X^{wh}}(t+1) = \overrightarrow{X^{wh}_{rand}}(t+1) - \overrightarrow{A^{wh}}.\overrightarrow{D^{wh}}. \tag{18.2.12}$$

Here, $\overrightarrow{X^{wh}_{rand}}$ denotes a random vector selected from the current population.

18.2.3.2 Grey Wolf Optimization (GWO)

Mirjalili et al. presented grey wolf optimization (GWO) recently with another promising BIC-based optimization technique adopted by several CPS domains for the ML models' parameter optimization [19]. Two key survival aspects of grey wolves are mathematically modeled in GWO: the leadership hierarchy and the hunting technique. Four different levels of grey wolves, namely the α, β, δ, and ω wolves are used for simulating the leadership hierarchy. The GWO encompasses three phases: looking for prey, encircling prey, and attacking prey. Multiple candidate solutions are considered for mathematically modeling the social hierarchy of wolves, where α is the best observed solution and β, δ, and ω are the second, third, and fourth best solutions accordingly. In the GWO algorithm, α, β, and δ wolves/agents/solutions guide the hunting scheme, which is followed by the ω.

During the hunting process, the grey wolves encircle the target prey as in the WO technique. The encircling trait of grey wolves can be modeled as follows:

$$\overrightarrow{D^{grw}} = \left| \overrightarrow{L^{grw}} . \overrightarrow{X_{Pr}}(t) - \vec{X}(t) \right|, \tag{2.13}$$

$$\overrightarrow{X^{grw}}(t+1) = \overrightarrow{X_{Pr}}(t) - \overrightarrow{K^{grw}} . \overrightarrow{D^{grw}}. \tag{2.14}$$

Here, t indicates the current iteration, $\overrightarrow{K^{grw}}$ and $\overrightarrow{L^{grw}}$ denotes the coefficient vectors, $\overrightarrow{X_{Pr}}$ corresponds to the position vector of the prey's location, $\vec{X}$ signifies the current position vector, and, finally, using $\overrightarrow{X^{grw}}$, GWO calculates the new position vector of a grey wolf agent. The coefficient vectors $\overrightarrow{K^{grw}}$ and $\overrightarrow{L^{grw}}$ can be determined using the following equations:

$$\overrightarrow{K^{grw}} = 2\vec{k}.\vec{r_1} - \vec{k}. \tag{18.2.15}$$

$$\overrightarrow{L^{grw}} = 2.\vec{r_2}. \tag{18.2.16}$$

18.2.3.3 Firefly Optimization (FO)

Xin-She Yang developed FO, which is another bio-inspired optimization algorithm for solving complex optimization problems [20]. The algorithm is designed based on the behavior of the fireflies, since they approach toward the direction based on the luminosity of other fireflies. The FO algorithm follows three basic principles for movement.

- All fireflies are unisex. Hence, the progression is irrelative to sex of the fireflies and solely dependent on the brightness of other fireflies.
- The brightness of the fireflies is determined by an encoded objective function.
- The attractiveness for progression is directly related to brightness. Consequently, brightness and the attractiveness decrease with the increment in the distance

In summary, a firefly moves toward brighter fireflies and moves randomly provided the unavailability of brighter fireflies in visible range. The firefly position update maintains the following equation:

$$\overrightarrow{X_i^{ff}}(t+1) = \overrightarrow{X_i^{ff}}(t)\gamma^{ff}e^{-\beta r_{ij}^2} + \overrightarrow{X_j^{ff}}(t) - \overrightarrow{X_i^{ff}}(t)\alpha^{ff}(t)\epsilon_i^{ff}(t). \tag{18.2.17}$$

The right-hand side of the expression expresses the interest of $\overrightarrow{X_i^{ff}}$ toward brighter firefly $\overrightarrow{X_j^{ff}}$. The last expression is a randomization term with $\alpha^{ff}(t)$ (i.e., $0 \le \alpha^{ff}(t) \le 1$). The term $\epsilon_i^{ff}(t)$ denotes a vector of several random values selected from a normal or some other distribution during the time, t. The exploitation parameter, $\alpha^{ff}(t)$ can be expressed as:

$$\alpha^{ff}(t) = \beta^{ff}(t)\delta^{ff}(t) \tag{18.2.18}$$

Here, $0 < \delta^{ff}(t) < 1$.

TABLE 18.2 Performance Comparison of Various Optimization Algorithms for Hyperparameter Optimization of NN-based PSCM Using UQVS Dataset

ALGORITHM	NUMBER OF LAYERS	NUMBER OF NEURONS IN HIDDEN LAYERS	ACCURACY	PRECISION	RECALL	F1-SCORE	CONVERGENCE EPOCHS
GWO	8	4	0.78	0.83	0.78	0.80	13
WO	12	6	0.88	0.89	0.88	0.89	19
FO	10	6	0.82	0.85	0.87	0.86	25

18.2.3.4 Evaluation Results

Table 18.2 shows the performance evaluation based various metric, which signifies that WO performs better than other two algorithms. The results indicate that the GWO shows the best performance in the case of convergence speed. However, the performance analysis also shows that GWO performs the worst, which suggests that the GWO optimization algorithm is being trapped in the local optima faster than the WO and FO algorithms. The convergence speed of WO is faster than the FO and not too much slower than the GWO algorithm. Hence, based on the performance analysis and the convergence speed, WO seems to be the better choice for both SVM and OCSVM hyperparameter optimization.

18.3 FORMAL ATTACK MODELING OF SHS

In our work, we use a formal method-based attack analyzer to figure out potential vulnerabilities since the tool allows us to explore the search space of possible system behaviors. Our attack analyzer takes several constraints like attacker's capability, accessibility, and goal along with controller ML models as an input and formally models them using satisfiability modulo theory (SMT)–based solvers. Depending on the satisfiability outcome of the SMT solvers, the proposed analyzer can figure out potential threats. Our proposed framework attempts to find the attack vector that can attain the attacker's goal compromising a minimum number of sensors within the attacker's capability. We consider the sensor measurement alteration attacks and following attack models throughout the write-up.

18.3.1 Attacks in SHS

The smart medical devices (both WBSDs and IMDs) are getting exposed to several vulnerabilities due to the advent of SHSs [21]. As discussed earlier, the healthcare industry is experiencing rapid growth in cyber attacks. Most popular cyber attacks in the healthcare systems include hardware Trojan [22], malware (e.g., Medjack [23]), Sybil attacks using either hijacked IoMT [24] or a single malicious node [25], DoS attacks [26], and man in the middle (MITM) attacks [27]. So it is imperative to study the vulnerabilities of an SHS before deploying it.

18.3.2 Attacker's Knowledge

In our attack model, we consider that attacker has blackbox access to the PSCM and ADM. The attacker also has an idea of data distribution of benign patients' sensor measurements. Based on that, the attacker can create or replicate the SHS underlying PSCMs and ADMs. The framework comes up with all possible attack vectors. The attack vector reveals the complete path of launching an attack.

18.3.3 Attacker's Capability

An attacker can break the lightweight encryption and obtain the sensor measurements if they have accessibility to that sensor device. The attacker needs to have access to multiple sensor measurements to launch attacks successfully. Since the PSCM model uses ML models to learn the patterns of sensor measurements from historical data, they cannot launch stealthy attacks targeting one/two measurements. For instance, it would not suffice to alter the blood pressure measurements only to misclassify a normal patient to be a high blood pressure patient. Our PSCM-based models learn the dependency of the sensor measurements for a particular label. Moreover, the ADM model detects whether unwanted manipulation has compromised the consistency of the sensor measurements as a whole. Our formal model specifies how many resources can be accessed and tempered by the attacker.

$$\sum_{s \in S} a_s \leq \text{Maxsensors,} \tag{18.3.1}$$

$$\forall_{s \in S} \, abs\left(\frac{\Delta P_s}{P_s}\right) < \text{Threshold.} \tag{18.3.2}$$

Our attack model assumes that the attacker cannot attack more than Maxsensors number of sensor measurements. Moreover, the attacker is also unable to alter measurements more than Threshold.

18.3.4 Attacker's Accessibility

An attacker can modify sensor measurements based on its accessibility.

$$\forall_{s \in S} \, a_s \rightarrow \left(\Delta P_s \neq 0\right). \tag{18.3.3}$$

Here, we consider a_s to be an integer value, where false is thought to be 0 and true to be 1.

18.3.5 Attacker's Goal

In our attack model, we assume that the attacker's goal is to minimally modify the sensor measurement to let the SHS control system misclassify a patient's status and thus provide wrong medication/treatment.

The PSCM model misclassifies a patient's label from j to $\bar{j}$, and the ADM also supports this inference if an adversary can alter sensor measurements based on the following constraints:

$$\text{alter}\left(j,\bar{j}\right) \rightarrow \forall_{s \in S} \left(\overline{P_s} \rightarrow P_s + \Delta P_s\right) \wedge \text{inference}\left(P,j\right) \wedge \text{consistent}\left(P,j\right) \wedge \text{inference}\left(\overline{P},\bar{j}\right) \wedge \text{consistent}\left(\overline{P},\bar{j}\right). \tag{18.3.4}$$

In equation (18.3.4), P_s shows the actual measurements, ΔP_s denotes the necessary alteration, and $\overline{P}_s$ represents the altered measurement value of sensor S.

18.4 ANOMALY DETECTION MODELS (ADMS) IN SHS

This section provides an overview of various ML-based ADMs of SHS along with a short overview of the proposed ensemble of unsupervised ML model.

18.4.1 ML-based Anomaly Detection Model (ADM) in SHS

The unsupervised clustering algorithms are mostly chosen for their notable performance in the case of anomaly detection in several CPS domains. We consider an ML-based ADM in our framework that groups the good data points based on labels provided by DCM and finds the optimal values for the parameters to cluster all the good data points. The formal constraint acquisition from clustering-based ADM give rise to some challenges, as follows.

18.4.1.1 Density-based Spatial Clustering of Applications with Noise (DBSCAN)

DBSCAN is a nonparametric unsupervised clustering algorithm that groups similar data points. The algorithm labels the data samples as noisy points provided they do not fit into any clusters. Two hyperparameters – epsilon and minpoints (minimum points for cluster creation) – need to be tuned for optimal clustering. However, explicit decision boundaries cannot be obtained from DBSCAN. In our ongoing work, we developed a concave-hull-based novel approach for extracting constraints from the DBSCAN algorithm [28]. For constraint acquisition simplicity, we modeled the DBSCAN-based consistency check by drawing boundaries considering the relationship among the combination of all pairs of features instead of all features together. Moreover, considering all features together presents another problem in the high-dimensional feature space since most of the clusters do not have sufficient data points for drawing boundaries in the high-dimensional space. Our constraints acquisition follows logical functions for checking the consistency of the PSCM-provided label. Our boundary acquisition technique is explained flowingly with a simple example scenario to facilitate the understanding of the process.

Figure 18.5 demonstrates two clusters C_1 and C_2 in a 2-dimensional data space, where C_1 comprises seven-line segments $(Ls_1, Ls_2, \ldots, Ls_7)$, and C_2 incudes three line segments $(Ls_8, Ls_9, and\ Ls_{10})$. The points of a particular line segment (Ls_i) are denoted as (x_a^i, y_a^i) and (x_b^i, y_b^i), where $y_b^i \geq y_a^i$. We use following logical functions to validate the consistency of data point (x, y) with the PSCM-provided label, j. Note that to check the consistency, we only use the historical data points of label, j, for a cluster creation and consistency check. If the data point in consideration falls within any of the clusters, we consider that point as a legitimate sample; otherwise, we label that as an anomaly.

- in Range Of Line Segment (x, y, Ls_i): Checks whether the y-coordinates of the data point are within the two end points of the line segment, Ls_i.

$$\text{inRangeOfLineSegment}(x, y, Ls_i) \Leftrightarrow y_a^i < y \leq y_b^i. \tag{18.4.1}$$

From the Figure 18.5, it can be said that inRangeOfLineSegment (x_t, y_t, Ls_1) returns true for the point (x_t, y_t) since $y_1^1 \geq y_t \geq y_2^1$. However, inRangeOfLineSegment (x_t, y_t, Ls_4) returns false for line segment Ls_4.

- leftOfLineSegment (x, y, Ls_i): Checks whether the x-coordinates of the data point are on the left side of the line segment, Ls_i.

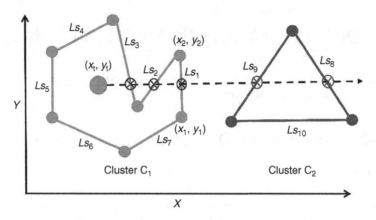

FIGURE 18.5   Logic behind checking whether a point is inside a polygon cluster in DBSCAN algorithm.

$$\text{leftOfLineSegment}\left(x, y, Ls_i\right) \Leftrightarrow \left(x\left(y_a^i - y_b^i\right) - y\left(x_a^i - x_b^i\right) - \left(x_a^i y_b^i - x_b^i y_a^i\right)\right) < 0. \quad (18.4.2)$$

It is apparent from Figure 18.5 that the leftOfLineSegment$\left(x_t, y_t, Ls_1\right)$ returns true for the data point (x,y). However, it returns false for the leftOfLineSegment$\left(x_t, y_t, Ls_5\right)$.

- intersect$\left(x, y, Ls_i\right)$: We draw an imaginary line parallel to the x-axis from the point of interest (x,y). The function checks whether the imaginary line segment Ls_i intersects the imaginary line. A data point is said to intersect with a line segment if the data point is both inRangeOfLineSegment and leftOfLineSegment, as shown in equation (18.4.3):

$$\text{intersect}\left(x, y, Ls_i\right) \Leftrightarrow \text{inRangeOfLineSegment}\left(x, y, Ls_i\right) \wedge \text{leftOfLineSegment}\left(x, y, Ls_i\right) \quad (18.4.3)$$

It is obvious from Figure 18.5 that intersect$\left(x_t, y_t, Ls_i\right)$ returns true for line segments 1, 2, 3, 8, and 9, while false for the rest.

- within Cluster$\left(x, y, C_k\right)$: Checks whether a data point (x,y) is within a cluster, C_k. The function performs intersect$\left(x, y, Ls_i\right)$ on all boundary line segments of a particular cluster and XOR them. The function outcome is true for a cluster, provided the imaginary line parallel to x-axis from data point (x,y) intersects with am odd number of line segments for that cluster. Equation (18.4.4) shows the function definition.

$$\text{withinCluster}\left(x, y, C_k\right) \Leftrightarrow \otimes_{1 \leq i \leq \text{numofLs}_{C_k}} \left(\text{intersect}\left(x, y, Ls_i\right) \wedge In\left(C_k, Ls_i\right)\right) \quad (18.4.4)$$

Here, $In\left(C_k, Ls_i\right)$ determines whether the cluster c_k includes the line segment, Ls_i. For instance, from Figure 18.5, it is clear that $\left(x_t, y_t\right)$ is inside cluster C_1 but not in C_2 since the imaginary line paralleled to the x-axis from it intersects three line segments (am odd number) of the cluster.

From Table 18.1, we can see that n_s denotes the total sensor measurements, and the overall possible patient health status is denoted by n_l. In this work, we assume each measurement is recorded/reported by a sensor measurement. From the table, we can also see that S represents the set of all the sensors, P is the measurement values of those sensor measurements, and L is the set of all possible patient statuses. Let us say the status of the patient in consideration is denoted by j, where $j \in L$. For simplicity and other reasons, we consider 2-dimensional clusters, and hence we think of two sensor relationships at a time. Thus, for the label $j \in L$ and sensor measurement pair (a,b), where $(a,b) \in S$, we obtain one or more clusters, $C_k^{a,b,j}$, which represents the relationship between the two measurements for that specific label, k. These clusters comprise a few line segments, which are represented as $Ls_l^{a,b,j}$. For a pair of sensor measurements from all measurements $P_{a,b} = (P_a, P_b) \in P$, DBSCAN algorithm checks consistencies using (18.4.5), (18.4.6), and (18.4.7).

$$\text{consistent}(P, j) \Leftrightarrow \text{forall}_{(a,b) \in S \wedge (a!=b)} \exists_{k \in C \text{withinCluster}} \left(P_{a,b}, C_k^{a,b,j} \right), \tag{18.4.5}$$

where

$$\text{withinCluster}\left(P_{a,b}, C_k^{a,b,j} \right) \Leftrightarrow \oplus_1 \left(\text{intersect}\left(P_a, \& P_b, Ls_l^{a,b,j} \right) \wedge \text{In}\left(C_k^{a,b,j}, Ls_l^{a,b,j} \right) \right) \tag{18.4.6}$$

$$\text{intersect}\left(P_a, P_b, Ls_l^{a,b,j} \right) \Leftrightarrow \text{inRangeOfLineSegment}\left(P_a, P_b, Ls_l^{a,b,j} \right) \wedge$$
$$\forall_i \left\{ \text{leftOfLineSegment} \right\} (P_a, \forall c\{P_b\}, \forall_i \{Ls_l^{a,b,j}) \tag{18.4.7}$$

18.4.1.2 K-means

The K-means algorithm produces almost similar constraints to those of DBSCAN with a variation in the number of clusters, noise points, etc. (Likas et al., 2003). One of the main challenges in the K-means algorithm is to determine the optimal value of k, which can be optimally selected with various techniques like the elbow method, the silhouette method, etc. [29].

18.4.1.3 One-class SVM (OCSVM)

The OCSVM model is a bit different from the state-of-the-art support vector machine (SVM) model for novel pattern detection [30]. The SVM is a supervised ML for drawing hyperplanes between different classes. But the SVM model is dependent on the labels for the training samples. For zero-day/novel attack detection, the OCSVM model can be leveraged, which separates the trained patterns from the origin using a decision boundary. The OCSVM model's decision boundary can be modified by tuning two hyperparameters: gamma and nu.

18.4.1.4 Autoencoder (AE)

The AE model is mainly an NN model that intends to regenerate the features. Section 18.2.2 shows the modeling notations of NN models [31]. The primary difference between the regular NN model and AE model is that the AE model contains some strict constraints. For instance, the number of nodes in the encoded hidden layer of the AE model needs to be less than the number of features of the model. Although the AE model mainly gets used for data compression applications, recently, the model has been used for abnormality detection as well.

18.4.2 Ensemble-based ADMs in SHS

The efficacy of the DBSCAN and K-means algorithm as ADM in SHS has been analyzed in our ongoing work [7]. Moreover, we propose an unsupervised ensembling of the AE and OCSVM models, which has shown significant performance in the smart building domain. The proposed ensembled model can be used as SHS ADM as well. The workflow of the algorithm is explained as follows.

18.4.2.1 Data Collection and Preprocessing

The primary stage of ensembled-based ADM is to create/collect benign sensor measurements from trusted devices. The dataset stores the SHS sensor measurements and corresponding actuation. The duplicate entries are removed from the dataset as the first step of preprocessing. Then the features are normalized, followed by the dimensionality of the feature space being reduced by the principal component analysis technique. The preprocessing steps allow faster model training. The dimensionality reduction of features is performed only for OCSVM model training. However, for AE model training, all features (sensor measurements) are used. The benign dataset is split into 75% training and 25% testing samples. Additionally, a dataset comprised of attack samples (generated based on the attack model as discussed in section 18.3) is also prepared for assessing the model's performance.

18.4.2.2 Model Training

The model training uses preprocessed data to train both AE and OCSVM models. The AE model is directly trained with $model^{AE}$ on historical benign samples, while the PCA components of training samples are used to train $model^{OCSVM}$. The models are trained separately, and after the training's completion, the prediction models are then used to calculate benign and anomalous sample separation thresholds.

18.4.2.3 Threshold Calculation

The threshold calculations for both AE and OCSVM are an essential step of the proposed ADM. The threshold value determines the decision plane for attacked and benign samples. The OCSVM decision function outcomes a score in $[-1,1]$, which can be used as the OCSVM model's weight without further processing. The threshold for the OCSVM model is 0, which signifies that a sample prediction of more than 0 will be counted as a benign sample and vice versa. In the case of the AE model, the difference between actual features (sensor measurements) and the model prediction is termed an error. The AE model's threshold calculation, threshAE, can be performed by taking the maximum error amid the training prediction and the corresponding features as shown in Algorithm 1. However, the threshold values of the models need to be normalized as they are on different scales.

18.4.2.4 Anomaly Detection

The proposed ensembled ADM uses both the trained models, $model^{AE}$, and $model^{OCSVM}$ and their thresholds, threshAE and threshOCSVM to detect anomaly. For a particular sample, the AE and OCSVM models are weighted with weightAE and weightOCSVM, respectively, depending on the distance from threshAE and threshOCSVM. Moreover, the sign of the distance is used to figure out the model's prediction, predAE and predOCSVM. If the AE and OCSVM models' outcomes disagree with each other, the normalized weights of the models (i.e., normWeightAE and normWeightOCSVM) are multiplied with the prediction of models to obtain the ensembled outcome.

Algorithm 1: Ensembled Unsupervised ML Model

Input: X,T
Output: $predSample$
Train autoencoder-based ADM, $model^{AE}$ on X
Train OCSVM-based ADS model, $model^{OCSVM}$ on $pca(X)$
$threshAE := \max(abs(X-model^{AE}(X)))$
$threshOCSVM := 0$
For each sample in T :
  $weightAE := abs(sample - model^{AE}(sample))$
  $weightOCSVM := model^{OCSVM}(pca(sample))$

$$normWeightAE := abs\big(normalize(weightAE) - normalize(threshAE)\big)$$

$$normWeightOCSVM := abs\big(normalize(weightOCSVM) - normalize(threshOCSVM)\big)$$

if $weightAE < threshAE$, **then** $predAE = 1$
else $predAE = -1$
if $weightOCSVM < threshOCSVM$, **then**
 $predOCSVM = -1$
else $predOCSVM = 1$

$$weightSample = \frac{1}{2} \times \big(normWeightAE \times predAE + normWeightOCSVM \times predOCSVM\big)$$

if $weightSample < 0$, **then** $predSample[sample] =$ **Anomaly**
else predSample[sample] = **Benign**
End For

18.4.2.5 Example Case Studies

This section provides two example scenarios from the COD dataset with numeric data to clarify how the proposed ADM works and to illustrate the need for the proposed ensemble model [32]. The detailed description of the dataset can be found in our published work [7].

Case Study 1
The first case study was performed to check the ADM performance on a benign sample. The prediction of the OCSVM model produces the weightOCSVM to be –0.038 – and labels the sample as an anomaly since the output is negative. The sample is labeled as benign by the AE model since the calculated weightAE (=0.41) does not exceed threshold, threshAE(=3.38). Moreover, the normalized OCSVM threshold, normalize(threshOCSVM), was found to be 0.84, and normalized weight, normalize(weightOCSVM), is calculated to be 0.28, which makes the normalized weight of OCSVM normWeightOCSVM to be 0.28. Similarly, the value of normWeightAE was found to be 0.46. Since normWeightAE is higher than normWeightOCSVM, the ensembled model prediction corresponds to the prediction of AE model. Thus the ensembled model correctly helps to reduce the false anomaly rate.

Case Study 2
Case study 2 is performed on two attack samples. The attack samples are obtained from a benign sample considering two different cost increment attacks.

For the attacked sample, the OCSVM model predicted weightOCSVM to be 0.06 and labeled the sample as an anomalous sample. However, the weightAE was found to be 1.23, which does not exceed threshAE. Hence, the sample is labeled as benign by the AE model. The measured normWeightOCSVM is 0.42. Similarly, the value of normWeightAE was determined as 0.39. The value of normWeightAE being higher than the normWeightOCSVM makes the ensembled decision correspond to the OCSVM prediction. Thus the ensembled model lowers the false benign rate for critical attack samples.

18.4.2.6 Evaluation Result

We evaluate the effectiveness of the proposed ADM using several metrics. Figure 18.6 shows the ADM's performance evaluation based on the cost consumption attack. The cost consumption attack attempts to measure CO_2, temperature, and occupancy sensor measurement of a smart building with an intent to increase the overall energy consumption of the building. We can observe a lot of such points in Figure 18.6, where both AE and OCSVM models differ in their opinions. However, the proposed ADM ensembles both models and can produce correct labeling in almost all cases. The OCSVM model does not provide any

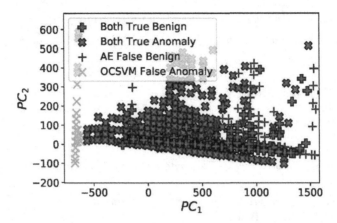

FIGURE 18.6   Ensembled ADM performance against cost consumption attack in smart building HVAC control system.

false benign labeling, while the AE model does not generate any false anomaly labeling in all considered attacks. That's why those legends are missing from the figure. The proposed model performs better than the individual OCSVM and AE models based on accuracy, precision, recall, and F1-score performance metrics for all considered attacks [7].

18.4.2.7 Hyperparameter Optimization of ADMs in SHS

The fitness function choice for the ADMs is more challenging as it includes only a dataset with no anomalous samples. Like the PSCM, ADM hyperparameter optimization uses the training loss as the fitness function. But using only the training loss/accuracy for the fitness function is not sufficient for all ADMs because, in most of the cases, the hyperparameter comes up with a flexible boundary around the benign samples of the training dataset. However, the flexible boundary performs poorly on the anomalous data samples. Hence a tightnessFunction is introduced for choosing the fitness function of the ADMs. In the case of OCSVM ADM, we use the sum of the distance of the support vectors from the boundary as the tightnessFunction, which overcomes the problem of poor performance on the anomalous samples. For the DBSCAN algorithm, the tightness is ensured by further minimizing the radius of the clusters, ε.

Our comprehensive evaluation demonstrated the effectiveness of hyperparameter optimization based on the ML model's performance, convergence, and scalability [7].

18.5 FORMAL ATTACK ANALYSIS OF SMART HEALTHCARE SYSTEMS

The formal attack analyzer takes the SHS dataset and trains both ADM and PSCM models. To recall, the PSCM is responsible for labeling patients' status, while the ADM checks the consistency of the sensor measurements of the PSCM-provided label. The attack analyzer leverages the learned model parameters from both ML models and converts them into decision boundaries to generate constraints from them. The constraints are formulated as a constraints satisfaction problem (CSP) and are fed into an SMT solver. The solver takes the constraints of the attacker's capability, accessibility, attack goal along with the ML-model constraints and assesses the threat associated with a patient's sensor measurements.

The solver returns a satisfiable (SAT) outcome provided the constraints are nonconflicting based on the measurements of the patient in consideration. The solver of the analyzer can report an attack vector indicating required alteration for accessible sensor measurements to misclassify the patient's measurement through PSCM and evade ADM and thus achieve the attack goal. The analyzer is capable of determining the most critical threats, which signify the minimal attacker's capability to obtain the attack goal. It can be said that an attack vector also implies that an attacker cannot successfully launch an attack to accomplish the attack goal if its capability is less than analyzer-provided capability. Hence the system is said to be threat resilient up to that capability. If the solver produces unsatisfiable (UNSAT) results, it means that, based on the attacker's capability, the attack goal cannot be attained. In this case, the analyzer gradually increases the attacker's capability and redetermines the attack feasibility until reaching an SAT point. The SMT solver uses several background theories to solve CSP problems, and the solver outcome possesses a formal guarantee.

18.5.1 Example Case Studies

Two case studies have been added in this section to understand the working principles of the attack analyzer. For the case studies, a DT-based PSCM- and DBSCAN-based ADM have been considered due to their significant performance in the case of both real and synthetic datasets.

Case Study 1
The first case study is on the synthetic dataset that resembles the futuristic healthcare system we are considering. The dataset comprised of various vital signs like heart rate, blood pressure, blood alcohol, respiratory rate, etc. was collected from several datasets like Fetal ECG synthetic dataset [33], UCI ML diabetes dataset [34], oxygen saturation variability dataset [35], BIDMC PPG respiration dataset [36], and StatCrunch dataset [7]. Those datasets contain over 17,000 samples each with eight sensor measurements. The reference datasets of Table 18.3 were used to generate the data. The dataset contains sensor measurements of six different patient statuses. Four different sensor measurements of a patient are demonstrated in Table 18.3. Our attack analyzer reports that the attacker cannot be successful with the attack intent of letting the SHS PSCM to incorrectly classify a high blood cholesterol–labeled patient as a high blood pressure one, compromising a single sensor measurement only. An attacker having access to two sensor measurements (i.e., systolic blood pressure and blood oxygen) might attain the attack goal and mislabel the patients so that caregivers deliver the wrong medication/treatment. However, the ADM can detect the attack because altering two sensors' measurements made for a consistency violation in terms of the learned patient sensor measurement distribution derived from historic data. Our attack analyzer identified a possible attack path by altering heart rate measurement, systolic blood pressure, diastolic blood pressure, and blood oxygen sensor measurements by 2%, 7%, 2.5%, and 5.7%, respectively. Sample DT and DBSCAN constraints from PSCM and ADM are shown in Table 18.4 and Table 18.5.

TABLE 18.3 Attack Scenario Under Different Attack Models

CURRENT STATE	HEART RATE	SYSTOLIC	DIASTOLIC	BLOOD OXYGEN
Actual measurements	122.94	75.98	153.56	93/2
Attacked measurements (DT constraints only)	122.94	73.46	153.56	98.5
Attacked measurements (both DT and DBSCAN constraints)	120.332	81.855	149.67	98.5

TABLE 18.4   Example of DT Algorithm-driven Constraints

$$\text{inference}(P,0) \rightarrow \big((P\_5 \leq 20.5) \wedge \big(((P\_7 \leq 8.5) \wedge (P\_4 \leq 94.005 \vee))$$
$$((P\_7 \leq 8.5) \wedge (P\_2 \leq 140.46) \wedge (P\_0 \leq 100.51)) \vee$$
$$((P\}\_5 > 20.5) \wedge (P\_3 > 130.495)))$$

$$\text{inference}(P,1) \rightarrow \big((P_5 > 20.5) \wedge (P_3 \leq 130.495) \wedge (P_2 \leq 140.46)\big)$$

$$\text{inference}(P,2) \rightarrow \big((P_5 > 20.5) \wedge (P_3 \leq 130.495) \wedge (P_2 > 140.46)\big)$$

$$\text{inference}(P,3) \rightarrow (P_5 \leq 20.5) \wedge (P_7 > 8.5) \wedge (P_5 \leq 20.5)$$

$$\text{inference}(P,4) \rightarrow (P_5 \leq 20.5) \wedge (P_7 \leq 8.5) \wedge (P_4 > 94.005) \wedge (P_2 \leq 140.46)) \wedge (P_0 > 100.51)$$

$$\text{inference}(P,5) \rightarrow (P_5 \leq 20.5) \wedge (P_7 \leq 8.5) \wedge (P_4 > 94.005) \wedge (P_2 > 140.46)$$

TABLE 18.5   Example of DBSCAN Algorithm-driven Constraints

$$...\big(\big((-0.75P_0^0 + 0.35P_1^0 + 79.929 \geq 0) \wedge (70.11 < P_1^0 \leq 70.86)\big)$$
$$\oplus \big((-1.5P_0^0 + 0.53P_1^0 + 171.7767 \geq 0) \wedge (68.61 < P_1^0 \leq 70.11)\big)$$
$$\oplus \big((-0.08P_0^0 + 0.53P_1^0 + 171.7767 \geq 0) \wedge (65.21 < P_1^0 \leq 67.44)\big)$$
$$\vee \big((0.08P_0^0 + 0.39P_1^0 + 1059.126 \geq 0) \wedge (85.27 < P_1^0 \leq 87.36)\big)$$
$$\oplus \big((0.04P_0^0 + 0.43P_1^0 + 21.929 \geq 0) \wedge (88.22 < P_1^0 \leq 92.33)\big)$$
$$\oplus \big((0.04P_0^0 + 0.41P_1^0 + 21.21 \geq 0) \wedge (87.35 < P_1^0 \leq 91.15)\big)\big)$$

Case Study 2

We have also verified the performance of our proposed framework using real test bed data collected by the University of Queensland Vital Signs dataset [8]. The dataset collects 49 sensor measurements from more than 30 anesthesia patients undergoing surery at Royal Adelaide Hospital and being monitored using Philips intellivue monitors and Datex-Ohmeda anesthesia machine. The prepossessed dataset possesses 209,115 samples with 26 sensor measurements and 58 patient status (labels) with 28 different alarms. Our attack analyzer found the feasibility of different attacks by raising wrong alarms instead of the intended ones. For instance, an adversary can activate alarms (APNEA, low blood pressure, low end-tidal carbon dioxide, high inspired concentration of sevoflurane) instead of other (APNEA, high minute volume) alarms by changing measurement values in artery diastolic pressure, artery mean pressure, effective end-tidal decreased hemoglobin oxygen saturation label, inspired decreased hemoglobin oxygen saturation label, end-tidal isoelectric point, inspired isoelectric point, effective end-tidal concentration of sevoflurane, and inspired concentration of sevoflurane sensors by 9%, 8%, 8.4%, 2.3%, 6%, 10%, 2%, 4%, respectively.

18.5.2 Performance with Respect to Attacker Capability

The performance of our attack analyzer is primarily evaluated by enumerating the total identified attack vectors in terms of attacker's capability. Figure 18.7 demonstrates the total number of attack vectors found by our analyzers w.r.t. variable number of compromised sensor measurements and different thresholds for FDI attack in the case of a synthetic test bed. The attacker can attain an attack goal by compromising one sensor only as depicted from the figure. There are at most three possible attack vectors (when the attacker has the capability of compromising eight sensor measurements), provided the attacker is restricted to compromise the sensor measurements within the 10% threshold. The increment of threshold can figure

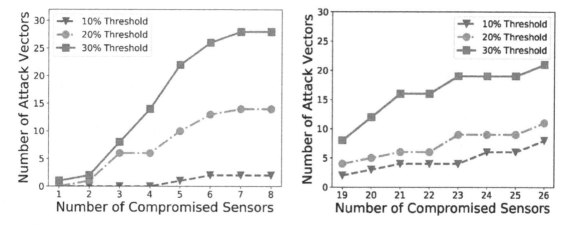

FIGURE 18.7 Performance analysis of SHChecker framework for identifying attack vectors w.r.t. attacker's capability for (Left) synthetic dataset and (Right) UQVS dataset.

out more attack vectors. For instance, our analyzer can identify 28 possible attack vectors, while it can compromise eight sensor measurements with a 30% threshold. A similar scenario can be seen from the real test bed, as in Figure 18.7.

18.5.3 Frequency of Sensors in the Attack Vectors

The proposed attack analyzer inspects the set of all possible attack vectors and figure out the involvement of the individual sensor measurements in the attack vectors. Figure 18.8 shows the sensor measurements' frequency with respective attack vectors for both datasets. The evaluation figures that almost all sensor measurements contribute similarly (with a few exceptions like systolic sensor measurement) in all attack vectors, while the attack threshold is around 30%. The study provides an insight into prioritizing the protection of sensor measurements for designing a defense tool based on their contributions in attack vectors. If the frequency of a sensor measurement exceeds that of the others, the sensor associated with that certain measurement should be enhanced for securing the SHS architecture. Thus our analyzer can effectively provide a design guide.

18.5.4 Scalability Analysis

The scalability analysis enables us to assess the feasibility of implementation of the proposed attack analyzer for the large-scale system. The scalability of the analyzer is evaluated by inspecting required time with variable size of the SHS. The size of SHS is varied by altering the number of sensors to build our SHS model. Figure 18.9 and Figure 18.10 show that the DBSCAN cluster creation time takes lot less time compared to the boundary creation time, and there is a linear growth for the sensor measurements. The figures suggest that the constraints generation for the ADM (DBSCAN) requires significantly more time than the DT clusters. However, the cluster, boundary creation, and constraints generation are performed before deploying the analyzer in real time. Hence the time requirements for these processes are insignificant. From the figures, it can also be seen that the execution time for the solver increases remarkably w.r.t. the attacker's capability due to the need for checking more constraints. In the case of real-time threat analysis, we can see an exponential increment of the execution time requirement, which creates scalability concerns for large SHSs.

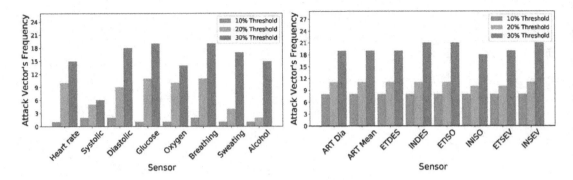

FIGURE 18.8 Frequency of the different sensors in the attack vectors for (Left) synthetic dataset and (Right) UQVS dataset.

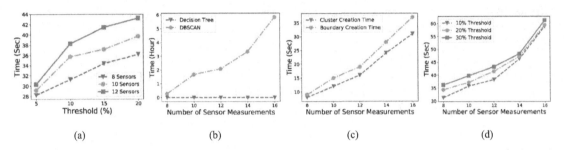

FIGURE 18.9 Execution time for the (a) cluster and boundary creation time, based on number of sensor measurements; (b) ML constraints generation, based on number of sensor measurements; (c) threat analysis based on threshold for data injection; and (d) threat analysis based on the number of sensor measurements measured from the synthetic dataset.

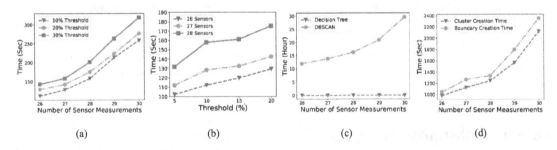

FIGURE 18.10 Execution time for the (a) cluster and boundary creation time, based on number of sensor measurements, (b) ML constraints generation, based on number of sensor measurements, (c) threat analysis based on threshold for data injection, and (d) threat analysis based on the number of sensor measurements measured from the UQVS dataset.

TABLE 18.6   Sensor Measurements to Compromise to Achieve a Targeted Attack Goal

CURRENT STATE	TARGET STATE	# SENSOR MEASUREMENTS TO COMPROMISE	SENSOR MEASUREMENTS
Normal	High cholesterol	3	Heart rate, alcohol, glucose
High blood pressure	Abnormal blood oxygen level	4	Systolic, diastolic, breathing, blood oxygen
High cholesterol	Excessive sweating	2	Heart rate, breathing

18.6 RESILIENCY ANALYSIS OF SMART HEALTHCARE SYSTEM

We consider a system to be resilient based to the degree to which it effectively and speedily saves its critical capabilities from disruption and disturbance carried out by adverse conditions and events. Our proposed analyzer can figure out the resiliency of a system for a targeted attack condition. A system is said to be k-resilient if it can perform smoothly even k of its components are faulty or nonresponding. The resiliency analysis for the synthetic dataset is shown in Table 18.6, which signifies that an attacker cannot be successful in its attack goal of misclassifying a normal patient into a high cholesterol patient provided it cannot modify more than two devices, and hence the system is denoted as 2-resilient for the attack goal. An attacker having alteration access to more than one sensor measurement can misclassify a high blood sugar patient into an abnormal oxygen level patient or an excessive sweating state patient into a normal one. Similarly, the UQVS dataset in which the alteration of the normal status patient to show a decrease in hemoglobin oxygen saturation (DESAT) is 20-resilient. The resiliency analysis capability of the proposed analyzer provides a design guide indicating the relationship between the number of protected sensors and associated risk.

18.7 CONCLUSION AND FUTURE WORKS

We present a comprehensive framework of SHS with both patient status/disease classification and anomaly detection capability. The framework uses ML-based models for accomplishing the classification and anomaly detection tasks. Moreover, the ML model hyperparameters are tuned with bio-inspired computing-based optimization algorithms. Additionally, the proposed framework shows significantly high performance in reducing the false alarm rate of the ADMs. Our framework also uses an ML and formal modeling-based attack analyzer to model and assess the security of ML-based SHS. The tool can analyze the potential threats that satisfy the attacker's goal. The attack analyzer extracts ML model constraints to understand the relationships between the sensor measurements and health states and the consistency among the measures. Moreover, the attack analyzer can synthesize potential attack vectors for a given attack model, where the attacker can change the health state (the actual one) to a wrong one (the targeted state). Our evaluation results on both of our datasets in consideration show that diverse attacks are possible by compromising different numbers and types of sensor values, even compromising only one sensor measurement. Though our current model considers only a single-label attack, we plan to expand it to multilabel goals in future work. Besides, we plan to study other efficient ML algorithms to improve the performance of our proposed framework. The proposed framework assumes that the underlying datasets contain sufficient data to capture the data patterns, which might not be valid for the emerging CPS domains like SHS. Hence, we will use the generative adversarial network–based data augmentation technique.

REFERENCES

[1] National Health Expenditure Data – NHE Fact Sheet. (2020). Opgehaal 21 September 2021, van www. cms.gov/Research-Statistics-Data-and-Systems/Statistics-Trends-and-Reports/NationalHealthExpendD ata/NHE-Fact-Sheet

[2] Chen, M., Li, W., Hao, Y., Qian, Y., & Humar, I. (2018). Edge cognitive computing based smart healthcare system. *Future Generation Computer Systems*, 86, 403–411.

[3]   Wiens, J., & Shenoy, E. S. (2018). Machine learning for healthcare: on the verge of a major shift in healthcare epidemiology. *Clinical Infectious Diseases*, 66(1), 149–153.

[4]   Bracken, B. (2021). Cyberattacks on Healthcare Spike 45% Since November. Opgehaal 16 Maart 2021, van https://threatpost.com/cyberattacks-healthcare-spike-ransomware/162770/

[5]   Hawkins, L. (2021). Cyberattacks increase in healthcare, but sector unprepared. Opgehaal 16 Maart 2021, van www.healthcareglobal.com/technology-and-ai-3/cyberattacks-increase-healthcare-sector-unprepared

[6]   Dyrda, L. (2020). The 5 most significant cyberattacks in healthcare for 2020. Opgehaal 16 Maart 2021, van www.beckershospitalreview.com/cybersecurity/the-5-most-significant-cyberattacks-in-healthcare-for-2020.html

[7]   Haque, N. I., Rahman, M. A., & Shahriar, H. (2021, July). Ensemble-based Efficient Anomaly Detection for Smart Building Control Systems. In *2021 IEEE 45th Annual Computers, Software, and Applications Conference (COMPSAC)* (pp. 504–513). IEEE.

[8]   Liu, D., Görges, M., & Jenkins, S. A. (2012). University of Queensland vital signs dataset: development of an accessible repository of anesthesia patient monitoring data for research. *Anesthesia & Analgesia*, 114(3), 584–589.

[9]   Demirkan, H. (2013). A smart healthcare systems framework. *It Professional*, 15(5), 38–45.

[10] Fell, J. C., & Voas, R. B. (2014). The effectiveness of a 0.05 blood alcohol concentration (bac) limit for driving in the United States. *Addiction*, 109(6), 869–874.

[11] Petersen, H., Baccelli, E., & Wählisch, M. (2014, June). Interoperable services on constrained devices in the internet of things. In *W3C Workshop on the Web of Things*.

[12] Pimentel, M. A., Johnson, A. E., Charlton, P. H., Birrenkott, D., Watkinson, P. J., Tarassenko, L., & Clifton, D. A. (2016). Toward a robust estimation of respiratory rate from pulse oximeters. *IEEE Transactions on Biomedical Engineering*, 64(8), 1914–1923.

[13] Wang, S. C. (2003). Artificial neural network. In *Interdisciplinary computing in java programming* (pp. 81–100). Springer, Boston, MA.

[14] Quinlan, J. R. (1986). Induction of decision trees. *Machine learning*, 1(1), 81–106.

[15] Haque, N. I., Khalil, A. A., Rahman, M. A., Amini, M. H., & Ahamed, S. I. (2021, September). BIOCAD: Bio-Inspired Optimization for Classification and Anomaly Detection in Digital Healthcare Systems. *IEEE International Conference on Digital Health*.

[16] Saputro, D. R. S., & Widyaningsih, P. (2017, August). Limited memory Broyden-Fletcher-Goldfarb-Shanno (L-BFGS) method for the parameter estimation on geographically weighted ordinal logistic regression model (GWOLR). In *AIP Conference Proceedings* (Vol. 1868, No. 1, p. 040009). AIP Publishing LLC.

[17] Hagan, M. T., Demuth, H. B., & Beale, M. (1997). *Neural network design*. PWS Publishing Co.

[18] Mirjalili, S., & Lewis, A. (2016). The whale optimization algorithm. *Advances in engineering software*, 95, 51–67.

[19] Mirjalili, S., Mirjalili, S. M., & Lewis, A. (2014). Grey wolf optimizer. *Advances in engineering software*, 69, 46–61.

[20] Yang, X. S. (2009, October). Firefly algorithms for multimodal optimization. In *International symposium on stochastic algorithms* (pp. 169–178). Springer, Berlin, Heidelberg.

[21] Haque, N. I., Rahman, M. A., Shahriar, M. H., Khalil, A. A., & Uluagac, S. (2021, March). A Novel Framework for Threat Analysis of Machine Learning-based Smart Healthcare Systems. *arXiv preprint* arXiv:2103.03472.

[22] Wehbe, T., Mooney, V. J., Javaid, A. Q., & Inan, O. T. (2017, May). A novel physiological features-assisted architecture for rapidly distinguishing health problems from hardware Trojan attacks and errors in medical devices. In *2017 IEEE International Symposium on Hardware Oriented Security and Trust (HOST)* (pp. 106–109). IEEE.

[23] Storm, D. (2015). MEDJACK: Hackers hijacking medical devices to create backdoors in hospital networks. Opgehaal 08 Januarie 2020, van www.computerworld.com/article/2932371/medjack-hackers-hijacking-medical-devices-to-create-backdoors-in-hospital-networks.html

[24] Almogren, A., Mohiuddin, I., Din, I. U., Almajed, H., & Guizani, N. (2020). Ftm-iomt: Fuzzy-based trust management for preventing sybil attacks in internet of medical things. *IEEE Internet of Things Journal*, 8(6), 4485–4497.

[25] Bapuji, V., & Reddy, D. S. (2018). Internet of Things interoperability using embedded Web technologies. *International Journal of Pure and Applied Mathematics*, 120(6), 7321–7331.

[26]   Deshmukh, R. V., & Devadkar, K. K. (2015). Understanding DDoS attack & its effect in cloud environment. *Procedia Computer Science*, 49, 202–210.

[27]   Pournaghshband, V., Sarrafzadeh, M., & Reiher, P. (2012, November). Securing legacy mobile medical devices. In *International Conference on Wireless Mobile Communication and Healthcare* (pp. 163–172). Springer, Berlin, Heidelberg.

[28]   Asaeedi, S., Didehvar, F., & Mohades, A. (2017). α-Concave hull, a generalization of convex hull. *Theoretical Computer Science*, 702, 48–59.

[29]   Patel, P., Sivaiah, B., & Patel, R. (2022). Approaches for finding optimal number of clusters using K-means and agglomerative hierarchical clustering techniques. *2022 International Conference on Intelligent Controller and Computing for Smart Power (ICICCSP)*, pp. 1–6, doi: 10.1109/ICICCSP53532.2022.9862439

[30]   Chen, Y., Zhou, X. S., & Huang, T. S. (2001, October). One-class SVM for learning in image retrieval. In *Proceedings 2001 International Conference on Image Processing (Cat. No. 01CH37205)* (Vol. 1, pp. 34–37). IEEE.

[31]   Ng, A. (2011). Sparse autoencoder. CS294A Lecture notes, 72(2011), 1–19.

[32]   Liu, K. S., Pinto, E. V., Munir, S., Francis, J., Shelton, C., Berges, M., & Lin, S. (2017, November). Cod: A dataset of commercial building occupancy traces. 1–2).

[33]   Hypertension. (2018, April). Opgehaal van https://catalog.data.gov/dataset/hypertension/

[34]   Martin, R. J., Ratan, R. R., Reding, M. J., & Olsen, T. S. (2012). Higher blood glucose within the normal range is associated with more severe strokes. *Stroke research and treatment*, 2012.

[35]   Bhogal, A. S., & Mani, A. R. (2017). Pattern analysis of oxygen saturation variability in healthy individuals: Entropy of pulse oximetry signals carries information about mean oxygen saturation. *Frontiers in physiology*, 8, 555.

[36]   Pimentel, M., Johnson, A., Charlton, P., & Clifton, D. (2017). BIDMC PPG and Respiration Dataset. *Physionet*, https://physionet.org/content/bidmc/1.0.0/.

A User-centric Focus for Detecting Phishing Emails

19

Regina Eckhardt and Sikha Bagui

Department of Computer Science, University of West Florida,
Pensacola, Florida, USA

Contents

DOI: 10.1201/9781003187158-23

313

19.1 INTRODUCTION

Cyber attacks are becoming ever more widespread, and the majority of cyber attacks are phishing attacks.[1] A study conducted by Cofense, formerly Phishme, found that 91% of all cyber attacks are phishing attacks.[2] In a phishing attack, an email is disguised in such a way that recipients are led to believe that the message is something that they want to read; hence the recipients click on the link or attachment and ends up revealing sensitive information.[3] Since phishing emails have no specific characteristic, they are difficult to detect, and little research has been done on the detection of phishing emails from a user's perspective.

For successful defense against phishing attacks, the ability to detect phishing emails is of utmost necessity. Measures to detect phishing can be classified into technical as well as user-centric [30, 39]. Technical measures include detecting phishing emails before the user receives the email, warning the email recipient through technical means [30]. Other technical measures include up-to-date security software that prevent users from following links to malicious websites by sending out warnings, asking for a second confirmation, or directly blocking potentially fraudulent websites [15]. Most research to date has focused on perfecting the technical aspects by trying to perfect the classification of phishing emails through machine learning (ML) or deep learning (DL) algorithms [51, 52].

To date, little emphasis has been placed on the user-centric measures as a line of defense again phishing attacks. User-centric measures focus on an email recipient and focus on increasing user awareness [30, 39]. Antiphishing training is often used to reduce susceptibility to phishing, but such training often focuses on specific types of phishing [12]. New forms of phishing attacks are continuously taking form, and it is difficult to keep up with new attacks with antiphishing training [31]. Moreover, user-centric measures need to be understood in the context of behavioral models. Software might detect an email as phishing and warn the user, but the attack may still be successful. This is because it is shown that users have to not only trust but also understand the behavior of technical phishing detectors to feel involved in the detection process [30].

As the second line of defense, this work develops an artifact that would be an interpreter for the ML/DL phishing detector. LIME and anchor explanations are used to take a closer look at cues of phishing emails and how they can be extracted by the means of explainable AI (XAI) methods [28]. Being able to interpret ML/DL algorithms enables an effective interaction between the technical approach of using a ML/DL phishing detector and a user-focused approach.

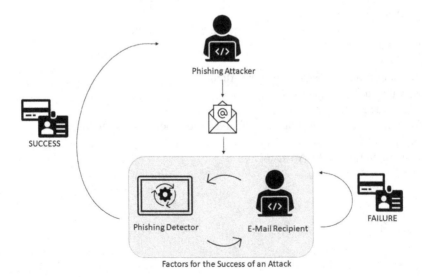

FIGURE 19.1 Success of a phishing attack.

Figure 19.1 demonstrates dependence of the success of a phishing attack on the interaction between technical phishing detectors and user involvement. For effective defense, both technical phishing detectors and active user involvement is necessary.

The rest of this paper is organized as follows: section 19.2 presents the background and related works; section 9.3 presents a summary of the dataset; section 19.4 presents the decision behavior of ML/DL models; section 19.5 presents the design of the artifact. The overarching goal of the artifact is to prevent users from falling for phishing attempts. Essential technical as well as theoretical fundamentals for the development of an artifact are provided. Section 19.6 presents the conclusions of this study.

19.2 BACKGROUND AND RELATED WORK

A comprehensive view of user-focused behavior cannot be understood without understanding the behavioral models that influence phishing susceptibility. This section presents a discussion of the behavioral models that form the backdrop of understanding phishing susceptibility. The latter part of this section presents previous research on user-focused measures and technical antiphishing measures, and the section concludes by discussing the gap in the research.

19.2.1 Behavioral Models Related to Phishing Susceptibility

The question of how and why phishing is so successful has been a major topic of research [12, 54]. To address this issue, what might affect a user's phishing susceptibility has to be analyzed. Research in phishing susceptibility ranges from empirical studies to behavioral models. Empirical studies showed that people read their emails in an automated way and make use of heuristics, so suspicious elements might go unnoticed [12, 38, 55, 58]. These heuristics include instincts developed over time [39]. Hence the question arises as to how we can increase users' attention. A 600 participant study found that frequent internet users were more likely to fall for phishing [36]. These users felt less scared to click on links in emails. This study discussed the need for knowledge on how to control the use of heuristics and the level of attention.

Behavioral models have also been introduced to investigate a user's phishing susceptibility. The rational actor theory assumes that humans think about and plan their behavior before making a thoughtful decision [5]. A coexistence of different human behaviors is covered in the dual-process theory. The dual-process theory focuses on influence techniques [58], with the state of mind being of utmost importance [12]. The dual-process theory was brought to a broad audience by [25] and introduces two behavioral systems, System 1 and System 2. System 1 behavior can be described as automated, intuitive, and unconscious, while System 2 behavior can be described as systematic, rational, and deliberate [59]. To reduce phishing susceptibility, the goal is to get users from System 1 behavior to System 2 behavior, that is, think when reading emails. The two systems were analyzed in phishing susceptibility research in different models such as the elaboration likelihood model (ELM) [21], the heuristic systematic model (HSM) [55], and directly [58, 9]. Both ELM and HSM explain how cognitive processing influences phishing deception by distinguishing between two information processing routes: the central route of careful consideration of information and the peripheral route of relying on simple cues only to evaluate a message [21]. HSM also follows an economic minded approach, that is, HSM assumes that a person's mind usually processes something with the least amount of effort, and more effortful processing is used only if more detailed information processing is needed [8].

Inattentive behavior relying on simple cues in phishing emails is not in line with the rational actor theory. What is needed is to investigate why people have inattentive behavior and methods for triggering thoughtful action when dealing with emails. Approaches to managing cognitive processing can help investigate ways for affecting the level of attention and effort one puts into reading emails. In addition to findings from the fundamentals of dual-process theory, the success of an efficient technique for determining phishing attacks depends on how well this technique takes advantage of a user's tendency to behave in System 1 [58]. It has also been shown that traditional approaches to security education do not have any effect when behaving in System 1 [9] and that users need to understand their increased vulnerability to fall for phishing in System 1 [12].

19.2.2 User-centric Antiphishing Measures

Existing antiphishing measures show effectiveness in System 2 behavior but are not able to shift behavior from System 1 to System 2. This is mainly due to the absence of thoughtful cognitive processing in the handling of emails. User-focused antiphishing measures can include technical measures as well as antiphishing training [1]. In this situation, technical measures would be tool tips helping users to evaluate embedded links in emails [56] or warning messages. Tool tips can be deployed in the form of highlighted elements such as the domain of a link in that email or a delay when being forwarded to a website after clicking on a link. This aims at drawing attention to possibly fraudulent elements [56]. Warning messages need to be recognized by the user and not left ignored [19].

Antiphishing training can be carried out by self-study, in classroom-style trainings, as well as in interventional trainings [30, 38]. Interventional measures [30], for example, can be used for training users to apply caution when opening a phishing email. Educational content for the self-study of antiphishing training is often provided in the form of texts [39], such as blogs with up-to-date information on current trends in phishing attacks [50], videos on how to detect phishing [26], and the use of games. Using games, users can, for example, test their own ability to detect phishing [47, 7]. A study done by Jagatic et al. [23] simulated phishing emails sent to students at Indiana University to test their phishing susceptibility. If the user fell for an attack, for example, by clicking on a link, the system would teach the participant about the attack [3].

Upon evaluating the various forms of antiphishing trainings, weaknesses in the various approaches become apparent, such as users not understanding cues provided by toolbars. If cues do not interrupt users from their usual actions, they go unnoticed and are easy to ignore [30]. This can be attributed to System 1 behavior of users.

Self-study material was found to be effective in increasing awareness, but it was found to be rarely actually used. Kumaraguru et al. [30] showed that emails directing users to study material are often ignored, and users usually do not seek this material on their own. According to Sheng et al. [57], users tend to ignore study material mainly because they cannot believe that they themselves can be victims of phishing attacks. Studies have shown that though classroom-style training, that is, in-class lectures as well as lab sessions, can actually increase one's ability to detect phishing emails, the difficulty of offering classroom-style training for large numbers of users is not very practical; hence this is considered to have only a minor impact on the whole [30]. Sending out simulated email tests is also problematic. Embrey and Kaivanto [12] state that, depending on the variety and frequency of repetitions, such tests might only cover some phishing types.

Nguyen [39] found that if people are not interested in learning about how they can detect phishing emails, the effectiveness of the training is low, especially in the long term. Moreover, most antiphishing training does not address recent research in phishing susceptibility [9], and users often do not follow the rules in a training. Trainings tend to address only certain phishing scenarios leaving the user unprepared for attacks carried out in different forms [12]. Hence different forms of antiphishing training can be effective only if specific requirements are satisfied [30].

19.2.3 Technical Antiphishing Measures

Technical antiphishing measures include rule-based filters, traditional machine learning methods, document representation, and deep learning techniques [22]. Rule-based filters [60] assess emails following a given rule. They can include blacklists, which are specific filters, where, for example, email and IP addresses of malicious senders or keywords are collected in a list, and incoming emails matching at least one list entry are automatically blocked [53]. One of the disadvantages of blacklists is that these lists have to be created and continuously updated [52]. Also, blacklists are not able to handle new patterns well. With blacklists, a new attacker is added to the blacklist only after an attack has succeeded [53].

Machine learning (ML) is based on predefined features of an email [52]. The ML model is fed labeled data, i.e., previously classified phishing emails as well as nonphishing emails. By iteratively classifying an email, checking the correctness of the classification, the set of features, and their outcomes, the model identifies the importance of the different features for the classification of an email [6]. The identified importance is reflected in weights assigned to the features, and, from this, the optimal set of features characteristic for a phishing email is found [2]. For the classification of a new email, the outcome of each feature is extracted and fed into the ML algorithm [53]. Unlike rule-based classifiers [51], features in ML models can capture more complex characteristics of a text by analyzing the underlying structure of phishing emails. Traditional ML classifiers can only find the best from given features and do not have the ability to learn unknown features [40].

Deep learning (DL), a subfield of ML, automatically learns complex patterns for effective and task-specific representations from raw data and uses them as features [40, 52]. No manual feature selection or feature engineering is necessary [14, 35] in DL, and DL can classify emails efficiently without human intervention [53]. It can handle large amounts of data, improving the performance of a network with an increasing amount of data [17]. DL has shown to achieve state-of-the-art performance in natural language processing [40, 53] and in many cases performs better than other ML techniques [35]. DL methods have shown better performance in spam email detection [35] and good performance in phishing detection [51].

The highest performance can be achieved by finding the best combination of classifiers as well as the best combination of classifiers and document representations [51, 52].

19.2.4 Research Gap

Although a substantial amount of research has focused on either user-centric or technical antiphishing measures (taken independently) to identify phishing emails, little work has been done on designing an architecture that incorporates the user-centric method in the backdrop of behavioral methods. With XAI, the aim is to improve the thoughtful cognitive handling of emails, moving a user's behavior from System 1 to System 2.

19.3 THE DATASET

The dataset, provided by AppRiver,[4] a company headquartered in Gulf Breeze, Florida, contains 18,305 emails, stored as a comma-separated-values (csv) file. Every email is categorized as either phishing or

TABLE 19.1 Data Used for Analysis

	TOTAL	PERCENTAGE
Whole data set	18,305	100.00%
Phishing emails	3,416	18.66%
Spam emails	14,889	81.34%
Training set	12,813	70.00%
Test set	5,492	30.00%

spam, and there are 3,416 emails in the phishing category. The information provided in each email is in the body of the email. Each email also contains a subject, location, industry, the number of recipients, as well as some facts about the size and structure of each email. For this analysis, only the body of the email was used. The entire dataset was split into two parts. Thirty percent of the data was used as a test, set and 70% was used as a training set. Table 19.1 provides an overview of the different samples used:

19.4 UNDERSTANDING THE DECISION BEHAVIOR OF MACHINE LEARNING MODELS

Since it has been shown that even the best DL/ML phishing detectors should be complemented by attentive email users to improve the detection rate, two different frameworks for interpreting ML classification models are presented.

19.4.1 Interpreter for Machine Learning Algorithms

In the paper " 'Why Should I Trust You?' Explaining the Predictions of Any Classifier," Ribeiro et al. [44] introduced methods to explain the decision behavior of ML models. People encounter ML in their daily lives, for example, for unlocking their smartphone by face recognition. Without knowledge about how these tools and applications work, it is difficult to trust their behavior and decisions [10], especially when they behave in unexpected ways, such as not recognizing the face or providing unusual recommendations. People do not use applications they do not trust, and they do not trust applications they do not understand [44]. To solve this problem in classification tasks, Ribeiro et al. [44] developed local interpretable model-agnostic explanations (LIME) and anchor explanations to provide easily understandable representations of the decision behavior of any classifier [44, 42].

Explaining a prediction means presenting artifacts to understand the relationship between the components of the instance and the prediction of the model [49]. The goal is to gain more trust and more effective use by providing explanations [45]. A first step is to understand why the model was chosen [63]. Explanations need to be interpretable; that is, the connection between the input and the response must be clear [44]. Since the degree of interpretability of a model can vary from person to person depending on personal background, a model must incorporate possible limitations for an individual. Explanations should be easy to understand [37]. Another option to guarantee a high level of interpretability is to characterize the audience of interest [61]. The explanations should have local fidelity; that is, the features used for explanation must be important in the local context [44]. Besides the local context, the explanations need to provide a global perspective to increase trust in the model as a whole [43]. Some models are easier to interpret while others are more difficult, but an interpreter should be able to explain any model independent of its level of complexity [42]. Local interpretable model-agnostic explanations (LIME) and anchor explanations cover the criteria for an explanatory model.

19.4.2 Local Interpretable Model-Agnostic Explanations (LIME)

LIME explains individual predictions of a classifier, making it interpretable [43]. It explains any black-box classifier with at least two classes. The algorithm works by approximating a prediction locally using an interpretable model. It first takes the prediction to be explained, where the prediction is an outcome of a model that is not interpretable in itself. Then two significant elements are needed: a model that is interpretable and an instance that is similar to the original prediction that can be fed to the interpretable model. With this model, an interpretable explanation of the original output can be found. In order to find an appropriate interpretable model, a proximity measure as well as a measure of complexity need to be incorporated. The proximity measure ensures taking an instance that is similar to the original prediction. The measure of complexity describes the level of complexity of the interpretable model aiming at a model of lower complexity. The functioning of LIME is mathematically described next.

Let $x \in R^d$ denote original representation, i.e., an email, and $x^0 \in \{0,1\}^{d0}$ be its binary representation. The explanation is a model $g \in G$, where G describes the class of interpretable models. This means a model g in this set can be presented to the user with artifacts. Since not every model in this set might be simple enough to be interpretable by anyone, the complexity of an explanation $g \in G$ is measured by $\omega(g)$. The actual model to be explained is denoted by a function:

$$f: R^d \rightarrow R$$

whereas $f(x)$ is the probability that x is a phishing email. Additionally, there is a proximity measure $\pi_x(z)$ between z to x in order to define locality around x. Aiming at an interpretable model imitating the original model as well as possible, the measure $L(f,g,\pi_x)$ describes how unfaithful g is in approximating f in the locality defined by π_x. The goal is to have both interpretability and local fidelity so that $L(f,g,\pi_x)$ needs to be minimized while $\omega(g)$ needs to be kept low in order for the model to still be interpretable by humans. LIME then produces an explanation of the following form:

$$\xi(x) = \underbrace{\arg\min}_{g \in G} L(f, g, \pi_x) + \omega(g)$$

This can then be used for different explanation families G, fidelity functions L, and complexity measures ω. The goal is to minimize L without making any assumptions about f to keep the resulting model-agnostic. The proximity of a classifying instance to the original prediction can be guaranteed by the proximity measure. In order to guarantee a stable algorithm, LIME iterates over N samples and uses K-Lasso to find the K features that best relate to the prediction. The features found are highlighted in the email and listed with their respective probability of being phishing or spam. A LIME explanation is presented in Figure 19.2.

As can be observed from Figure 19.2, the major advantage of LIME is that it is visually attractive and easy to follow. But its major drawback is that LIME explanations are locally faithful; that is, for example, the features "iCloud" and "Account" found to be relevant for the classification of this email into "Phishing" do not necessarily have to be features categorizing another email as phishing too. To overcome this disadvantage, anchor explanations are introduced.

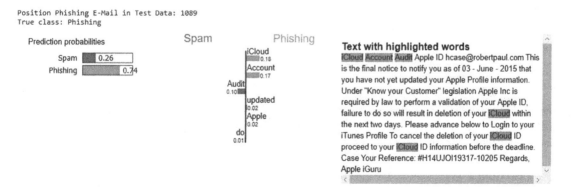

FIGURE 19.2   Example LIME explanation.

19.4.3 Anchor Explanations

Anchor explanation are a set of elements that "anchor" the prediction in a form that changes to the rest of the elements in the input instance but that do not change the prediction [16]. It is able to explain any black-box classifier with two or more classes and provide model-agnostic, interpretable explanations for classification models.[5] Based on an approach similar to LIME explanations, anchor explanations address the drawback of LIME; that is, anchors are not limited to approximating the local behavior of a model in a linear way. This kind of approximation does not allow any conclusion on the faithfulness of the explanations and their coverage region.

The idea of anchor explanations is to explain the behavior of classification models by anchors that are specific rules ensuring a certain prediction with a high level of confidence. In Goerke and Lang [16], a rule is said to anchor a prediction if changes in other feature values do not affect the prediction. In order to find an anchor, an anchor candidate is chosen while the rest of the words in the email are replaced with a specific probability. This replacement is done by either unknown token (UNK) or words similar to the original email content. An anchor will always anchor a specific prediction; that is, a set of words found to be an anchor for the classification of one email into phishing cannot be an anchor for another email to be classified as spam. This is the major improvement when comparing anchor explanations to LIME. For text classification, anchors consist of the words that need to be present to ensure a prediction with a high probability independent of the other words in the input. Starting with a black-box model:

$$f: X \rightarrow Y$$

and an email $x \in X$, the goal is to explain the outcome $f(x)$ by rules. X describes the original dataset, the set of emails, while Y represents the classification into phishing or spam, that is, the respective probabilities for each category. A rule can be defined as follows:

Definition 1: A rule R is a set of words with $R(x) = 1$ if every element of this set occurs in the instance x.

R needs to act on an interpretable representation since its features are used for the explanations that need to be interpretable. In the case of the phishing detector, a rule consists of words or phrases and can be examined in every email. The rule also needs to be able to explain the prediction. The ability to serve as an anchor implies the ability to classify an email into phishing or spam with the existence of this rule only. This is true if, for another email z, a perturbing distribution D and a confidence level τ it holds that:

$$E_{D(z|R)}[I_{f(x)=f(z)}] \geq \tau$$

with $R(x) = 1$, that is, whenever the probability for another email z satisfies the rule R to produce the same prediction is at least τ, τ needs to be fixed whereby a higher value of τ represents more confidence that a rule might be related to the classification of an email. The preceding formula can also be transformed into:

$$P(E_{D(z|R)}[I_{f(x)=f(z)}] \geq \tau) \geq 1 - \delta,$$

that is, the probability for a rule being an anchor is at least $1 - \delta$. Here, the smaller the value for δ, the more difficult it is for a rule to be called an anchor. For the usage of rules, the term coverage is of importance and can be defined as follows:

Definition 2: For each rule the coverage,

$$cov(R) = E_{D(z)}[R(z)].$$

19.4.3.1 Share of Emails in the Data for Which the Rule Holds

Once several rules that satisfy the anchor condition have been generated, the one with the largest coverage is preferred. The question left unanswered is how to get an anchor from a set of rules, and this is addressed with the application of the beam-search method. This method represents an approach to finding the optimal anchor with the largest coverage [46].

Beam-search is used to find the optimal number as well as set of anchors. It is an improvement of the standard bottom-up construction [57]. The standard bottom-up construction starts with an empty rule $R = \varnothing$ as an anchor R. The empty rule is complemented by the feature candidate with the highest precision r_1 so that the new anchor is:

$$R = \{r_1\}$$

This iteration is repeated with the remaining feature candidates until the probability of the constructed anchor hits $1 - \delta$ [46]. The final anchor found might not be optimal since, a feature, once added, cannot be removed from the set in a later iteration [57]. The beam-size method addresses this shortcoming and considers sets of B candidates instead of single candidates in each step. This approach can provide anchors with larger coverage [46]. The value for τ as well as settings for the other parameters mentioned can be fixed during the implementation of the anchor explanations. One variable for which a value needs to be assigned is the confidence threshold. It is defined as follows:

Definition 3: The share c of instances satisfying the rule that lead to the same prediction can be described by

$$c = \frac{\Sigma_{z \in X} I_{R(z)=1 \cap f(z)=f(x)}}{\Sigma_{z \in X} I_{R(z)=1}}.$$

The minimum of this share is called the confidence threshold c_t [27] and will be set to $c_t = 0.95$.

The higher the confidence threshold, the higher the probability that another set of words will satisfy this rule and lead to the same prediction. This means that the respective anchor candidate is more stable in guaranteeing a specific prediction the higher the threshold.

These functionalities of the anchor explanations were implemented with the help of the alibi packages for python. Just like the output of LIME, this algorithm outputs the anchor, i.e., the set of words that characterize the classification of this email. It also provides the emails with the word replacements that led to the classification. The implementation was done with the default settings of this algorithm, except that the anchor candidate was replaced by UNK with a certain probability. An exemplary output for a phishing email can be seen in Figure 19.3.

For this email, Figure 19.3, it can be noted that a precision of 0.75 was set and that no anchor could be found that satisfied this precision. Nevertheless, the algorithm output the best anchor from all candidate anchors. The elements in this anchor are "a," "us," "1999," "prevent," "account," "are," "." and the anchor has a precision of about 0.007. Some of the elements of the anchor seem to be useful in terms of presenting it to a user such as "prevent" or "account," but the remaining are less meaningful. The computation time for finding anchor explanations for an email varies significantly between the emails, from a few minutes up to several hours. Taking this into account, in addition to not always being able to find anchors satisfying the precision level and the existence of elements in the anchor that lack in their content, there is indeed an upside potential for the anchor explanations. The outputs of LIME and anchor explanations both seem to provide reasonable and understandable explanations, highlighting the relevant words or phrases in the email, making it more user-friendly than providing all examples of "successful" replacements. With LIME being only locally faithful, anchor explanations provide more stable explanations for predictions. For the design of the artifact, the advantages of both LIME and anchor explanations were incorporated for an optimal solution.

Wells Fargo: Your Privacy and security is our priority. It's our responsibility to maintain and secure your account. We are implimenting a new
security devices that we enable us bind your account to your location to prevent illegal login from an unknown device or location. Update your
profile to continue with online access. Your security is important to us. 1999 - 2015 Wells Fargo. All rights reserved.
Position Phishing E-Mail in Test Data 16
Could not find an anchor satisfying the 0.75 precision constraint. Now returning the best non-eligible anchor.
Anchor: a AND us AND 1999 AND prevent AND account AND are AND .
Precision: 0.01
{'names': ['a', 'us', '1999', 'prevent', 'account', 'are', '.'], 'precision': 0.007462686567164179, 'coverage': 0.0063, 'raw': {'feature':
[25, 32, 66, 40, 20, 23, 49], 'mean': [0.001110494169905608, 0.0023094688221709007, 0.004807692307692308, 0.00684931506849315,
0.005154639175257732, 0.003333333333333335, 0.007462686567164179], 'precision': [0.001110494169905608, 0.0023094688221709007,
0.004807692307692308, 0.00684931506849315, 0.005154639175257732, 0.0033333333333333335, 0.007462686567164179], 'coverage': [0.4951,
0.2512, 0.1235, 0.0622, 0.0311, 0.0146, 0.0063],'examples':
[{'covered': array([['UNK UNK UNK UNK Privacy and UNK is UNK priority . It UNK our UNK to maintain and secure UNK account . We are
            implimenting a UNK UNK UNK that UNK UNK UNK UNK your UNK UNK your location to prevent UNK login UNK an UNK UNK or
            UNK UNK UNK your UNK UNK UNK with UNK access UNK Your security UNK UNK UNK UNK . 1999 - 2015 UNK UNK UNK UNK rights UNK UNK'],
           ['UNK UNK : Your UNK and UNK is UNK UNK UNK It 's our UNK to UNK and UNK your UNK . We UNK UNK a UNK security devices UNK we UNK
            us bind UNK account UNK your UNK UNK prevent UNK UNK UNK unknown UNK UNK UNK UNK your UNK to UNK with online access .
            Your security is important to UNK UNK 1999 - UNK Wells Fargo . UNK UNK UNK UNK"], ["Wells UNK UNK UNK UNK UNK security UNK our
            UNK UNK It 's UNK UNK to UNK and secure your UNK UNK We UNK UNK a UNK security devices UNK we UNK us UNK your account to UNK UNK
            UNK UNK illegal login UNK an UNK device UNK location UNK UNK your profile UNK UNK with online access UNK Your UNK UNK UNK important
            UNK UNK. 1999 UNK UNK UNK UNK . UNK UNK UNK ."],
           ["Wells UNK : UNK Privacy UNK UNK is our priority UNK UNK 's our responsibility UNK UNK UNK secure UNK account UNK We UNK
            implimenting a UNK security devices that we enable UNK bind UNK UNK UNK UNK UNK to prevent UNK login from an unknown UNK UNK UNK
            . Update UNK UNK UNK UNK with online UNK . Your security is UNK to us UNK 1999 - 2015 Wells UNK . UNK UNK UNK UNK"],
```

**FIGURE 19.3**   Example anchor explanation.

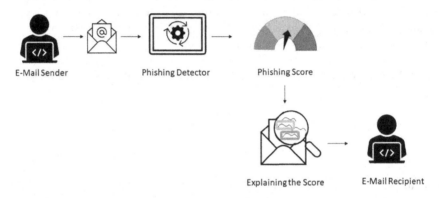

E-Mail Sender          Phishing Detector          Phishing Score

Explaining the Score          E-Mail Recipient

**FIGURE 19.4**   Design of the artifact.

# 19.5 DESIGNING THE ARTIFACT

## 19.5.1 Background

The goal of the artifact is to prevent users from falling victim to phishing attempts. It achieves this objective by drawing the user's attention to cues in emails. This artifact enables users to carefully examine the parts of the email most relevant to classifying it as a phishing attempt as well as alerts the email recipient, with the objective of moving the user from System 1 behavior to System 2 behavior.

The artifact is designed, as shown in Figure 19.4, based on research in the areas of phishing susceptibility, antiphishing training, and the generation of explanations for document classification based on methods of AI [32, 34, 46]. An AI system takes an email as the input and assigns a score that represents the likelihood that the particular email is a phishing attempt. Based on the score, the emails are assigned to one of three classes: "phishing," "suspicious," and "legitimate." Emails classified as "phishing" are immediately discarded, and those classified as "legitimate" are passed through. For "suspicious" email, an XAI system determines which property of the email's text, that is, the presence of certain words, phrases, or typographical features, led to the score assigned by the phishing detector. For this, an efficient search-based
```

XAI algorithm is employed that only requires query access to the phishing detector. The parts of the text that contributed to the email's classification as "suspicious" are highlighted when presented to the user.

19.5.2 Identifying Suspected Phishing Attempts

As described in [62], the starting point is an instantiated ML-based phishing detector. It is assumed that the phishing detector takes an email as input and outputs a bounded score s that reflects the likelihood that the email is a phishing attempt. Otherwise, the phishing detector is treated as a black-box, and no particular properties are assumed. Without loss of generality, it is assumed that $s \in [0,1]$. To determine which emails to discard, explain, or pass through, two threshold values t_{phish} and $t_{suspicious}$, are defined. All emails for which $s > t_{phish}$ are classified as phishing. This threshold is chosen such that the rate of false positives in this class is minimal; that is, the least legitimate emails as possible are mistakenly discarded. Accordingly, the lower threshold $t_{suspicious}$ is set such that practically no phishing email is assigned a score: $s^1 t_{suspicious}$. Hence, only emails are considered for which $t_{phish} \geq s \geq t_{suspicious}$, that is, suspected phishing attempts.

19.5.3 Cues in Phishing Emails

With the rise of more advanced phishing attacks, phishing emails have become more versatile and increasingly personalized. Nevertheless, there are common traits in phishing emails which have been reported in the scientific literature. These are used in antiphishing trainings. Since the technical implementation exclusively processes the text of an email, the focus is on traits in the text. Elements beyond plain text, e.g., links or logos, are not considered, as is technical information such as server addresses and electronic signatures. The latter is highly specific to organizations and typically dealt with through rule-based antiphishing measures. The textual traits of phishing emails, summarized in Table 19.2, can broadly be classified into syntactic, stylistic, and user-related features[4, 24, 29, 36]. Syntactic features include the style of the email and the order of information in a message. Stylistic features can further be partitioned into prompting features, the group of senders, and elements beyond the text. Prompting features would include things like a question for the confirmation of banking details, message quality, elements in the form of appeals, creation of time pressure, and an unusually strong emphasis on security [24]. The group of senders can provide a first hint to a possible fraudulent goal of an email. For example, financial services are the most targeted industry in terms of phishing attacks [11]. User-related features include trust vs. distrust and the lack of focus of the user [36]. Since the focus is on traits in the emails text, user-related features are not discussed in further detail.

Different groups and different aspects within these groups of cues necessitate the assignment of appropriate weights for their respective importance. The importance of an aspect cannot be generalized and varies depending on the email and its content. Therefore, the goal is to highlight and extract the relevant cues for each email individually. This has to be done in a way that users' System 2 behavior is triggered by, on the one hand, their increasing phishing awareness and, on the other hand, decreasing phishing susceptibility, as described in the related works section.

ML models are known to pick up on stylistic and syntactic features. For stylistic features, words or groups of words can be detected that express a prompting or demand, and the sentiment and purpose for these prompting elements can be investigated [4, 24, 29]. An example is a prompt to a recipient to update his or her bank account where specific information can be expressed through different keywords such as "bank," "account," "update," or "details." The phishing detector can scan incoming emails for words or groups of words that are proven to be characteristic for phishing emails. Elements beyond plain text can be picked up by ML as long as they are in a form that is recognizable by ML. The category of the sender can be found through keywords that represent the industry, whereby the industry can be revealed by analyzing the prompting elements. Syntactic features such as the style of an email in terms of length or tone can be analyzed in an indirect way [4]. The length can be considered by counting words while the analysis of the

TABLE 19.2   Textual Traits of Phishing Emails

CUE	CAPTURED BY XAI	HOW? OR WHY NOT?
1 Logos, links, or link types (Jakobsson, 2007; Blythe et al., 2011; Moody et al., 2017; Kim & Kim, 2013)	Yes	If distinguishable from regular text elements, they can be recognized by a detector.
2 Asking for confirmation of banking details (Blythe et al., 2011)	Yes	By words semantically related to the cue
3 Poor spelling and grammar (Jakobsson, 2007; Blythe et al., 2011)	Yes	Being treated as a regular text element they can be recognized by a detector
4 Specific group of senders and premises (Jakobsson, 2007; Blythe et al., 2011; Moody et al., 2017; Kim & Kim, 2013)	Partly	By being mentioned in the text to be analyzed
5 Style of the email (Blythe et al., 2011)	Yes	There exist ML methods that are able to analyze the style of a text.
6 Order in message (Kim and Kim, 2013)	Yes	Not only words but also the structure of a text can be analyzed.
7 Message quality elements (Kim & Kim, 2013)	Yes	By words semantically related to the cue
8 Trust and distrust, boredom proneness, lack of focus (Moody et al., 2017)	No	Depends on the email recipient and cannot be investigated by analyzing the email only
9 Time pressure (Kim & Kim, 2013)	Yes	Words indicating time pressure can also be detected as relevant for the classification of the email.
10 Semantic network analysis (Kim & Kim, 2013)	No	The algorithm does not take into account word frequency, proximity, and co-occurrence
11 Too much emphasis on security (Jakobsson, 2007)	Partly	By words semantically related to the cue

tone is the same as for prompting elements. Generally, it can be stated that ML phishing detectors are able to capture the cues on a token-based approach, i.e., they catch relevant tokens that refer to specific groups of cue characteristics for the categorization of an email into phishing or nonphishing. Thus the score of a ML-based phishing detector depends on the presence or absence of traits.

19.5.4 Extracting Cues

To identify the cues and phrases in the email that best convey the suspicion to the user, an XAI system was designed that extracts this information (traits) from the underlying phishing detector. In terms of the XAI nomenclature, this can be considered a black-box outcome explanation problem [18]; that is, the aim is to explain the phishing detector's score, not the detector's inner workings [13, 34]. With the phishing detector assumed to be a black-box, the XAI algorithm has to be model-agnostic [33, 44]. This further allows for updates of the phishing detector and prevents compromise of its effectiveness [44]. Also, representation flexibility is required [44]; that is, parts of the actual email text are highlighted, which might be very different from the numerical machine learning model's input. Another flexibility can be achieved through the extraction of phrases, instead of tokens, by connecting tokens and assigning an appropriate length to the connected tokens. A typical example that illustrates the advantage of using phrases is the negation of words such as "not good" that will not be recognized as "bad" if treated separately. When thinking about the implementation of a phishing detector and explainer in a real-world scenario, the aspect of efficiency is of high importance since it needs to run on the fly for massive amounts of emails coming in at a high frequency. The usage of a search-based approach, i.e., searching for relevant elements, can satisfy both characteristics of the artifact as being model-agnostic and efficient.

19.5.5 Examples of the Application of XAI for Extracting Cues and Phrases

The first characteristic listed in Table 19.1 is having links in an email where there is a demand from the sender of that email to click on the link. While Blythe et al. [4] and Kim & Kim [29] only emphasize the existence of links in emails, Moody et al. [36] and Canova et al. [7] differentiate between the different types of links. LIME and anchor explanations are only looking for words or phrases. Since LIME and anchor explanations do not recognize links, they cannot differentiate one type of link from another. A link can be recognized only if the explanation algorithm assigns importance to this link as a word or phrase for the classification of the email. Also, words such "click" or "link" can provide information about the existence of a link. This is illustrated in Figure 19.5.

Figure 19.6 illustrates the second point in Table 19.1. Blythe et al. [4] state that a typical phishing email asks for a confirmation of banking details. This information can be captured by explanation algorithms, since this is often expressed by keywords such as "bank," "account," "update," "details," etc.

Another characteristic that can at least partly be found through explanation algorithms is poor spelling and grammar, an example of which is presented in Figure 19.7.

Most research on the detection of phishing emails mention the fifth characteristic listed in Table 19.2. This characteristic relates to phishing emails being exposed through a specific group of senders. For this

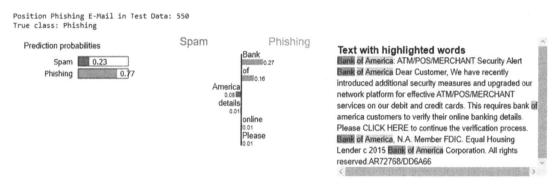

FIGURE 19.5   Example click on link.

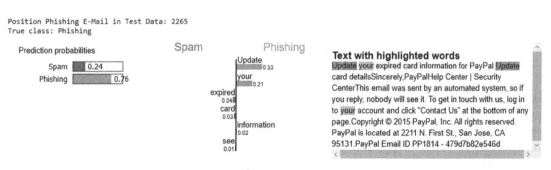

FIGURE 19.6   Example banking details.

Position Phishing E-Mail in Test Data: 2899
True class: Phishing

FIGURE 19.7   Example error.

Position Phishing E-Mail in Test Data: 67
True class: Phishing

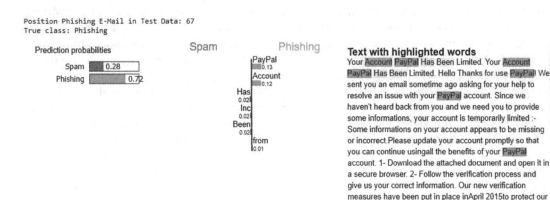

FIGURE 19.8   Example PayPal.

analysis, only the bodies of the emails were considered. Though a separate focus on the sender was not included, the sender can still be investigated as a significant factor in the classification of the emails. In this case, the sender is also mentioned in the body and the subject of the text. Blythe et al. [4] and Kim and Kim [29] state that phishers most often pretend to be financial institutions. An example is presented in Figure 19.8.

Logos of the sender as well as the style of an email can only be partly analyzed by the explanation algorithm. Just like links, the algorithm does not recognize the logo as a logo, but it can recognize it as a characteristic element in an email. Indirectly, the style of an email in terms of length or tone [4] can also be analyzed. The tone of an email can be categorized by finding relevant words. The length of an email, however, cannot be expressed through LIME or anchor explanations, though this might be important in neural networks. The email presented in Figure 19.9 is striking due to its short length and its being in mostly capital letters with a lot of exclamation marks. The capital letters and the exclamation marks evoke a tone of excitement in that email. The word "ignore" was used as an imputation method whenever the body of an email was empty so that the algorithm does not give an error message when working with an empty body, an empty subject, or both empty.

As with the tone of an email-specific message, quality elements can be captured with LIME and anchor explanations. These can be recognized in a direct or indirect way. Expressing specific feelings or appeals belong to this category. These would require an additional interpretation in a neural network rather than just providing a list of words that are relevant for the decision. Appeals can be grouped into rational, emotional, and motivational appeals [29]. Rational appeals deal with the reasoning in different forms such as "one event is a sign of another" [29]. This means that phishing attackers are trying to not just ask for sensitive information without any explanation. They feel that providing reasoning makes them appear less

FIGURE 19.9   Example tone.

FIGURE 19.10   Example reasoning appeal.

FIGURE 19.11   Example emotional appeal.

suspicious. In the example presented in Figure 19.10, significant changes in the account's activity are used as the reason for the update of the user's account information.

As shown in Figure 19.11, emotional appeals deal with different emotions and try to make the email user feel appreciated [29].

Figure 19.12 presents an example of a motivational appeal. Motivational appeals make use of the needs of humans, and, like emotional appeals, they are based on the specific message that is targeted through that email.

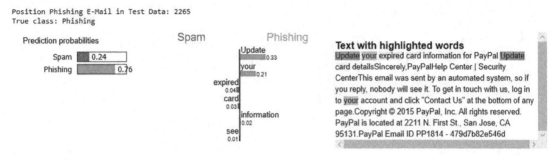

FIGURE 19.12   Example motivational appeal.

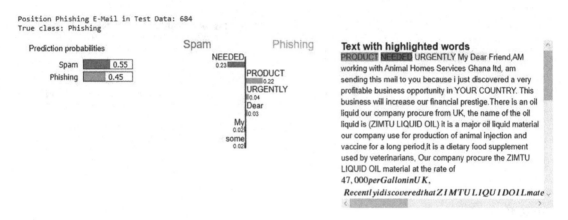

FIGURE 19.13   Example time pressure.

Another element that often comes up when the detection of phishing is being discussed is time pressure. Time pressure can be expressed by a lot of different words. These words can be part of the set that is significant for the decision of the network. Figure 19.13 presents an example of a time pressure.

According to Jakobsson [24], along with time pressure, an inordinate emphasis on security can be characteristic of phishing emails. While phishing attackers try to hide their intention by giving more attention to security, this can reveal their actual goal. Just like the time pressure aspect, security can be expressed by a variety of words, and therefore it can be identified by LIME as well as anchor explanations. An evaluation of whether too much emphasis was placed on security needs to be validated individually for each email.

Though each of the characteristics in Table 19.1 were explained separately, in reality many of the characteristics from Table 19.1 can be found in an email. Figures 19.14 and 19.15 are examples of phishing emails containing motivational, emotional, rational appeals, sent from a specific group of senders, asking for a confirmation of banking details, having a specific style (short, business email, advertising), containing message quality elements, and putting a strong emphasis on security.

Other factors are important in helping the email user detect a phishing email, and these are not included in the preceding methods, for example, the order of the words in a message with some key structural elements [4]. These structural elements can be viewed in the preceding methods in an indirect way, for example, through the tone of one of the emails just described. The structure itself cannot be displayed as a characteristic element in the decision of a neural network. However, a semantic network analysis can be helpful for an algorithm to detect patterns in a classification task by taking into account word frequencies, the proximity between words or group of words, and co-occurences. However, the order in a message, for example, cannot be expressed with the explanation algorithms used. The same holds for trust

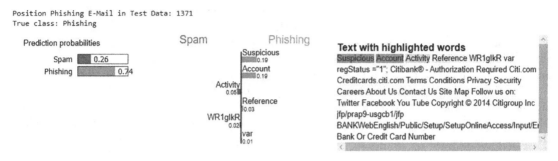

FIGURE 19.14   Examples of phishing emails from Citibank.

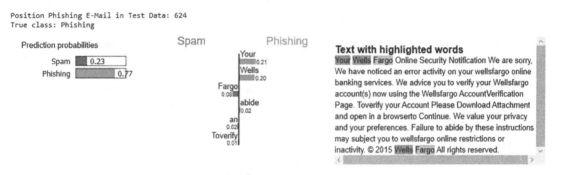

FIGURE 19.15   Examples of phishing emails from Wells Fargo bank.

or distrust of the user in an email, the boredom proneness and lack of focus. These depend solely on the email user and can only be affected by specific training on awareness. In summary, these findings can help improve the detection of phishing emails.

19.6 CONCLUSION AND FUTURE WORKS

The proposed model marks a significant milestone in detecting phishing attacks. The goal of this work was to develop a solution capturing the interaction of a technical phishing detector and user involvement. The objective of the XAI models was to move a user's behavior from System 1 to System 2. An artifact was designed, presenting its ability to improve phishing detection.

19.6.1 Completion of the Artifact

From the comparison of the interpreters, it can be concluded that both the interpreters leave space for further improvement for the completion of the artifact. For example, LIME might classify the same email differently in various runs due to its random properties. That is, the same words could be found as being characteristic for a phishing email as for a spam email [20].

Anchor explanations address the drawback of the local fidelity of LIME but can be improved in other aspects. For example, depending on the email, the computation of anchors for one email can take several hours. This does not provide the desired computation time for explanations in almost real time. Also, the implementation of the algorithm with the alibi package could not always find anchors for the specific precision level, and the anchors sometimes found elements of less content included. For future work, it is recommended that these drawbacks of anchors be fixed by fine-tuning the parameters. This, with the ability to highlight anchors with LIME explanations, can lead to a fast and easily understandable interpreter complementing the decisions of the corresponding phishing detector.

The objective was to get users to use System 2 thinking [9, 58 by raising their attention through highlighting suspicious parts of email text. Raising the user's attention should aid users in determining whether an email is genuine or phishing by drawing attention to important phrases in emails potentially classified as phishing [41, 55].

NOTES

1   www.csoonline.com/article/2117843/what-is-phishing-how-this-cyber-attack-works-and-how-to-prevent-it.html
2   https://cofense.com/
3   AppRiver. AppRiver, 2002.
4   Alibi anchors. https://docs.seldon.io/projects/alibi/en/stable/methods/ Anchors.html. Accessed: 2021-01-08.
5   Keras.io. callbacks–keras documentation. https://keras.io/callbacks/. Accessed: 2021-01-05.

REFERENCES

[1]   Alagha, A. A. A., Alkhalaf, M. N. I., & Taher, A. W. A. (2015). Phishing and Online Fraud Prevention. https://uspto.report/patent/app/20140165177
[2]   Akinyelu, A. & Adewumi, A. (2014). Classification of Phishing Email Using Random Forest Machine Learning Technique. *Journal of Applied Mathematics*. 6 pages. DOI: 10.1155/2014/425731
[3]   Aniket, B. & Sunil, B. M. (2018). State of Research on Phishing and Recent Trends of Attacks. i-manager's *Journal on Computer Science*, 5(4):14.
[4]   Blythe, M., Petrie, H., & Clark, J.A. (2011). F for Fake: Four Studies on How we Fall for Phish. Proceedings of the International Conference on Human Factors in Computing, CHI 2011, Vancouver, BC, Canada, May 7–12. DOI: 10.1145/1978942.1979459
[5]   Brighton, H. & Gigerenzer, G. (2012). Are Rational Actor Models "Rational" Outside Small Worlds? In Okasha, S., & Binmore, K. *Evolution and rationality*, Cambridge University Press, 84–109.
[6]   Burdisso, S. G., Errecalde, M., & Gomez, M. My. (2019). A Text Classification Framework for Simple and Effective Early Depression Detection Over Social Media Streams. *Expert Systems with Applications*, 133:182–197.
[7]   Canova, G., Volkamer, M., Bergmann, C., & Borza, R. (2014). Nophish: An Anti-phishing Education App. In: Mauw S., Jensen, C. D., (eds). Security and Trust Management. STE 2014. *Lecture Notes in Computer Science,* vol 8743, Springer, Cham., 188–192. https://doi.org/10.1007/978-3-319-11851-2_14
[8]   Chaiken, S. & Trope, Y. (1999). *Dual-process theories in social psychology.* Guilford Press, New York.
[9]   Dennis, A. R. & Minas, R. K. (2018). Security on Autopilot: Why Current Security Theories Hijack our Thinking and Lead us Astray. *Data Base for Advances in Information Systems*, 49(s1):15–37.
[10]   Dhurandhar, A., Chen, P-Y. Luss, R., Tu, C-C., Ting, P., Shanmugam , & Das, P. (2018). Explanations Based on the Missing: Towards Contrastive Explanations with Pertinent Negatives. *Proceedings of the 32nd International Conference on Neural Information Processing Systems*, 590–601.
[11]   Drake, C. E., Oliver, J. J. & Koontz, E. J. (2004). Anatomy of a Phishing Email. *Proceedings of the First Conference on Email and Anti-Spam (CEAS)*, 1–8.
[12]   Embrey, I. & Kaivanto, K. (2018). Many Phish in the C: A Coexisting-Choice-Criteria Model of Security Behavior, *Economic Working Paper Series*, Lancaster University Management School, 1–23.

[13] Fernandez, C. & Provost, F. (2019). Counterfactual Explanations for Data-Driven Decisions. *40th International Conference on Information Systems, ICIS 2019,* Munich, Germany, 1–9.

[14] Ganesh, H. B., Vinayakumar, R., Anand Kumar, M. & Kp, S. (2018). Distributed Representation using Target Classes: Bag of Tricks for Security and Privacy Analytics. *Proceedings of the 1st Anti-Phishing Shared Task Pilot at 4th ACM IWSPA co-located with 8th ACM Conference on Data and Application Security and Privacy (CODASPY 2018),*Tempe, Arizona, USA, March 21, 2018.

[15] Gerber. S. (2018). 11 Security Strategies to Protect Your Company From Phishing Attacks, TNW News. https://thenextweb.com/news/11-security-strategies-protect-company-phishing-attacks

[16] Goerke, T. & Lang, M. (2019). Scoped Rules (anchors). In *Interpretable Machine Learning.* https://christo phm.github.io/interpretable-ml-book/anchors.html

[17] Goodfellow, I., Bengio, Y. & Courville, A. (2017). Deep Learning. MIT Press, USA.

[18] Guidotti, R., Monreale, A., Ruggieri, S., Turini, F., Pedreschi, D., & Giannotti, F. (2019). A Survey Of Methods For Explaining Black Box Models. *ACM Computing Surveys,* 51(5), 1–42.

[19] Gupta, B. B., Arachchilage, N. A., & Psannis, K. E. (2018). Defending Against Phishing Attacks: Taxonomy of Methods, Current Issues and Future Directions. *Telecommunication Systems,* 67(2):247–267.

[20] Harikrishnan, N.B., Vinayakumar, R, & Kp, S. (2018). A Machine Learning approach towards Phishing Email Detection. *Proceedings of the 1st Anti-Phishing Shared Task Pilot at 4th ACM IWSPA co-located with 8th ACM Conference on Data and Application Security and Privacy (CODASPY 2018),* Tempe, Arizona, USA, March 21, 2018.

[21] Harrison, B., Svetieva, E. & Vishwanath, A. (2016). Individual Processing of Phishing Emails. *Online Information Review,* 40(2):265–281.

[22] Hiransha, M., Unnithan, N. A., Vidyapeetham, R., & Kp, S. (2018). Deep Learning Based Phishing Email Detection. In: R. Verma, A. Das (eds.): *Proceedings of the 1st AntiPhishing Shared Pilot at 4th ACM International Workshop on Security and Privacy Analytics (IWSPA 2018),* Tempe, Arizona, USA.

[23] Jagatic, B. T. N., Johnson, N. A., Jakobsson, M. & Menczer, F. (2007). Social Phishing, *Communications of the ACM,* 50(10), 94–100.

[24] Jakobsson, M. (2007). The Human Factor in Phishing. In *Priv. Secur. Consum. Inf.* 7, 1–19.

[25] Kahneman, D. (2012). Thinking, Fast and Slow. Penguin Books.

[26] Karumbaiah, S., Wright, R. T., Durcikova, A. & Jensen, M. L. (2016). Phishing Training: A Preliminary Look at the Effects of Different Types of Training. In *Proceedings of the 11th Pre-ICIS Workshop on Information Security and Privacy,* Dublin.

[27] Klaise, J., Looveren, A. V., Vacanti, G., & Coca, A. (2021). Alibi Explain: Algorithms for Explaining Machine Learning Models. *Journal of Machine Learning Research,* 22, 1–7.

[28] Kluge, K., & Eckhardt, R. (2020). Explaining the Suspicion: Design of an XAI-based User-Focused Anti-Phishing Measure, *16th International Conference on Wirtschaftsinformatik,* March 2021, Essen, Germany.

[29] Kim, D. & Kim, J. H. (2013). Understanding Persuasive Elements in Phishing Emails: A Categorical Content and Semantic Network Analysis. *Online Information Review,* 37:835–850.

[30] Kumaraguru, P. Sheng, S. Acquisti, A. Cranor, L. F. & Hong, J. I. (2010). Teaching Johnny Not To Fall For Phish. *ACM Trans. Internet Techn.,* 10:7:1–7:31.

[31] LaCour, J. (2019). Phishing Trends and Intelligence Report – The Growing Social Engineering Threat. *Technical report,* PhishLabs Team, 2019.

[32] Lei, T., Barzilay, R., & Jaakkola, T. (2016). Rationalizing Neural Predictions. *EMNLP 2016 – Conference on Empirical Methods in Natural Language Processing,* 107–117.

[33] Lipton, Z. C. (2018). The Mythos of Model Interpretability. *Queue,* 16(3):1–27.

[34] Martens, D., & Provost, F. (2014). Explaining Data-Driven Document Classifications. *MIS Quarterly: Management Information Systems,* 38(1):73–99.

[35] Mohan, V. S., Naveen. J., Vinayakumar, R., Suman, K., Das, A., & Verma, R. M. (2018). A. R. E. S: Automatic Rogue Email Spotter Crypt Coyotes. In: R. Verma, A. Das (eds.): *Proceedings of the 1st AntiPhishing Shared Pilot at 4th ACM International Workshop on Security and Privacy Analytics (IWSPA 2018),* Tempe, Arizona, USA, 21-03-2018, published at http://ceur-ws.org

[36] Moody, G., Galletta, D. F. & Dunn, B. K. (2017). Which Phish get Caught? An Exploratory Study of Individuals' Susceptibility to Phishing. *European Journal of Information Systems,* 26:564–584.

[37] Narayanan, M., Chen, E., He, J., Kim, B., Gershman, S., & Doshi-Velez, F. (2018). How do Humans Understand Explanations from Machine Learning Systems? An Evaluation of the Human-Interpretability of Explanation. arXiv, 1802.00682.

[38]   Neumann, L. (2017). Human Factors in IT Security. In F. Abolhassan, editor, *Cyber Security. Simply. Make it Happen.*, 75–86. Springer International Publishing.

[39]   Nguyen. C. (2018). *Learning Not To Take the Bait: An Examination of Training Methods and Overlerarning on Phishing Susceptibility.* PhD thesis.

[40]   Nguyen, M., Nguyen, T. & Nguyen, T. H. (2018). A Deep Learning Model with Hierarchical LSTMs and Supervised Attention for Anti-Phishing. In: R. Verma, A. Das. (eds.): *Proceedings of the 1st AntiPhishing Shared Pilot at 4th ACM International Workshop on Security and Privacy Analytics (IWSPA 2018),* Tempe, Arizona, USA, 21-03-2018, published at http://ceur-ws.org

[41]   Parsons, K., Butavicius, M., Pattinson, M., McCormac, A., Calic, D., & Jerram, C. (2015). Do Users Focus on the Correct Cues to Differentiate between Phishing and Genuine Emails? *ACIS 2015. Proceedings – 26th Australasian Conference on Information Systems,* 1–10.

[42]   Peltola, T. (2018). Local Interpretable Model-agnostic Explanations of Bayesian Predictive Models via Kullback-Leibler Projections. *Proceedings of the 2nd Workshop on Explainable Artificial Intelligence (XAI) at IJCAI/ECAI.* arXiv:1810.02678v1

[43]   Ribeiro, M. T. (2016). LIME – Local Interpretable Model-Agnostic Explanations Update. https://homes. cs.washington.edu/~marcotcr/blog/lime/

[44]   Ribeiro, M. T. Singh, S., & Guestrin, C (2016a). "Why Should I Trust You?": Explaining the Predictions of Any Classifier. *Proceedings of the 22nd ACM SIGKDD International Conference on Knowledge Discovery and Data Mining,* 1135–1144. https://doi.org/10.1145/2939672.2939778

[45]   Ribeiro, M. T., Singh, S., & Guestrin, C. (2016b). Model-Agnostic Interpretability of Machine Learning. *ICLM Workshop on Human Interpretability in Machine Learning,* New York, NY. 91–95.

[46]   Ribeiro, M. T., Singh, S., & Guestrin, C (2018). Anchors: High-Precision Model-Agnostic Explanations. In *32nd AAAI Conference on Artificial Intelligence,* New Orleans, 1527–1535.

[47]   Sheng, S., Magnien, B., Kumaraguru, P., Acquisti, A., Cranor, L. F., Hong, J. & Nunge, E. (2007). Anti-Phishing Phil. *Proceedings of the 3rd symposium on Usable Privacy and Security – SOUPS '07,* 229, 88. ACM Press.

[48]   Silic, M. & Lowry, P. B. (2020). Using Design-Science Based Gamification to Improve Organizational Security Training and Compliance. *Journal of Management Information Systems,* 37(1): 129–161. DOI: 10.1080/07421222.2019.1705512

[49]   Shmueli, G. (2010). To Explain or to Predict? *Statist. Sci.* 25 (3) 289–310, https://doi.org/10.1214/10-STS330.

[50]   Sorensen, M. (2018). The New Face of Phishing. APWG. https://apwg.org/the-new-face-of-phishing/

[51]   Unnithan, N. A, Harikrishnan, N. B, Ravi, V., & Kp, S. (2018). Machine Learning Based Phishing Email detection. *Proceedings of the 1st Anti-Phishing Shared Task Pilot at 4th ACM IWSPA co-located with 8th ACM Conference on Data and Application Security and Privacy (CODASPY 2018),* Tempe, Arizona, USA, March 21, 2018.

[52]   Vazhayil, A., Harikrishnan, N. B, Vinayakumar, R, & Kp, S. (2018). PED-ML: Phishing Email Detection Using Classical Machine Learning Techniques. In: R. Verma, A. Das (eds.): *Proceedings of the 1st AntiPhishing Shared Pilot at 4th ACM International Workshop on Security and Privacy Analytics (IWSPA 2018),* Tempe, Arizona, USA, 21-03-2018, published at http://ceur-ws.org

[53]   Vidyapeetham, R., Ganesh, H. B., AnandKumar, M., Kp, S. (2018). DeepAnti-PhishNet:Applying Deep Neural Networks for Phishing Email Detection. In: R. Verma, A. Das (eds.): *Proceedings of the 1st AntiPhishing Shared Pilot at 4th ACM International Workshop on Security and Privacy Analytics (IWSPA 2018),* Tempe, Arizona, USA, 21-03-2018, published at http://ceur-ws.org

[54]   Vishwanath, A., Herath, T., Chen, R., Wang, J. & Rao, H. R. (2011). Why do People get Phished? Testing Individual Differences in Phishing Vulnerability Within an Integrated, Information Processing Model. *Decision Support Systems,* 51(3):576–586.

[55]   Vishwanath, A., Harrison, B., & Ng. Y. J. (2018). Suspicion, Cognition, and Automaticity Model of Phishing Susceptibility. *Communication Research,* 45(8):1146–1166.

[56]   Volkamer, M., Renaud, K. &Reinheimer, B. (2016). TORPEDO: TOoltip-poweRed Phishing Email DetectiOn. In *IFIP Advances in Information and Communication Technology,* 471, 161–175.

[57]   Wiseman, S., & Rush, A. M. (2016). Sequence-to-Sequence Learning as Beam-Search Optimization. *Proceedings of the 2016 Conference on Empirical Methods in Natural Language Processing,* Austin, Texas, 1296–1306. 10.18653/v1/D16-1137

[58] Wright, R. T., Jensen, M. L., Thatcher, J. B., Dinger, M. & Marett, K. (2014). Research Note – Influence Techniques in Phishing Attacks: An Examination of Vulnerability and Resistance. *Information Systems Research*, 25(2):385–400.

[59] Yu, K., Taib, R. Butavicius, M. A., Parsons, K. & Chen, F. (2019). Mouse Behavior as an Index of Phishing Awareness. In D. Lamas, F. Loizides, L. Nacke, H. Petrie, M. Winckler, and P. Zaphiris (Eds.). *Interact.* Springer International Publishing, 539–548.

[60] Suzuki, Y.E., & Monroy, S.A.S. (2022). Prevention and mitigation measures against phishing emails: a sequential schema model. *Security Journal.* 35(4):1162–1182, doi:10.1057/s41284-021-00318-x.

[61] Linardatos, P., Papastefanopoulos, V., & Kotsiantis, S. (2021). Explainable AI: a review of machine learning interpretability methods, *Entropy* 23(1): 18, https://doi.org/10.3390/e23010018.

[62] Kluge, K., & Eckhardt, R. (2021). Explaining the Suspicion: Design of an XAI-based User-focused Anti-phishing Measure, Lecture Notes in Information Systems and Organization. In: Frederik Ahlemann & Reinhard Schütte & Stefan Stieglitz (ed.), *Innovation Through Information Systems*, Springer, pp. 247–261.

[63] Zhang, Z., Al Hamai, H., Damiani, E., Yeun, C.Y., and Taher, F. (2022). Explainable artificial intelligence applications in cyber security: state-of-the-art in research. *arXiv*, 16:06, doi.org/10.48550/arXiv.2208.14937.

Printed in the United States
by Baker & Taylor Publisher Services